Behavioral
Mechanisms
in Ecology

Behavioral Mechanisms in Ecology

Douglass H. Morse

Harvard University Press
Cambridge, Massachusetts
and London, England   1980

Library of Congress Cataloging in Publication Data

Morse, Douglass H 1938–
    Behavioral mechanisms in ecology.

    Bibliography; p.
    Includes index.
    1. Animals, Habits and behavior of. 2. Animal
ecology. I. Title.
QL751.M876   591.51   80–12130
ISBN 0–674–06460–7

To John Corliss

Acknowledgments

DURING THE LONG gestation period of this book I have profited tremendously from the assistance and insight of many people. I commenced this project during a sabbatical year at the Edward Grey Institute, Oxford. David Lack and Christopher M. Perrins made facilities there available to me, including the splendid Alexander and Elton libraries.

The book was completed at the University of Maryland, where my colleagues and graduate students were a continuing source of stimulation, ideas, and encouragement. I discussed most of the material presented here in my graduate behavioral ecology course, which provided a valuable forum. John O. Corliss, chairman of the Department of Zoology, was wonderfully supportive of this venture, and he continually fostered an atmosphere that encouraged scientific endeavor and ferment. The adept editorial pen of Nancy E. Stamp helped greatly in bringing the manuscript toward a readable form at a time when that seemed impossible.

The following colleagues at Maryland and other institutions generously critiqued one or more chapters or provided other useful advice: J. David Allan, Edward R. Buchler, R. Michael Erwin, Robert S. Fritz, Douglas E. Gill, Timothy G. Halverson, David W. Inouye, Murray Itzkowitz, Eugene S. Morton, Marjorie L. Reaka, Wolfgang M. Schleidt, Michael D. Shalter, Nancy E. Stamp, Geerat J. Vermeij, Frederick E. Wasserman, and Paul W. Woodward.

The artwork was done by Jaquin B. Schulz and the graphics by

Martha L. Tabor. Several staff members of the National Museum of Natural History, Smithsonian Institution, helped Ms. Schulz and made collections available to her. Gregory Paul assisted in the reconstruction of the hadrosaur, and Wayne Hoffman and Storrs Olson provided useful information on the bird illustrations.

I am indebted to the following publishers and individuals for permission to use materials belonging to them: Academic Press (*Advances in Ecological Research, Theoretical Population Biology*), American Association for the Advancement of Science (*Science*), American Ornithologists' Union (*Auk*), American Society of Mammalogists (*Journal of Mammalogy*), American Society of Zoologists (*American Zoologist*), Baillière Tindall (*Animal Behaviour*), *Biological Bulletin,* Blackwell Scientific Publications (*Journal of Animal Ecology, Journal of Applied Ecology*), E. J. Brill (*Behaviour*), British Ornithologists' Union (*Ibis*), Cooper Ornithological Society (*Condor*), Cornell Laboratory of Ornithology (*Living Bird*), Duke University Press (*Ecology, Ecological Monographs*), I. Eibl-Eibesfeldt, Entomological Society of Canada (*Canadian Entomologist*), Geological Society of America, MacMillen Journals (*British Birds, Nature*), Methuen and Company, National Research Council of Canada (*Canadian Journal of Zoology*), New York Entomological Society, Oxford University Press, Verlag Paul Parey (*Zeitschrift für Tierpsychologie*), Society for the Study of Evolution (*Evolution*), Springer-Verlag (*Oecologia*), United States Geological Survey, University of California Press, University of Chicago Press (*American Naturalist*), University of Notre Dame Press (*American Midland Naturalist*), The Wildlife Society (*Journal of Wildlife Management*), Wilson Ornithological Society (*Wilson Bulletin*), Zoological Society of London (*Journal of Zoology, London, Symposia of the Zoological Society of London*).

This book evolved out of a desire to expand my research on the foraging patterns of birds and to place that work in a broader perspective. Therefore I have drawn heavily on my own work, and in turn my research has been influenced by what I have learned in writing the book. The research has been generously supported by several grants from the National Science Foundation.

Contents

Behavioral Mechanisms in Ecology

1 *Introduction*

THIS BOOK IS ABOUT the relationships of animals to their resources and about their relationships to one another through those resources. It brings together parts of two flourishing biological disciplines, ecology and ethology. In recent time these disciplines have developed in almost complete isolation, but this was not always the case. Charles Darwin's *Expression of Emotions in Man and Animals* (1872) is still required reading for students of behavior, and his *On the Origin of Species* (1859) rests almost everywhere on a deep and sophisticated appreciation of ecological principles. Chapter 3 of the *Origin*, which treats the mechanism of competition and its implications, shows how thoroughly Darwin had integrated his understanding of ethology and ecology. Competition is a major underpinning of the theory of evolution through natural selection, and in several critical passages Darwin bases his argument on a combination of ethological and ecological considerations.

Natural history, the description of the lives of animals, is rich in observations useful to both the ethologist and the ecologist. But the theoretical questions addressed by natural historical studies have usually been those of systematics (the phylogenetic relationships between different groups of organisms), not those of ethology or ecology, much less those linking them. The ties between the two fields received little explicit attention before the late 1960s, in part because two very different sorts of questions had been asked.

Traditional ethology has been concerned with the way particular behavioral patterns function. For example, studies of communication attempt first to provide a descriptive analysis of the behavior, and then to infer why it takes the form it does. This behavior may have important implications for the way in which the participants subsequently exploit resources in their environment, and the ecological context (competitors, predators, arrays of resources) may restrict the ways in which a behavioral repertoire develops and the conditions under which its different parts are used. For the most part, such problems fall outside the domain of ethology.

Traditional ecology has been more concerned with high-order processes such as demography, energy flow, and the species composition of communities than it has with the proximate mechanisms that underlie these "numbers games." The behavior of the animals in question tends to be treated as a black box, an approach that is encouraged by the fact that the animals or plants most convenient for traditional ecological studies are often not ones convenient for behavioral study. Conversely, the forms typically favored by ethologists (birds and fish) are seldom well suited for quantitative ecological observation.

Pioneering departures from these trends can be found in the studies of Crook on birds and primates (for instance, his 1965, 1970a), in which he explores the role of ecological factors, primarily resources and predators, as determinants of social systems. Group size, composition, range, and defense may all differ as a function of these ecological variables, which therefore bear importantly on the development of behavioral patterns. Crook's basic hypotheses have been refined and extended in subsequent work (Eisenberg, Muckenhirn, and Rudran, 1972; E. O. Wilson, 1975; Clutton-Brock and Harvey, 1978). However, the topics treated by these authors are not the only interesting ones to be found in the interface between social structure and environmental variables. The behavioral repertoires resulting from varying patterns of social structure may have important implications for the types of exploitation patterns that can be developed. In turn, the spacing patterns resulting from different social systems may affect the carrying capacity of the habitat and the species composition of the animals and plants associated with it.

Behavioral ecology is not, strictly speaking, a discrete area of ecological theory, as is population ecology or community ecology. Rather it is a way (and I believe a profitable way) of approaching many topics of ecological importance. For this reason I shall not attempt to outline a theory of behavioral ecology. It seems to me that the promise of this approach is that deeper understanding, and new questions, will arise from a more inclusive attack on the old questions.

In chapters 2 and 3 I discuss first the fundamental problem of foraging efficiency; then—from a somewhat broader point of view—the overall economics of foraging. This provides the background for a consideration of other constraints. Animals often vie directly for space, rather than for food itself, so I next discuss habitat selection (chapter 4). Many factors may prevent animals from foraging at the theoretical maximum level of efficiency. One of the most important of these factors is predation (chapter 5). In order to survive, it may be necessary to compromise foraging efficiency, but the cost of predator avoidance may differ markedly from species to species, from one avoidance technique to another, and from one geographic region to another.

High temperatures interfere with the foraging abilities of many animals, and low temperatures greatly increase the outflow of heat for animals that operate at high body temperature. If an animal is exposed to a wide temperature range, a behavioral strategy that minimizes heat loss may be favored over one that maximizes the rate of energy gain (chapter 6). Many animals take advantage of heat sources in the environment. Heliothermic forms, the sun worshippers of the animal kingdom, are the most conspicuous examples. Animals can never feed directly on radiant energy the way plants do, but they can greatly reduce their needs for food by using a certain amount of radiant energy and thermoregulation. Ectothermic forms that operate most efficiently at relatively high body temperatures (lizards, for example) often have no other sources of body heat; but even endotherms, which have a choice, often make considerable use of "passive solar."

Factors such as these determine the size of the reproductive effort an animal can make at a given time (chapter 7). Reproduction not only demands an increased *quantity* of resources, it may also demand *types* of resources different from those used by adults for their own maintenance, thus compromising foraging efficiency. The need to protect young or to bring food to them may also constrain foraging repertoires; parental care is an important (and probably underestimated) determinant of resource exploitation patterns.

Chapters 8 through 11 discuss competition, a pervasive and important influence on foraging patterns. Individuals may attempt to secure resources before others can get to them, or they may directly interfere with one another's attempts to gather resources. Such interference may be actively aggressive, or it may be manifested merely in passive avoidance. In either case the foraging patterns of one or both individuals will be modified.

Competition for mates (chapter 8) may be even more intense than for food or space. Mate selection is not a major theme of this book, but I treat it because competition for mates has important implications for resource exploitation, and ecological constraints in turn may strongly influence sexual selection.

Chapters 9 and 10 consider how the problem of defending a site against both conspecifics and predators may affect spacing patterns. Territorial systems are compared with hierarchical systems in which no fixed space is defended, but in which members of a group have unequal rights, depending on their rank. The strongest competition usually occurs among members of the same species, but interspecific competition (chapter 11) can be surprisingly intense; its effects on patterns of resource exploitation are generally similar to those of intraspecific competition but may differ in some interesting ways.

Chapter 12 explores the advantages and liabilities of sociality, mainly as it affects foraging and defense against predators. In chapter 13 I attempt to delineate some of the major unsolved problems in behavioral ecology, and I suggest ways they might profitably be approached.

Most chapters contain a concluding section entitled "Synthesis," in which I attempt to draw out some of the logical consequences of material in the preceding sections, and here I also offer some untested speculations.

2 *Variability in Foraging Patterns*

THE WAY ANIMALS EAT depends on the variety of foods available to them and on the characteristics of those foods. Potential foods differ with respect to the efficiency with which they can be found and consumed, the intensity with which other animals compete for them, the vulnerability to predation entailed by their use, and the way climate and other environmental factors interfere with their use (fig. 2.1). Only those individuals who can accumulate adequate energy stores in spite of these demands are able to reproduce. Even if no food limitations exist, efficient consumers should be more successful because they will have more time for other activities such as rasing young and avoiding predators.

The interactions among these factors are only poorly understood and are frequently ignored. For example, ecologists have often treated interactions between pairs of species as if they were independent of the rest of the community; this need not be the case, as Neill (1974) has clearly shown in his experimental studies on aquatic microcosms. The range of resource variability encountered over a life span will further affect feeding repertoires through natural selection. Individuals exposed to a constant supply of resources are in a much better position to specialize on them than are those that experience changes in their kind or abundance. Resource variability itself can be strongly affected by relationships among consumers. For example, individuals who rank low on either an intraspecific or interspecific dominance hierarchy may find resources less predictable than high-ranking individuals, sim-

5

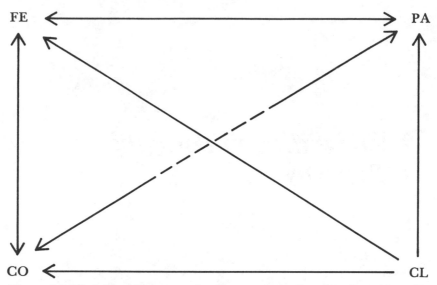

Figure 2.1 Relationships among key factors that affect foraging. CL = climates and other environmental factors; CO = competition; FE = foraging efficiency; PA = predator avoidance.

ply because high-ranking individuals may prevent orderly exploitation of resources by subordinates (Morse, 1974).

Given the large number of factors that influence the way animals exploit their foods, it is reasonable to ask whether foraging patterns are amenable to analysis. Fortunately, animals do appear to respond to certain factors in predictable ways.

Generalists and specialists

Foragers can be divided into *generalists* and *specialists,* according to the number of different resources they use. These two categories are best thought of as the ends of a continuum describing the proportion of available resources taken. The concept is most useful in comparing individuals (or species) with one another under similar conditions. The members of most species fall comfortably between the boundaries defining the ultimate generalist and the ultimate specialist (fig. 2.2). Opportunities for being a specialist necessarily depend on the number of resources available. If only one or two resources are available, there is no meaningful distinction between the two categories. Most animals, however, have real options.

It is generally believed that a compromise exists between the number of resources an animal uses and the proficiency with which it exploits them; that is, a jack-of-all-trades is a master of none. Levins

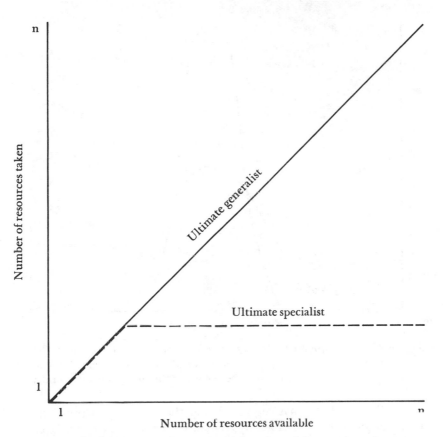

Figure 2.2 Options open to foragers as a function of the number of resources available. Most foragers to be discussed fall well within the area delineated by the diagonal line and the dashed horizontal line. Monophagous species fall along the dashed line.

(1968) suggested that high proficiency in resource use is attained only if an animal exploits few resources. Ideally, then, the area under any two resource utilization curves (fig. 2.3) should be equal. Selective pressure for efficient exploitation should narrow the range of resources used, whereas interspecific competition and environmental fluctuation should broaden it. An optimal exploitation curve should reflect the environmental variability and biotic interactions experienced by the species. Within a species, differences in the status and activities of individuals may give them different optimal curves.

Although Levins' hypothesis is a useful concept, few data exist on the relation of feeding efficiency to the range of resources exploited (see Schoener, 1974). On the basis of a few studies in the literature,

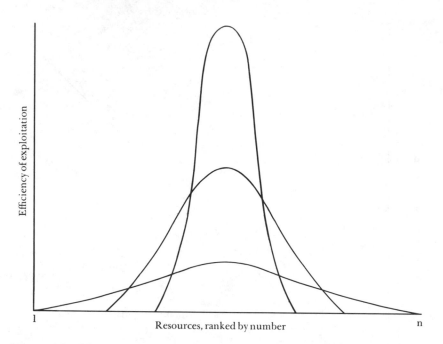

Figure 2.3 The proposed trade-off between the number of resources that an animal may exploit and its overall efficiency of exploiting them. Efficiency of exploitation of any food item may be measured in energy gain per unit time.

D. S. Wilson (1975) has suggested that the efficiency of prey capture by predators is not described by a smooth curve, as implied above, but rather by one that increases rapidly from minimum prey size, reaches a plateau, and then descends slowly with increasing prey size (fig. 2.4). Extrapolating from this curve, Wilson argues that although large predators can use foods unavailable to small predators, small predators are able to exploit considerably fewer foods that are unavailable to large predators, so that large predators have a competitive advantage. If this is a general relationship, it could explain why some kinds of animals are more successful as predators than others. As Wilson observes, however, the advantage can easily be negated by characteristics of the prey-size distribution or by the relative positions of different predators on this curve.

Several workers have asked how many resources are required to support a species. Most investigations have centered on the hypothesis that one species requires one resource (see, for example, MacArthur and Levins, 1964; Rescigno and Richardson, 1965). To decide how many resources are needed to support a species, it is necessary to determine how many resources an animal recognizes. Does a single

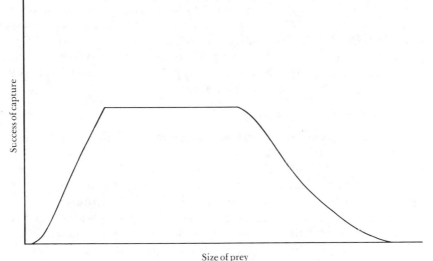

Figure 2.4 Shape of the success-of-capture curve. As the predator proceeds from the smallest to the largest prey, it quickly reaches its peak efficiency, followed by a plateau and a gradual decline. (From D. S. Wilson, 1975, copyright 1975 by The University of Chicago.)

herbivorous species represent one resource to a carnivore, or are several similar herbivorous species (perhaps congeners) treated as one resource by the carnivore? Perhaps the young of a given herbivorous species are considered a single resource and the adults are treated as another. It is important to define resources as the consumer views them when distinguishing generalists from specialists (MacArthur, 1968). Patterns of choice may depend on abilities of consumers to discriminate, and a supposedly generalist population may be made up of individuals that all exploit the same resources, of specialists exploiting different resources, or of any intermediate combination (Morse, 1971a; Roughgarden, 1974). Even within a habitat, individuals may differ in their amount of specialization, a matter often ignored in discussions of this subject (Morse, 1971a). For example, some individual trout (*Salmo gairdneri*) regularly exploit a considerably wider range of prey than others (Bryan and Larkin, 1972). Haigh and Maynard Smith (1972) provide additional examples where the one-species-one-resource argument seems to fail.

Generalists often occur where they have few competitors. This may result from low resource availability or from the limitation of a species' range to an area such as an island. A single resource or small number of resources may be inadequate to support a population, with the result that any species present will necessarily exploit a wide variety of resources.

In practice, few if any studies permit an appropriate definition of a resource under varying levels of availability. This makes it difficult to decide whether the generalist habit is a consequence of the abundance of different resources or some other factor.

Stereotypy and plasticity

Resource patterns may also be characterized in terms of their variability over time. Whether members of a population adopt a *stereotyped* or *plastic* repertoire of food items should depend primarily on the availability of those items over time. I define stereotypy as the tendency to exploit different (or identical) resources in the same way regardless of conditions experienced, and plasticity as the tendency to exploit different (or identical) resources in different ways under changing conditions (Morse, 1971a). As in the case of generalists and specialists, it is convenient to consider these terms relative to each other.

Although the two terms somewhat parallel those of generalist and specialist, generalist is not synonymous with plastic individuals as I have defined them here, nor is specialist with stereotyped individuals. Specialists may be either stereotyped or plastic; the former will not use any resource outside a narrow range of choices, whereas the latter will do so occasionally. Generalists might also be either stereotyped or plastic; the former will continually use a broad range of resources with high predictability, whereas the latter will use a wide range of resources with little long-term predictability (Morse, 1971a).

Under what conditions do stereotyped and plastic individuals naturally occur? Klopfer and his colleagues have provided some possible answers in their attempt to relate feeding repertoires and habitat selection to the distribution, abundance, and phylogenetic relationships of species in natural communities. Although parts of their argument extend beyond the topic of stereotypy and plasticity per se, I shall include some of them briefly, since they tie stereotypy and plasticity to several other subjects.

Klopfer and MacArthur (1960) noted that nonpasserine birds (perching birds unrelated to songbirds and their allies) are proportionately more common in the more or less predictable lowland tropical forests, whereas passerine birds (songbirds) are proportionately more common at higher latitudes, where conditions are more variable. They argued that this distribution occurs because nonpasserine birds are more stereotyped than passerine birds, a legacy of low neurological plasticity in the phylogenetically old nonpasserines, as compared with the phylogenetically younger passerines. Unfortunately, the authors presented no direct evidence in support of this assumption about stereotypy and neurological differences. Klopfer and MacArthur also assumed that a high level of stereotypy is advantageous under relatively constant and predictable conditions and that a high level of plasticity is advantageous in the greatly varying climates and fluctuating food supplies typically experienced at high latitudes.

In 1961 the same authors reported that bill sizes differ less between closely related (and potentially competing) tropical birds than they do between closely related birds of temperate regions, and they argued from this that tropical species have narrower niches (ranges of resource or space use) than do temperate species. Narrow niches may (but need not necessarily) be correlated with a high degree of stereotypy. This assumes that bill morphology is an adequate measure of foraging behavior; the assumption is plausible, but has been questioned by Schoener (1965) and by Willson, Karr, and Roth (1975).

Klopfer and MacArthur considered these tropical birds to be what I call stereotyped specialists, that is, individuals that exploit the same narrow range of resources at all times (feeding type 1, fig. 2.5). Yet it is doubtful that small differences between feeding structures are confined to stereotyped specialists; small differences could also characterize species pairs that have mean bill sizes moderately well adapted to a variety of foods (feeding type 2, fig. 2.5). In each case broad niche dimensions could exist, accompanied by appreciable niche overlap. Thus either broad niches with high between-species overlap, or narrow, largely exclusive niches, could account for the great number of bird species in tropical lowland forests. Croxall (1977) has demonstrated that birds found in the lowland forests of New Guinea have greater niche overlap, but not necessarily narrower niches, than do their high-altitude ecological counterparts. His conclusion is important to this discussion, since medium- and high-altitude tropical birds show many ecological similarities to temperate populations (Orians, 1969a).

Klopfer (1965, 1967) subsequently compared tropical and temperate passerine birds in laboratory experiments, to determine whether their purported differences in stereotypy had a phylogenetic or ecological basis. Hand-reared tropical tanagers showed a somewhat higher degree of stereotypy than temperate sparrows, although these tanagers had a higher level of plasticity than predicted. Klopfer concluded that the high degree of stereotypy normally attributed to these birds prob-

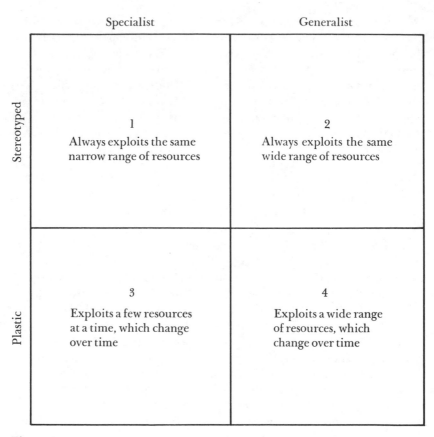

Specialist   Generalist

Stereotyped

1

Always exploits the same narrow range of resources

2

Always exploits the same wide range of resources

Plastic

3

Exploits a few resources at a time, which change over time

4

Exploits a wide range of resources, which change over time

Figure 2.5 Basic feeding types of animals.

ably was an effect, not a cause, of the large number of species present in the tropics. However, the small samples used did not permit statistical analysis, so the question of differences in stereotypy remains open. If his conclusions are supported by future studies, it will be clear that an animal's experiences may be of considerable importance in determining how it exploits resources.

A correlation between the degree of stereotypy and the number of species would suggest that individuals belonging to species on islands with few other species should be less stereotyped than individuals of the same species in mainland areas where there are many other species. Sheppard, Klopfer, and Oelke (1968) found this to be the case for catbirds (*Dumetella carolinensis*) and cardinals (*Cardinalis cardinalis*) from Bermuda and the adjacent North American mainland. Crowell's data (1962) on free-ranging catbirds and cardinals from the same two areas, as well as Klopfer's data (1967) on bananaquits (*Coroeba flaveola*) from Puerto Rico and Panama, also support this conclusion.

Although these studies have not established unequivocal cause-effect relationships among behavioral patterns, environmental factors, and phylogenetic relationships, they have pointed out possible correlations among them, which may reflect the potential range of responses that an organism can make under a variety of conditions. Similar species should reduce each other's foraging opportunities, regardless of how they interact, if they are resource-limited. This restriction should hold whether stereotypy precedes high diversity or high diversity precedes stereotypy.

Invading new communities

The discussion of stereotypy implies that some individuals can adjust to unusual conditions more readily than others. This is important in the case of displaced animals and could largely determine which ones colonize and what interactions occur between them and other residents. Frequently such displaced colonizing animals undergo ecological release (expansion of one or more niche dimensions) in the absence of competitors (MacArthur and Wilson, 1967; Grant, 1972). Most studies of ecological release have primarily been concerned with documenting that release occurs (Schoener, 1967, 1968a), and evaluating its implications for the community as a whole (MacArthur and Wilson, 1967; MacArthur, 1972). Unfortunately, little information exists about the behavior of individuals during the initial period of colonization. Comparisons are usually made between mainland and island populations, although the island forms are often sufficiently distinct to be accorded separate taxonomic status (Crowell, 1962; Sheppard et al., 1968). Prospects of successful colonization will depend very much on whether the invader can respond immediately to its new situation.

Changes in resource exploitation patterns might be of either a primarily behavioral (type I) or a primarily genetic (type II) kind. These characteristics parallel those of plasticity and stereotypy, respectively. I assume that changes in type I depend on a genetic foundation and that populations exhibiting these changes may in time exhibit type II changes. Type I colonizers can quickly alter their foraging patterns in response to different conditions. These highly plastic individuals will make predictably better colonizers than type II individuals. Type I individuals will probably use a broader range of resources of habitats than type II individuals, although this depends on the niche size in precolonization situations. Type II colonizers will initially exploit resources in much the same way that they did at their point of origin, even in the absence of competitors, and only change later in evolutionary time, if at all. Chances of extinction will be relatively high during this extended period (see MacArthur and Wilson, 1967; Ricklefs and Cox, 1972). Thus the contribution of type II individuals to successful colonization should be limited, particularly if their favored resources are in short supply, and their opportunities should differ directly with the range of resources that they typically exploit. Diamond (1970) reported that some bird populations on South Pacific islands differ little from their mainland counterparts, thus matching the characteristics of type II colonizers. However, it is unclear how long these populations have been present on the islands.

The importance of the distinction between type I and type II individuals may be illustrated by comparing the foraging patterns of individuals from the same population in species-rich and species-poor communities. Several species of New World warblers (Parulidae) coexist along the coast of Maine (MacArthur, 1958; Morse, 1971b, 1977a). As many as five of them have been found together in mainland spruce forests, although adjacent islands of one hectare or less do not support more than three of these species. I focused my attention on several of these islands, located from a few hundred meters to a few kilometers from the five-species mainland populations. The birds from the small islands are part of the mainland populations; therefore, the genetic basis of their behavioral patterns should not differ from that of individuals in large populations (Morse, 1970a, 1971b). The species are all migratory and must reclaim territories each year, a habit that provides opportunities for mixing. Occasional movement occurs between the islands and mainland during the breeding season. These birds occupy the small islands as early in the spring as they do the adjacent large forests, and known individuals often use the same island in successive years. All of these factors indicate that the island-inhabiting birds are typical of the population as a whole.

An inverse relationship occurs between the stereotypy of the common inhabitants of the small islands (measured by differences in foraging diversity) and the minimum size of the islands they occupy. If one species is present, it is the parula warbler (*Parula americana*). If two species are present, they are the parula and the yellow-rumped (*Dendroica coronata*) warblers. Black-throated green warblers (*D. virens*) occur only where the other two species are also present. Of these species, black-throated green warblers exhibited the least plasticity, which accords well with their restriction to certain of the islands. These birds were not particularly successful colonists. They used the habitat on the smallest islands similarly to the manner in which they used it on the mainland and also with a much lower breeding success than the other two species (Morse, 1971b) (fig. 2.6). The other two commonly occurring species (yellow-rumped and parula warblers) showed ecological release on the islands where black-throated green warblers were absent, partially filling the space occupied elsewhere by the latter species. In the absence of yellow-rumped warblers, the parula demonstrated even greater release. Thus, black-throated green warblers fit the criterion of a type II colonizer (little immediate behavioral modification), and yellow-rumped and parula warblers fit type I (immediate behavioral modification). The latter two species' opportunities for colonizing new areas should thus be greater than those of the black-throated green warbler. However, it may be dangerous to extrapolate from these populations to others. Some populations of black-throated green warblers frequent deciduous forests, and others occur in coniferous forests of the southern Appalachian Mountains in the absence of other members of this group (Bent, 1953).

Ricklefs and Cox (1972) suggested that there are considerable differences in the adaptability of prospective colonists. Species with low levels of adaptability presumably have high extinction rates as colonists. As seen in fig. 2.6, dispersal rates during the breeding season appear to be greatest in parula and yellow-rumped warblers, for more wandering individuals of these species are seen than of others, relative to their overall population density (Morse, 1971b, 1977a). The parula warbler has the largest average clutch size of the three species (Palmer, 1949), and the yellow-rumped warbler has slightly larger clutches than the black-throated green warbler (MacArthur, 1958). The correlation between degree of adaptability and likelihood of successful colonization is reflected in the nesting success of these species on the islands, and also in the frequency with which the species occupied the islands (Morse, 1971b, 1977a).

Thus marked differences exist in the responses of even closely related animals to new surroundings, and these should be of major importance in determining their success as colonizers. The success of individuals in newly occupied areas may also depend on the location from which they came. My study (1971b) of colonizing warblers illustrates that it may be dangerous to draw conclusions about stereotypy or plasticity solely from phylogenetic considerations. Congeners may display far greater differences than Klopfer and MacArthur's initial arguments (1960, 1961) would suggest.

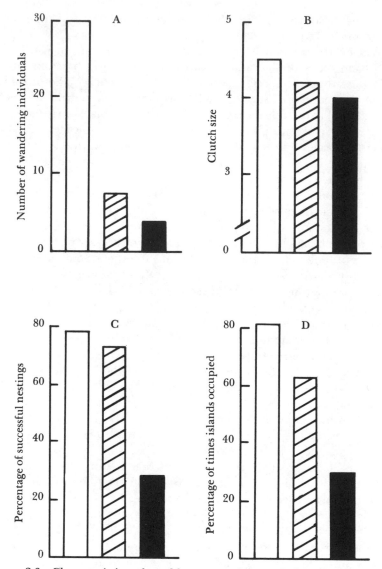

Figure 2.6 Characteristics of warblers occupying small islands. *A,* number of wandering, nonbreeding individuals, corrected for the abundance of the different species on the mainland staging areas; *B,* mean clutch size; *C,* percentage of successful nestings on islands; *D,* percentage of times that islands were occupied. White bars = parula warbler; hatched bars = yellow-rumped warbler; black bars = black-throated green warbler. (Modified from Mac-Arthur, 1958, copyright 1958 by the Ecological Society of America; Morse, 1971b, copyright 1971 by the Ecological Society of America; Morse, 1977a; and unpublished data.)

Foraging patterns associated with changing conditions

High plasticity is an obvious strategy for coping with a changeable (even if predictably changeable) environment, but there are others. These include dormancy, hoarding, and migration. The option adopted may differ with the conditions experienced and the basic attributes of an animal.

DORMANCY

Many animals, both endotherms and ectotherms, become dormant in response to marked changes from an optimal condition. This capability may permit occupation of a single area, typically one that periodically undergoes severe climatic fluctuations, without developing an extremely plastic foraging repertoire. Although the physiological details lie beyond the scope of this discussion, dormancy entails three problems of interest here: (1) obtaining food adequate for survival at decreased metabolic rates, (2) finding an appropriate site for this inactive period, and (3) obtaining food adequate for recovery after emergence. The unfavorable conditions may be winter cold, summer heat, or drought. In this chapter I discuss species that lay down energy stores in the form of fat and those that hoard food. Related subjects involving behavioral thermoregulation are discussed in chapter 6.

If obtaining food adequate for entering dormancy is an important consideration, one would predict that prior to this period individuals would increase their rate of food intake or gathering behavior (Lyman, 1954; Irving, 1972). Increases in time spent foraging have been noted in middle and late summer, prior to winter dormancy, in several species of mammals and are associated with marked increases of weight, amounting to as much as 50 percent of body weight in the case of lean and fat woodchucks (*Marmota monax*) (Davis, 1967). Much of the weight is gained well before an animal becomes inactive—for example, the Belding ground squirrel (*Spermophilus beldingi*) (Loehr and Risser, 1977) and the woodchuck (Davis, 1967). In fact, rate of food intake declines by the end of the active period in the ground squirrels (Loehr and Risser, 1977), woodchucks (Fall, 1971), and hoary marmots (*M. caligata*) (Barash, 1976). Barash notes that declines in food intake occur earlier in adult male and nonreproductive female hoary marmots than in reproductive females and young. He attributes this difference to the late start in building up reserves by the latter two categories. Few female yellow-bellied marmots (*M. flaviventris*) gain weight prior to weaning their young (Andersen, Armitage, and Hoffman, 1976).

Finding an adequate hibernaculum may also require searching time in animals unable to prepare such sites themselves. Those incapable of making their own hibernacula may find this a real problem, as may burrowers where satisfactory sites are rare. Limited supplies of hibernaria are believed to control the population sizes of both arctic ground squirrels (*S. undulatus*) (Carl, 1971) and yellow-bellied marmots (Andersen et al., 1976). Even where sites are abundant, their use may entail costs of preparation, as in ground-dwelling squirrels and marmots, or of migration, as in bats.

On emergence, individuals are likely to face a potentially severe food problem, although some come out of hibernation still carrying fat. This fat represents a reserve that can be used if the winter is particularly severe or that can be expended for other activities early in the spring. Winter mortality in yellow-bellied marmots appears to be associated with late and low snowpacks, which provide less insulation for the animals (Svendsen, 1974). In general, the period of dormancy appears to be a time of high mortality (Michener and Michener, 1977). Woodchucks, Belding ground squirrels, and yellow-bellied marmots may leave dormancy with some fat remaining, and their initial food intake may be relatively low (Fall, 1971; Loehr and Risser, 1977; Andersen et al., 1976). Ground squirrels and marmots tend to emerge before food in the environment is adequate to sustain them, and they immediately commence breeding. In their mountainous habitats it may be important to get young off as early as possible because of the short favorable season, and this early start will affect the survival of both the young and their mothers in the following winter (Andersen et al., 1976). Such strategies, however, increase the amount of fat reserves that must be laid on during the preceding summer. The amount that is laid on may dictate the time at which the adults can emerge in the spring.

Food Hoarding

Many mammals and virtually all birds cannot enter dormancy, and several mammals capable of becoming dormant cannot survive prolonged inclement periods without hoarding additional food. Food hoarding is highly developed in rodents that use mast or seed crops during the winter, but this behavior is not confined to temperate areas. Even rodents of tropical forests may hoard regularly—for example, agouti (*Dasyprocta pratti*) (Morris, 1962; Smythe, 1970a), green acouchi (*Myoprocta pratti*) (Morris, 1962), and African giant rat (*Cricetomys gambianus*) (Ewer, 1967). In the agouti, these hoards are used during times of extreme food shortage in the late dry season, when starvation may otherwise occur (Smythe, 1970a).

It is often stated that hoarding in mammals occurs in response to food deprivation. Although shortage of food no doubt is the ultimate stimulus associated with hoarding, there is no a priori reason why it should invariably act as the proximate stimulus as well. Most work on hoarding in mammals has been performed on laboratory rats (*Rattus norvegicus*) (Lanier, Estep, and Dewsbury, 1974), but considerable question exists about this species' hoarding tendencies under natural conditions (Beach, 1950; Steininger, 1950; Smith and Ross, 1953). Golden hamsters, which are tireless hoarders under natural conditions, show a strong tendency to cache long after they have become satiated (Smith and Ross, 1953), a trait that is also well developed in the red-tailed chipmunk (Lockner, 1972). Both hamsters and chipmunks may naturally encounter large fall seed crops; it is not surprising that they show innate tendencies to store food, regardless of their momentary energetic condition. Barry (1976) found that temperature and photoperiod affected hoarding responses in five subspecies of three species of deer mice (*Peromyscus*) and that differences among the species reflected the different conditions under which they lived (fig. 2.7). Two northern races, *P. leucopus novaboracensis* and *P. maniculatus bairdii,* increased their hoarding in response to both low temperatures and shortened photoperiod. The field-dwelling *P. m. bairdii* responded most strongly to photoperiod and the forest-dwelling *P. l. novaboracensis* to temperature. The desert-dwelling *P. m. blandus* hoarded more at high temperatures, which Barry interpreted as a response to the decreasing availability of food often associated with high temperatures. Another desert form, *P. eremicus eremicus,* hoarded with high frequency under all conditions. The tropical *P. l. castaneus* from lowland Mexico did not show strong differences in hoarding tendencies under the different regimes. This study strongly suggests that hoarding may be triggered by more than one cue and that these cues may be closely tied to the environmental conditions experienced.

In the several species of birds that store food, the trends seen among the titmice are of great interest. These birds show an increasing tendency to store food at progressively higher latitudes (Gibb, 1954, 1960; Haftorn, 1956), which is important to them for at least three reasons. First, temperatures at high latitudes are usually more severe than at lower latitudes. Second, cold seasons are longer, resulting in an extended period during which little or no food is available. Third, days are very short during the winter, curtailing the foraging time of diurnal organisms. Adoption of a nocturnal habit would seem to be a plausible response to the last contingency during the winter; however, in contrast to many mammals, small birds have not exploited this option (see Brooks, 1968). Small species of titmice in a given area

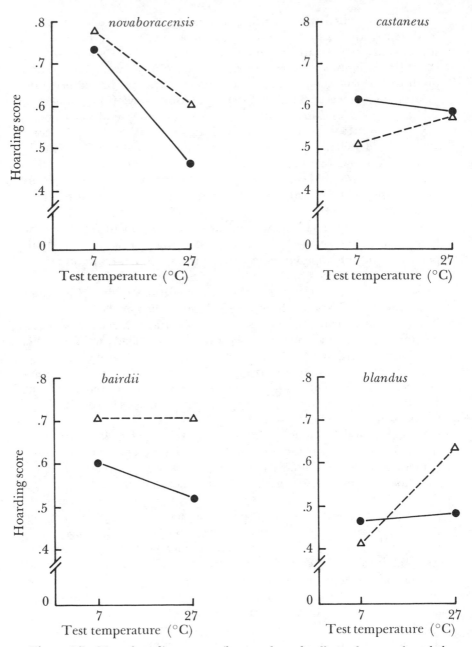

Figure 2.7 Mean hoarding scores (log number of pellets, plus one, hoarded per day) of four *Peromyscus* subspecies under various combinations of temperature and photoperiod. Dashed lines indicate short-day photoperiod, solid lines indicate long-day photoperiod. (Modified from Barry, 1976.)

cache resources with greater regularity and frequency than do larger species, which corresponds with the severity of energetic rigors that they experience as a result of their differing surface-area-to-volume ratios. Gibb (1954) and Krebs (1971) noted that of the tits frequenting Wytham Woods near Oxford, England, the smallest species, coal tit (*Parus ater*), and two of the three slightly larger ones, marsh tit (*P. palustris*) and willow tit (*P. montanus*), stored food regularly, whereas the third species of this medium-sized component, blue tit (*P. caeruleus*), and the largest species, great tit (*P. major*), were not observed to do so. In Switzerland, Bürkli (1973) also noted a tendency for coal tits to store food, while great tits did not do so under similar conditions. In Russia Bardin (1975) also found that great and blue tits did not store seeds, and Haftorn (1974) noted that great and blue tits at high latitudes are more or less dependent on man for food during winter. The other species of tits are probably dependent on stored food for survival at the northern end of their winter range (Haftorn, 1956; Grossman and West, 1977).

Storage of food is widespread among the Corvidae (crows, jays, nutcrackers, and related species) (Turcek and Kelso, 1968). This behavior also appears to be more prevalent among species living at high latitudes and altitudes, although aviary observations (Turcek and Kelso, 1968) have suggested that it occurs among some tropical species as well. Nutcrackers (*Nucifraga* spp.) have developed sublingual pouches that facilitate carrying seeds (Portenko, 1948; Bock, Balda, and Vander Wall, 1973), and the gray jay (*Perisoreus canadensis*) stores food in boluses held together by its especially sticky saliva (Bock, 1961; Dow, 1965). Turcek and Kelso (1968) discussed additional morphological adaptations for caching in this family.

Some food items are more apt to be stored than others. Gwinner (1965) noted that ravens (*Corvus corax*) store fat, which has relatively good storage capacities, preferentially to lean meat, which does not store well and is eaten immediately. Perhaps for similar reasons titmice cache insects as well as seeds, but seeds usually dominate (Haftorn, 1956).

Caching also occurs regularly among species that feed on prey items too large to be consumed at once. Shrikes (*Lanius* spp.) often cache the remains of partly eaten prey (Miller, 1931; Cade, 1967), as do hawks and owls (Mueller, 1974a; Collopy, 1977). American kestrels (*Falco sparverius*) in the laboratory showed a much stronger tendency to store intact mice than those with their skin removed, which would presumably desiccate and decay quickly (Mueller, 1974a). Caching is well developed in some mammalian carnivores, such as tigers (*Panthera tigris*) (Schaller, 1967) and pumas (*Felis concolor*)

(Hornocker, 1970). However, it is less frequently observed in lions (*Panthera leo*), perhaps because of the difficulty of concealing the remains of prey in open country (Schaller, 1972), and also because several individuals often feed on a single prey item, unlike the solitary tigers and pumas. If rare but large items are the major fare of a predator, caching behavior appears roughly analogous to hoarding by herbivores at a site of temporary food abundance.

Occasionally carnivores kill far more prey than they can consume on the spot, the familiar weasel-in-the-chicken-coop syndrome. Under natural conditions this phenomenon has been reported in foxes (*Vulpes vulpes*) preying on gulls (*Larus ridibundus*) in a nesting colony (Kruuk, 1964) and spotted hyenas (*Crocuta crocuta*) killing large numbers of Thomson's gazelles (*Gazella thomsonii*) (Kruuk, 1972a). MacDonald (1976) has established experimentally that foxes often, but not invariably, cache on the spot those prey they do not consume. Although the foxes are able to find the cached prey, they seldom exhume and eat them later. Kruuk (1972a) has suggested that surplus killing occurs when the predators come upon an unusually large number of prey that are for some reason vulnerable. This kind of bounty occurs so infrequently that feedback preventing such overkilling has not developed, nor has a counter-strategy on the part of the species victimized. A simple killing response would in fact probably serve the predator well most of the time. Inhibitions may act on searching, rather than killing—a pattern that occurs in sticklebacks (*Gasterosteus aculeatus*) (Beukema, 1968). Regardless of the basis for the initial killing, it is a point from which caching behavior would develop in carnivores, particularly where prey are very unevenly distributed.

In summary, food is often hoarded under natural conditions where its availability varies, particularly under severe climatic conditions. Some species spontaneously store food even when not hungry; this may be highly adaptive, because such surpluses often occur before inclement periods (for example, mast crops ripening in the fall). This behavior may indicate periodic food shortages even in situations that at first appear to be undemanding.

MIGRATION

Large numbers of birds and smaller numbers of other animals (bats, caribou, butterflies, dragonflies) avoid severe conditions by migrating. However, this alternative is available only to highly mobile species or to ones that inhabit areas such as mountainous regions that exhibit great local variation in weather conditions. Not surprisingly, flying forms exhibit large-scale migrations more often than others.

Liabilities as well as advantages accrue to migrants, in that they will almost invariably be exposed to conditions quite different from those that they would experience in any single area. Contingencies may arise on the wintering grounds, breeding grounds, or migration paths. When seasonal residents move into an area that already supports a population of permanent residents, they may be forced to compete with them. In some cases the invasion of migrants coincides with a season of plenty, but elsewhere conditions are less favorable when the migrants arrive (Moreau, 1966, 1973). For instance, although birds wintering just south of the Sahara Desert experience a period of reasonably abundant food on arrival, conditions subsequently deteriorate (Moreau, 1966, 1973). This may severely restrict the number of individuals that can be supported.

In addition to latitudinal migration, regular altitudinal movements occur in mountainous regions. Mountain sheep (*Ovis* spp.) (Geist, 1971) and rosy finches (*Leucosticte* spp.) (Austin, 1968) breed in high alpine areas and winter at lower elevations. Altitudinal movements of birds occur in the tropics as well, often in response to the availability of fruit (see Slud, 1960, 1964). Some species even undertake long-distance movements by swimming, as exemplified by many fishes, cetaceans (Brodie, 1975), pinnipeds (Bartholomew, 1970), and even those oceanic waterfowl in a flightless condition resulting from their fall molt (see Salomonsen, 1968).

For most migratory species it is usually unclear whether the breeding season, winter season, or migration is the most critical, and it is likely that the critical period even differs from population to population or from year to year (Morse, 1980). Although it has often been assumed that the breeding season is the limiting period of an animal's life, in several cases limitation appears to occur on the wintering grounds or in migration (Lack, 1954; Fretwell, 1969a, 1972; Morse, 1971a). If this is the case, foraging adaptations may be strongly related to those periods, as Fretwell (1969a, 1972) argued for several seed-eating sparrows and finches. Although winter residents may frequent habitats bearing considerable resemblance to their summer ones (MacArthur, 1958; Morse, 1971a), a close match often does not occur, particularly in species that summer in northern forests and winter in tropical lowlands. In many areas of tropical America (Willis, 1966a) and Africa (Morel and Boulière, 1962; Moreau, 1966, 1973; Thiollay, 1977; Sinclair, 1978), migrant birds frequent second-growth habitats, probably because they are largely excluded from the primary habitats by the permanent residents. In Panama, Willis (1966a) found that permanent residents are usually socially dominant over winter visitors, whom they exclude from favored resources. Most of the permanent residents appear to be well adapted to existing climax conditions, and many make little use of the second-growth areas used by migrants (Terborgh and Weske, 1969). However, Lack and Lack (1972) noted that North American migrants are widespread in relatively undisturbed forest in Jamaica, which may be a result of its small number of permanent residents.

If migrants are partly confined to second-growth areas, wintering space may be at a premium; in that case, adaptations for that season may also be at a premium. The recent destruction of large tracts of pristine forests in most tropical areas, which often regenerate into second-growth forests, has substantially increased the area available to migrants. This rapid change in habitats could even shift the limiting area from the wintering grounds to the breeding grounds (Morse, 1971a, 1980). The most recent studies suggest that migratory species are somewhat more prominent in primary forest than previously believed, especially at the northern edge of the neotropics (Keast and Morton, 1980), but this does not in any qualitative way alter the conclusions mentioned above.

Yearly changes in conditions can also decrease the predictability of resources available to migrants. Winstanley, Spencer, and Williamson (1974) argued that climatic deterioration south of the Sahara was responsible for a catastrophic decline, up to 77 percent in one year, in numbers of whitethroats (*Sylvia communis*) breeding in England. Morel (1973) and Morel and Morel (1974) demonstrated corresponding decreases in both wintering and breeding species of the sub-Saharan region. When conditions in the south of the Sahara eventually ameliorated, numbers of English whitethroats began to increase (Batten and Marchant, 1977).

Problems of resource exploitation in migration may also be formidable, with resources even less predictable than on summering and wintering grounds (Rappole and Warner, 1976). Migration itself is energetically very expensive, and in anticipation of it individuals often put on substantial amounts of fat, amounting to as much as half the normal weight of the bird (Odum, 1960). Even if an individual uses relatively similar habitats at both breeding and wintering sites, it may migrate through areas in which vegetation of similar structure is unavailable. Parnell (1969) found that migrating New World warblers used habitats rather similar to those in which they nested; however, in many situations it is impossible to do so. Forest-dwelling, New World warblers are regularly found in savanna grasslands on migration, where their ability to obtain food must be severely taxed at times (Morse, 1980). Thus, migratory individuals may be under strong selective pressure to maintain plasticity.

It is plausible that an individual might change from one stereo-typed condition to another at different seasons. Moss, Miller, and Allen (1972) have suggested that feeding preferences in a sedentary species, the red grouse (*Lagopus lagopus*), may change with physio-logical state. More specifically, Andrew (1972) and Andrew and Rogers (1972) have reported that the injection of testosterone markedly increased the persistence with which domestic chicks (*Gallus domesticus*) searched for different-colored grains. Migratory, free-ranging individuals could conceivably exhibit similar changes asso-ciated with hormonal fluctuations, although this possibility remains to be tested. If such changes occur, they should promote selection for individuals that are morphologically generalized in a way that per-mits them to perform these different activities with relatively high efficiency. However, it seems more likely that a relatively high level of plasticity will usually accompany a morphological compromise asso-ciated with a variety of conditions.

Although migratory movements may provide relief from seasonal fluctuations in climate or resources, few migratory animals are mobile enough to respond to daily changes in local conditions by moving away from them when they are poor and returning when they are good. However, a few striking exceptions do exist. Some swifts (Apodidae) move in front of unfavorable weather patterns or ex-ploit areas where they do not breed during periods when the weather is poor at their breeding sites (Svardson, 1950; Udvardy, 1954). This ability is particularly important to species that feed totally on winged insects, prey that remain inactive during periods of rain, extremely high wind, or cold. Swifts are among the most rapid of fliers, and their trait of feeding relatively huge boluses of insects to their young a few times during a day (Lack, 1956) may in part be an adaptation permitting them to bring back food from great distances. Further-more, the small clutch size characteristic of these birds means that the food brought to the young will only be split a few ways. Short-term movements are an alternative to the tendency of some related species (such as hummingbirds) to enter short-term torpor at such times. In fact, swift broods are capable of entering torpor.

Sea birds also carry out movements in response to weather patterns (Ashmole, 1971), as do red deer (*Cervus elaphus*), who may move up and down mountainsides as much as 600 m in a day (Darling, 1937). These movements represent the climatic equivalent of a horizontal migration of several hundred kilometers.

In summary, although migration removes an individual from an area during a time at which it would be unable to survive, it is a drastic strategy. Not only are conditions uncertain in the place to which it is moving, but it must often winter in an area that differs considerably from its summering area. The need to respond to a wide variety of conditions may limit the degree of specialization of migratory species. Baker (1978) discusses many of these problems in further detail.

The responses that animals make to highly changeable conditions thus differ with their own basic attributes. A technique of survival open to one group may be unavailable to another. These attributes are, however, strongly affected by the environmental regime experienced. Long-term dormancy is a solution unavailable to birds, but their ability to move from a temporarily unfavorable area is unexcelled. In spite of the difficulties experienced with any of the strategies discussed, they permit exploitation of resources and colonization of areas that would otherwise be unavailable. The solutions to these problems are of importance in determining the biotic makeup of communities that are subject to great fluctuation.

Ontogeny of foraging behavior

The problem of mastering a feeding repertoire confronts every individual first starting to forage for itself. The development and perfection of techniques used in finding, capturing, and processing food items are not easily attained by all species and may take considerable time to perfect. In species such as the great tit, extremely high mortality apparently occurs immediately after the rather brief period of parental care (Lack, 1966). Davies and Green (1976) and Davies (1976a) noted progressive changes in the foraging efficiency of young reed warblers (*Acrocephalus scirpaceus*) and spotted flycatchers (*Muscicapa striata*) (fig. 2.8). In both cases a decrease in pecking at inedible objects and an increase in complexity of capture techniques occurred. Flycatchers also flew longer distances to capture prey as they grew older, suggesting a change in effective perceptual capabilities.

Difficulty of foraging technique correlates positively with the length of the period of parental care. Often difficulty of foraging is directly related to the size of the prey item or the force required to take it. Thus extended, parental care is often associated with high trophic levels, probably because these species most frequently feed on prey that are capable of defense or escape or that occur in low densities. Large carnivorous mammals such as lions may remain dependent for several years and often attain maximum proficiency only after considerable predatory experience (Bertram, 1975; Schaller, 1967, 1972). But since these species also take a long time to reach adult size (which

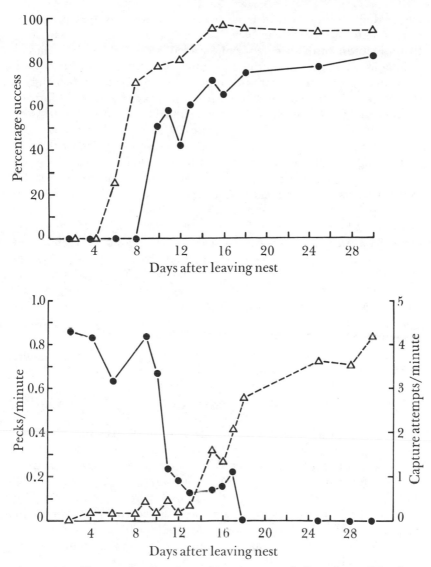

Figure 2.8 Changes in foraging of young spotted flycatchers (*Muscicapa striata*). *Left,* percentage success of capture attempts (dashed line) and percentage of capture attempts involving a sally of more than 1 m (solid line). *Right,* rate (pecks per minute) of pecking at inedible objects (solid line) and rate of attempting to capture live prey (per minute) (dashed line). (Modified from Davies, 1976a.)

affects their ability to take large prey) it is not easy to separate the factors determining proficiency. Fogden (1972) noted that insectivorous forest birds in equatorial Sarawak were fed by their parents as long as six to seven months, long after they reached adult size, and associated this period with the difficulties involved in learning to recognize and find the highly cryptic forms upon which they foraged.

Where it is possible to compare the hunting success of adults and full-grown juvenile birds, adults typically have greater success than their young. Because the birds have all reached adult size by the time of fledging, differences in success with particular prey must be attributed to experience or some other factor independent of size. For example, certain oystercatchers *(Haematopus ostralegus)*, which have a very specialized pattern for opening bivalves and crabs, do not approach maximum efficiency for many months, and in extreme cases adults feed the young for nearly a year (Norton-Griffiths, 1969). However, young reared on polychaetes may be fed by their parents no more than six or seven weeks. Royal terns *(Thalasseus maximus)*, which fish by diving from the air, may feed their young for six months (Ashmole and Tovar, 1968; Buckley and Buckley, 1974). Young brown pelicans *(Pelecanus occidentalis)* (Orians, 1969b) and little blue herons *(Florida caerulea)* (Recher and Recher, 1969) experience lower rates of success in capturing fish during their first winters than do adults. Difficult activities such as opening bivalves or capturing fish thus seem to require extensive practice.

Apparently juveniles in these species initially obtain enough food only by fishing more hours than their parents. This is known to be the case in sandwich terns *(Sterna sandvichensis)* (Dunn, 1972) and herring gulls *(Larus argentatus)* (Verbeek, 1977). The success of juveniles eventually seems to increase, although the apparent increase in their foraging efficiency might result solely from relatively high mortality rates among incompetent individuals (Orians, 1969b) or from the fact that adults dominate the best feeding sites. Relegation of juveniles to inferior sites might also be expected in species with territories, as in herons along the water's edge, but it seems unlikely where individuals feed in flight, as in terns.

Dunn (1972) noted that young sandwich terns feed in flocks with adults. By midwinter they are no longer taking food from their parents, but they may improve their feeding success by observing the techniques used by adults. Observational learning has frequently been demonstrated in foraging animals (Alcock, 1969; Krebs, MacRoberts, and Cullen, 1972).

Studies on individually marked animals are needed to separate these alternatives in most species. Species with distinctively colored age classes, such as the brown pelicans and little blue herons, are particularly suitable for this, but even in these species it is impossible to determine individual differences within classes. Where distinctions are possible, there are indications that differences in efficiency associated with age are widespread among forms feeding on difficult-to-capture prey (Cook, 1978; Groves, 1978; Searcy, 1978; Morrison, Slack, and Shanley, 1978).

In summary, the greater the difficulty of coping with a food item, the longer the period before proficiency is reached, and the greater the usual investment of parental care into the offspring. Many resources may initially seem poor choices for exploitation (bivalves rather than polychaetes in the case of oystercatchers) because of the considerable investment required to support offspring while they learn to feed on them. But the absence of strong competition for such resources may be a tremendous advantage to a persevering specialist that can learn to use them.

The origin of foraging adaptations

Maintenance of a broad range of foraging capabilities is naturally dependent on having access to a large number of inherited or learned behavioral patterns. Newly developed traits ultimately provide the basis for behavioral availability, but the rate of invention of new traits probably seldom limits a species' repertoire; the slow accumulation of new traits could easily provide the repertoire required to exploit most resources efficiently. Why does behavioral change eventually cease? This subject has two distinct aspects to it: the origin of new traits and the conditions under which they occur and are retained.

THE ORIGIN OF NEW FORAGING ADAPTATIONS

New foraging techniques might arise de novo, but it seems more likely that they would be based on existing behavioral patterns. Even techniques that appear to be totally new may represent capacities that were previously restrained by interspecific competition or by characteristics of the environment typically exploited.

Highly plastic individuals should develop new techniques of resource exploitation most frequently. Several interesting behavioral changes have been reported that may represent the first stages in the modification of overall patterns of resource exploitation. Regardless of their probability of eventual success, these behavioral changes warrant close study, since they provide insight into how new patterns of resource exploitation may have evolved in the past. First I shall provide examples that illustrate different origins. Usually it is un-

Figure 2.9 Novel foraging behavior by great tits. *Left,* opening milk bottle; *right,* pulling a string for a reward. (Drawings by Jaquin Schulz, based on illustrations in Fisher and Hinde, 1949, and Thorpe, 1963.)

clear how novel behavioral patterns have arisen. Fagen (1974) discussed the possibility that novel behavioral patterns (including foraging behavior) are a consequence of play by immature individuals. If true, this may provide some answers as to how new patterns develop, but it seems unlikely to account for all such cases, especially in animals not known to engage in play.

Existing behavioral patterns may be applied in new contexts. Tits (*Parus* spp.) have learned to open milk bottles and drink cream from

them over the past few decades (fig. 2.9). Subsequently this behavior spread rapidly throughout the British Isles (Fisher and Hinde, 1949). This trait bears close similarities to bark tearing, and since tits often forage on the bark of birch trees and use birch bark in their nests, bark tearing may have been a preadaptation that facilitated opening milk caps. Tits exhibit strong learning abilities in a variety of other situations (fig. 2.9); for instance, they can master several complex tasks associated with food finding (Smith and Dawkins, 1971; Krebs et al., 1972) and pull strings for rewards (Herter, 1940; Thorpe, 1943). Their high capacities for observational learning (Krebs et al., 1972) suggest how such behavior, once initially perfected, might be rapidly transmitted. Subsequent to the development of bottle opening by the tits, paper milk caps were replaced by aluminum ones, to which the tits readily adapted. Fisher and Hinde (1949) and Hinde and Fisher (1951) noted that several other species also open milk bottles occasionally, but whether they learned this habit by observing the tits is unknown. The opening of milk bottles fits all the criteria for the development of novel foraging behavior.

Changes in the distribution of oystercatchers associated with the use of new feeding techniques fit a similar pattern. In recent years oystercatchers have wintered in increasing numbers in the British Isles, a buildup accompanied by extensive foraging in grassy fields. It appears that they did not use these areas extensively before 1950, although Heppleston (1971) suggested that the habit may be somewhat older than this. Dare (1966) reported that the use of fields increased strikingly subsequent to the unusually cold period during the winter of 1962–63, when most of the intertidal invertebrates, which formed their traditional food source, died. The terrestrial foraging areas provide oystercatchers with additional food sources (earthworms, cranefly larvae), which allow them to survive during periods in which they may experience extreme difficulty in obtaining adequate food from their customary oceanside haunts. Hunting for earthworms and other soil invertebrates probably bears considerable resemblance to hunting for intertidal polychaetes. Polychaetes are taken by many individuals (Norton-Griffiths, 1967), but they are not an important item in all oystercatchers' diets (Heppleston, 1971; Dare and Mercer, 1973).

Although the oystercatcher has often been considered primarily a mollusk feeder that has extended its range of food to include a wide range of other items, Heppleston (1971) considers it to be a generalized feeder. If his view is correct, then this shift in feeding patterns supports the idea that most major changes in resource exploitation patterns involve species that already employ a broad range of resource exploitation techniques. Young oystercatchers learn feeding techniques directly from their parents (Norton-Griffiths, 1967, 1969), so they probably could have learned this trait from other individuals, even if they are no longer dependent on their parents by the time the fields are first frequented, in the winter. The populations discussed by Heppleston (1971) and Dare and Mercer (1973) are coastal, but other populations of oystercatchers have subsequently expanded their breeding ranges into agricultural lands considerable distances from the sea (Heppleston, 1972).

Hunting, meat-eating, and cannibalism have recently been reported in groups of wild chimpanzees (*Pan troglodytes*) (Kawabe, 1966; Suzuki, 1971; Bygott, 1972; Teleki, 1973). Bygott suggests that these are new behavioral patterns and that they must have arisen from certain elements of the aggressive repertoire, rather than from modifications of their typical herbivorous foraging techniques. He bases this proposal on the failure of chimpanzees to eat meat that they have not killed and the higher frequency of aggressive behavior in males than in females. The social content of the kill, described in detail by Teleki (1973), is consistent with this interpretation. However, the frequency of observations of chimpanzee hunting raises some question as to whether these behavioral patterns are in fact novel, or whether they represent normal but rarely used parts of their foraging repertoire. The detailed field observations made on chimpanzees over the past fifteen years may have simply revealed behavior not noted by earlier workers.

On the other hand, complicated hunting tactics associated with capturing large prey have been reported only three times in primates, twice in chimpanzees and once in baboons (Hamilton and Busse, 1978). These patterns may represent the expansion of an ancient primitive habit, insectivory. As noted above, the occasional capture of larger prey occurs in chimpanzees and baboons (Hamilton and Busse, 1978), and the development of more elaborate techniques of prey capture would be logical if large animals proved to be a profitable resource.

It is difficult to determine everything a foraging animal can do. Foraging repertoires may be restricted either by the presence of competing species (Morse, 1974) or by the variety of resources. For these reasons a titmouse would be unlikely to probe frequently in bark on tree trunks where large numbers of woodpeckers and nuthatches were present, and such behavior would also be uncommon on smooth-barked trees. Foraging patterns may sometimes appear in hand-reared individuals that seldom if ever are noted under natural circumstances. For instance, hand-reared Carolina wrens (*Thryothorus ludovicianus*) regularly hawk and glean for insects by hovering, behavior seldom if ever performed by this species in the field (J. H. Fellers, personal communication; E. S. Morton, personal communication).

The spread of foraging techniques within a population has been studied in great detail by primatologists examining the responses of free-ranging Japanese macaques (*Macaca fuscata*) to supplemental foods not encountered under natural circumstances (see Kawai, 1965). One young female learned to remove sand from sweet potatoes by washing them in water. Over a period of several years this habit spread through the troop. The same female later learned to separate kernels of wheat from sand by throwing both wheat and sand together into the water, a technique that also spread through the group. Peers and members of the family acquired these behavioral traits, but social superiors did not; this prevented the techniques from spreading even more rapidly. Beck (1976) found that in captive pig-tailed macaques (*M. nemastrina*) transmission of acquired behavior was not dependent on social status. Because the Japanese macaques are artificially provisioned, it is unlikely that their foraging patterns will spread, but this case does demonstrate that species may rapidly learn to exploit new resources.

CONDITIONS THAT FACILITATE NEW FORAGING ADAPTATIONS AND THEIR RETENTION

New foraging adaptations seem to occur most frequently where conditions have recently changed. During earlier times episodes of glaciation or crustal movement of the earth may have been major initiations of habitat change, but currently the most important force by far is human activity. The novel foraging patterns of great tits on milk bottles and oystercatchers in fields show dramatically how changed conditions can lead to an expansion of behavioral patterns. Where diversity is either artificially or actually low, novel foraging techniques may be retained for longer periods than under more competitive conditions.

Populations in areas of low species diversity such as an island enhanced opportunity to experiment with new types of behavior, simply because competition, even if intense, is exerted by fewer species than in high-diversity situations. These animals typically sample wider varieties of resources than their mainland counterparts. However, the restriction of such populations to islands reduces markedly the possibility that these behaviors will spread widely.

Galapagos or Darwin's finches (Geospizinae) provide a good illustration of novel foraging techniques in an island fauna. Because of the intense interest in these birds, information on their natural history probably exceeds that of most (or any) other island animals (Lack, 1947; Bowman, 1961).

Bowman and Billeb (1965) report that the sharp-beaked ground finch (*Geospiza difficilis*) on Wenman Island pecks at the blood-engorged pinfeathers of red-footed boobies (*Sula sula*) and masked boobies (*S. dactylactra*) attending their nests. This finch is the only known avian sanguivore! The boobies appear unable to repulse these finches effectively, although the loss in body reserves may be considerable, and serious problems of feather regeneration may occur. The boobies are known to experience severe energetic problems in rearing their young, so the additional stress provided by the finches could be significant (see Nelson, 1969). Bowman and Billeb hypothesize that this feeding behavior has arisen from the finches' habit of pecking at hippoboscid flies that attend the boobies. These flies are common ectoparasites that feed on bird blood. If perched on the boobies' white feathers, their dark coloration would present a conspicuous target for the finches.

Judging from the current widespread nature of the ground finches' parasitic feeding behavior on Wenman Island, the absence of similar reports from earlier dates, and the apparent absence of this behavior on other islands inhabited by this finch or any of its close relatives, it appears that feather picking is a new behavior, and one to which the hosts have not yet developed a response. Probably the next phase in this relationship will be for the boobies to commence resisting the attacks effectively, which will probably suffice to halt this interesting pattern.

DeBenedictis (1966) reports "bill-brace" feeding behavior for two other Darwin's finches, *Geospiza conirostris* and *G. fuliginosa*, on Hood Island. The birds are able to move stones by bracing their beaks or heads against immovable objects while pushing backward on the stone to be moved with both feet (fig. 2.10). *Geospiza conirostris*, which weighs bout 25 g, is able to move stones weighing over 350 g in this way. The birds then feed under the stones. This behavior has not been reported elsewhere.

Another unusual behavior is a cleaning symbiosis between Darwin's finches and reptiles. *Geospiza fuliginosa* (and to a lesser extent *G. fortis*) groom the Galapagos giant tortoise (*Geochelone elephantopus*) on at least three islands. The finches approach and sometimes even display to the tortoises. The tortoises in turn extend their legs and neck, which exposes ticks to the birds, who then pick them off (MacFarland and Reeder, 1974). *Geospiza fuliginosa*, *G. difficilis*, and *G. conirostris* remove ticks from marine iguanas (*Amblyrhynchus cristatus*) (Carpenter, 1966; Amadon, 1967).

Tool use is probably the best known novel foraging behavior of Darwin's finches (fig. 2.10). Tool use has been reported for several species of these finches (see Alcock, 1972), all of which use spines or twigs to dislodge prey from crevices. Interspecific observational learning may have followed a single development of this behavior, for there is some suggestion that tool use does not develop in finches reared in isolation from others (Millikan and Bowman, 1967). However, the reduced species diversity of the islands and the capabilities of Darwin's finches to perform a wide variety of other unusual behaviors leave open the alternative that the use of probes developed independently among these finches.

Additional evidence of the behavioral diversity to be found in island populations comes from unusual foraging patterns reported from other species on the Galapagos. The Galapagos mockingbirds (*Nesomimus* spp.) show a marked tendency toward egg eating and carnivory (Hatch, 1965; Harris, 1968; Bowman and Carter, 1971), habits not noted in mainland mockingbirds (*Mimus polyglottos*) (Bent, 1948). *Nesomimus macdonaldi* has even been observed peck-

Figure 2.10 Novel foraging behavior by Galapagos finches. *Left,* bill-brace feeding behavior of *Geospiza conirostris; right, Camarhynchus pallidus* using spine to dislodge prey from a crevice. (Drawings by Jaquin Schulz, based on an illustration in DeBenedictis, 1966, and a photograph by I. Eibl-Eibesfeldt.)

ing about the teeth of sea lions (*Zalophus californianus*) for food (Trimble, 1976).

The extent of novel foraging behavior in island populations is illustrated by the taste for meat, particularly the kidneys of sheep, displayed by the kea (*Nestor notabilis*), a parrot from New Zealand. Parrots seldom take animal matter under normal circumstances (Van Tyne and Berger, 1959; Dorst, in Thomson, 1964), although keas have been known to take insects and worms (Oliver, 1955). Captive parrots also readily take meat in some cases (Risdon, 1973), which suggests that they have the ability to digest such food. Attacks on sheep were not noted until several years after they had been introduced (Marriner, 1908). New Zealand is an area without large native mammals and their associated predators, although the moas probably once occupied an ecological niche similar to that now filled by sheep (see Storer, 1971; Morse, 1975a).

To test whether low diversity promotes novel foraging techniques I compared tool use in isolated areas with that in typical mainland situations. Because tool use is so conspicuous, the literature probably gives a fair impression of its relative frequency.

Recent reviews by van Lawick-Goodall (1970) and Alcock (1972) cite a total of twenty-seven instances of foraging-related tool use. Twelve of these (44 percent) occurred on islands or islandlike situations characterized by depauperate faunas. The proportion of tool-use

reports from isolated regions is particularly impressive both because relatively little field work has been done in such regions, and because the areas, numbers of species, and numbers of individuals are all small relative to those on mainlands. Most examples of tool use recorded from mainland areas (nine of fifteen) are in primates, and other studies indicate that several species of captive primates can readily learn to use tools (Beck, 1976). At least one avian example comes from a continental area of exceedingly low species diversity. It involves brown-headed nuthatches (*Sitta pusilla*) using flakes of longleaf pine bark as levers in knocking off other flakes, thus exposing insects hidden underneath (Morse, 1968a). Reports of such behavior appearing after van Lawick-Goodall (1970) and Alcock (1972) follow the same pattern. Tool use thus seems to bear out the predicted correlation between species diversity and frequency of novel foraging behavior, but other examples are needed to test the generality of this finding. Interestingly, tool use has not created major new foraging opportunities for any species other than our own, although it occurs regularly in some groups of animals. Tool use has recently been reported in other contexts, such as defense against predators (Hamilton, Buskirk, and Buskirk, 1975; Janes, 1976).

Variations in feeding adaptations within a population

When resources are scarce, the occurrence of different foraging techniques within a population may minimize competition. Whether significant variation exists should depend on the range of resources the population can exploit relative to that which any individual can exploit (see Levins, 1968), and on the profitability with which the different resources can be exploited. If making more effective use of one item requires making less efficient use of another, a basis for variability exists. Variability associated with resource exploitation may be either continuous (graded) or discontinuous (polymorphic).

CONTINUOUS VARIATION

Many anecdotal observations suggest that continuous differences in foraging are commonplace, but few studies have used known individuals. Thus it is not usually possible to eliminate other factors, such as differences in available food supplies and environmental conditions, as explanations of foraging differences. A good demonstration of intrapopulational differences in feeding is that of Bryan and Larkin (1972), who reported that brook trout (*Salvelinus fontinalis*), cutthroat trout (*Salmo clarki*), and rainbow trout (*Salmo gairdneri*) showed individual foraging differences under field conditions, and that these differences could not be attributed to size, growth rate, or area of capture. A field experiment on laboratory-reared rainbow trout, which controlled for absolute and relative abundance of food items, also demonstrated specialization (Bryan and Larkin, 1972).

Laboratory studies on wild-captured American crows (*Corvus brachyrhynchos*) demonstrate wide variation in feeding preferences in this highly omnivorous species (R. W. Powell, 1974). Members of free-ranging flocks of black-capped chickadees (*Parus atricapillus*) (Glase, 1973) and Carolina chickadees (*P. carolinensis*) (D. H. Morse, unpublished data) foraged differently from one another in ways influenced by age, sex, and position in a dominance hierarchy. Partridge (1976) believes that variation in foraging patterns of hand-reared great tits (*Parus major*) can be attributed to differences in experience and learning. Consistent differences often occur between individual New World warblers temporarily isolated on small islands (Morse, 1971b, 1973a), but it is impossible to rule out the possibility that these are caused by differences in available resources. Holmes, Sherry, and Bennett (1978) have found marked differences in the foraging of American redstarts (*Setophaga ruticilla*) holding adjacent territories. Although some of this variability may be attributed to differences in the territories or sexes, the within-sex variation was great enough to suggest that other factors were acting as well, as Partridge's data (1976) on great tits also suggest.

One might expect individuals with different-sized bills, claws, and other foraging implements to feed on quantitatively different foods. To date most speculation on this subject is based on measurements of beaks (Van Valen, 1965; Willson, 1969; Soulé and Stewart, 1970; Rothstein, 1973) or other feeding structures (Keast, 1965a, 1966; Schoener, 1967; Roughgarden, 1974), using preserved specimens for analysis. Behavior of the organisms themselves has been little studied in the field from this viewpoint (Morse, 1971a). One notable exception is a study on Galapagos finches, *Geospiza fortis* and *G. scandens,* by Grant et al. (1976), who found marked within-population variation in the foods selected and habitat patches frequented by these birds. Choices varied with the size of an individual's bill, consistent with their prediction. *Geospiza fortis* may have the greatest relative bill variation of any passerine bird (Grant et al., 1976); it remains to be seen whether less variable species will demonstrate comparable foraging differences.

DISCONTINUOUS VARIATION

Differences in foraging between individuals with different-sized bodies or feeding structures have been reported more frequently than individual foraging differences in structurally monomorphic populations, probably because the former group may be readily distinguished in the field. However, with rare exception foraging differences have been correlated with only two types of polymorphism, sexual dimorphism and age (or size) classes. The opportunity for foraging polymorphisms to arise independently of unrelated selective pressures seems low. Sexual selection and the associated separation of sexual roles in rearing young may produce discontinuous morphological variation, which may subsequently promote foraging differences. Foraging differences associated with sexual roles might also occur in the absence of significant morphological differences. Where indeterminate growth occurs, different-sized individuals often exploit different foods.

One foraging polymorphism that is not sex or size related has been reported under natural conditions. Norton-Griffiths (1969) found that European oystercatchers within a single population open mussels either by stabbing or by hammering, but not with both techniques. This behavior is learned from the parents and appears to be highly resistant to change. Stabbing is used to feed on gaping mussels (under water), and hammering is used to feed on closed mussels (often above the tide line). Thus, birds of the two types are spatially separated to some degree. Hartwick (1976) looked for similar differences in the closely related black oystercatcher (*Haematopus bachmani*) on the west coast of North America, but found only stabbers. It is thus possible that the pattern described by Norton-Griffiths is a local phenomenon.

Selander (1966) discusses the relationship between morphological differences and sexual foraging differences in birds, the factors associated with the extents of such differences, and their basis. Two congeneric woodpeckers occurring on the mainland of North America and Hispaniola illustrate his arguments well (fig. 2.11). The Hispaniolan woodpecker (*Centurus striatus*) is the only woodpecker present where it was studied, but several other woodpeckers coexist on the North American mainland in the range of the congeneric golden-fronted woodpecker (*C. aurifrons*). Male Hispaniolan woodpeckers have beaks over 21 percent longer and tongues over 34 percent longer than those of the females. The males are primarily probers; the females, gleaners. This difference constitutes a type of ecological release, largely by the females, into parts of the habitat characteristically occupied by other woodpeckers on the mainland. In contrast, male golden-fronted woodpeckers have beaks and tongues 9 percent longer than those of the females. Both male and female golden-fronted woodpeckers are primarily probers, and the mainland population as a whole occupies a narrower foraging niche than does the insular one (Selander, 1966).

Feeding differences related to sex occur in boat-tailed grackles (*Cassidix major*), in which males are strikingly larger than females. The large size and large tail of the male make him unable to feed by hovering over the water and dipping to pick up fishes or floating matter, a technique frequently used by the smaller female (Selander, 1966). Many birds of prey are strikingly dimorphic, with the female generally larger than the male. Differences in size of males and females are closely related to the sizes of prey they capture (Hoglund, 1964; Storer, 1966; van Beusekon, 1972; Brosset, 1973). Sexual dimorphism in foraging has also been reported in carrion crows (*Corvus corone*) (Holyoak, 1969) and European goldfinches (*Carduelis carduelis*) (Newton, 1967). Perhaps the most remarkable adaptation of this sort was that of the huia (*Neomorpha acutirostris*), an extinct bird from New Zealand. The male huia had a bill that was roughly woodpeckerlike, while the female had a long decurved bill. Males apparently dug in dead wood and females probed into crevices in search of insect food (Buller, 1888).

Differences in body size may be reinforced by social relationships among the individuals. For example, the foraging dimorphism between male and female Puerto Rican emerald hummingbirds (*Chlo-*

rostilbon maugaeus) is related to the behavioral dominance of the males (C. B. Kepler, personal communication). This factor also accentuates foraging differences in some woodpeckers (Ligon, 1968a; Hogstad, 1978a) and probably in hawks (see Storer, 1966; Mills, 1976).

Although sexual dimorphism in foraging has been reported most frequently in birds, it is not unique to them. Schoener (1967) has demonstrated differences in both food size and feeding site among anoles (*Anolis conspersus*) on Grand Cayman Island, West Indies (figs. 2.12, 2.13). In general, males take larger prey than females and also use larger and higher perches than females. These differences are predictable on the basis of differences in their sizes. The males' heads are disproportionately larger than those of the females, and this facilitates the taking of larger prey. Schoener (1974) cities several other cases where sexual dimorphism in diet correlates with body size. Bartholomew (1970) hypothesized that highly sexually dimorphic male and female pinnipeds secure quite different food, largely as a result of their strikingly disparate body sizes arising from intense sexual selection. Thus, where marked differences in foraging structures occur, they are often accompanied by somewhat different foraging habits, even if these differences are direct consequences of body size.

In certain cases sexual differences in foraging apparently do not occur, although morphological differences are present. Among woodpeckers of the genus *Dendrocopos* (Kilham, 1965; Ligon, 1968a, 1968b, 1973; Jackson, 1970; Kiesel, 1972; Morse, 1972; Austin, 1976), sexual foraging differences occur in some cases but not in others. Even within a single species, some populations show sexual differences in foraging while others do not (Ligon, 1968a, 1973; Koch, Courchesne, and Collins, 1970; Morse, 1972). These differences may depend on the number of other woodpeckers present (Morse, 1972).

Jehl (1970) argued that bill differences between male and female stilt sandpipers (*Micropalama himantopus*) and least sandpipers (*Calidris minutilla*) were not associated with foraging differences, because where he studied them on their breeding grounds food was abundant. Instead, he believed that these size differences were associated solely with pair formation. However, selection should still favor those individuals that take the most profitable items, which will often differ with the size of the implement used to gather the items. Also, food limitation for shorebirds may take place at other times of the year (Baker and Baker, 1973).

In other cases, marked sexual differences in foraging occur even in

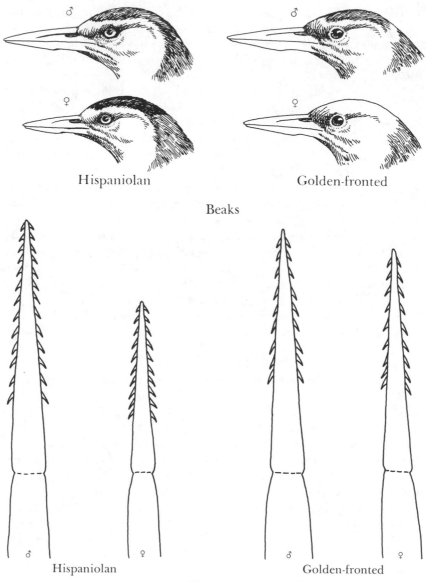

Beaks

Tongues

Figure 2.11 Beaks, tongues, and foraging patterns (*facing page*) of Hispaniolan (*Centurus striatus*) and golden-fronted (*C. aurifrons*) woodpeckers, illustrating the sexual differences in these parameters in the two species. White bars = males; hatched bars = females. (Heads drawn by Jaquin Schulz, based on line drawings in Selander, 1966; tongues redrawn from Selander, 1966; foraging motions modified from Selander, 1966.)

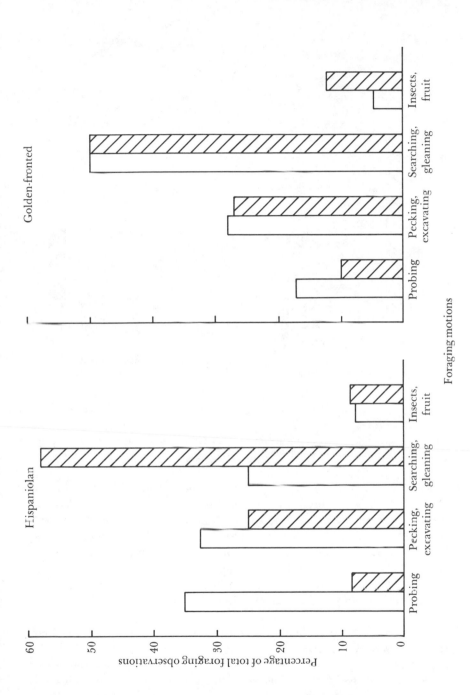

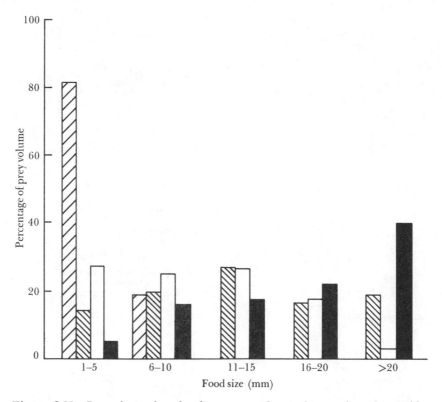

Figure 2.12 Prey sizes taken by four age and sex classes of anoles. Wide-striped bars = juveniles; narrow-striped bars = subadult males; white bars = adult females; black bars = adult males. (Modified from Schoener, 1967, copyright 1967 by the American Association for the Advancement of Science.)

the absence of strong morphological variation. Morse (1968b) found such differences in the foraging of male and female New World warblers (*Dendroica* spp.) nesting in spruce forests of northeastern North America. These birds live at high densities, and only the females incubate. Males spend much of their time singing in the tree-tops, an activity most effectively performed where they are most conspicuous, while the nest-tending activities conducted by the females are carried on lower in the trees. Males usually forage higher in the trees than do females (fig. 2.14), and part of the time they are displaying simultaneously. Females spend about 85 percent of their time incubating, and feed intermittently in the lower parts of the trees at an extremely rapid rate. By foraging away from the nest, the male spares the female's food supply near the site of incubation. Williamson (1971) has noted that several species of vireos (*Vireo* spp.) nest-

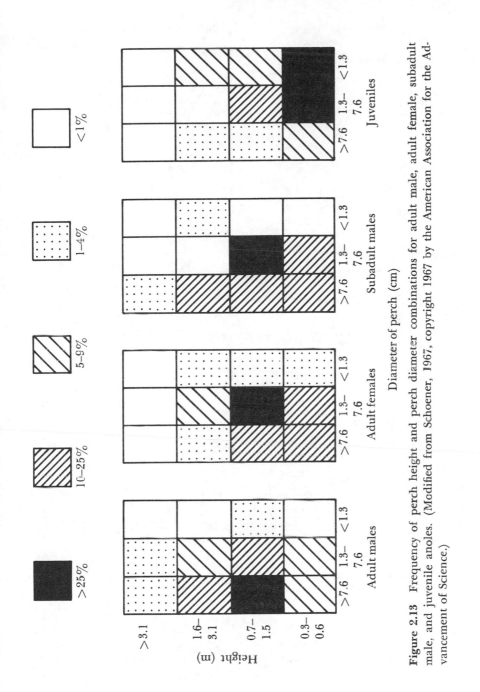

Figure 2.13 Frequency of perch height and perch diameter combinations for adult male, adult female, subadult male, and juvenile anoles. (Modified from Schoener, 1967, copyright 1967 by the American Association for the Advancement of Science.)

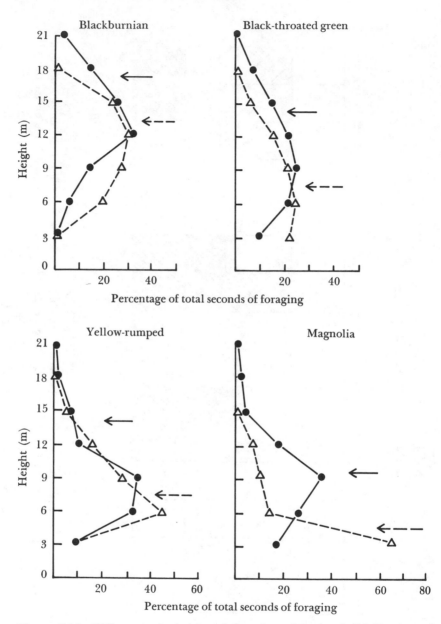

Figure 2.14 Differences in height of foraging of male (solid lines) and female (dashed lines) spruce-woods warblers. Solid arrows = mean singing heights of males, dashed arrows = mean nest heights. (Modified from Morse, 1968b, copyright 1968 by the Ecological Society of America.)

ing in the deciduous forests of the eastern United States partition their foraging in a similar way.

In a few instances the sexes are known to partition their feeding areas horizontally. Robins (1971) found that male and female Henslow's sparrows (*Passerherbulus henslowii*) fed in different parts of their territory (fig. 2.15), although no distinguishable differences existed between the parts. The advantage of this scheme may be similar to that described for warblers and vireos. Similar results have been reported for European wrens (*Troglodytes troglodytes*) (Harrisson and Buchan, 1934; Armstrong, 1955) and Bewick's wrens (*Thryomanes bewickii*) (Miller, 1941). Perhaps such differences are widespread among basically monomorphic species. Wyrwoll (1977) found that male goshawks (*Accipiter gentilis*) tend to hunt farther from the nest than females. Although females take food that is on the average larger than that taken by the males, considerable overlap between the two sexes occurs in food items taken (Storer, 1966).

In addition to the obvious examples associated with complete metamorphosis (which I do not discuss here), it has been clearly documented that different size classes in a population take different foods—for example, several fishes (Keast, 1965a, 1966), *Plethodon* salamanders (Fraser, 1976), and *Anolis* lizards (Schoener, 1967). Most of this information has come from examining the stomach contents of different-sized animals. However, Schoener (1967) has described the foraging sites of adult and juvenile anoles on Grand Cayman Island, West Indies. Adults occupy larger and higher perches in trees and secure larger prey than juveniles (figs. 2.12, 2.13). The microhabitats frequented might account for the differences in prey taken, although later studies (Schoener, 1968a; Andrews, 1971) suggested that differences in location do not completely account for the observed partitioning. For example, large males take some prey items too big for the juveniles.

Differences resulting from age are probably widespread, although they have seldom been documented. They presumably are correlated with the prey items that can be most readily exploited by foragers of different sizes, and with how wide a range of items can be manipulated. This factor has important implications for the range of food items that a population may exploit and also for the potential coexistence and diversity of species. Indeterminate growth is the rule in many major taxa of animals, with birds being a particularly striking exception. Thus, the great size range spanned by the members of a population with indeterminate growth may be of major significance in establishing the composition of communities, specifically in the competitive relationships between species and the consequent diversity of species.

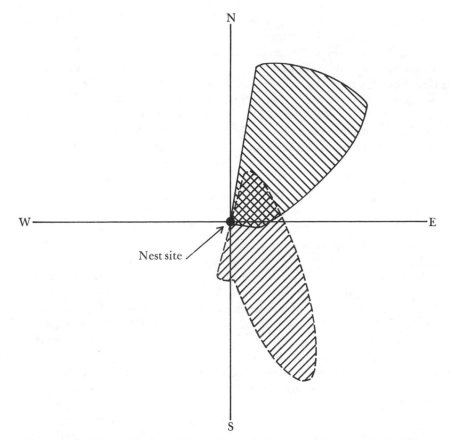

Figure 2.15 Directions in which male (solid line) and female (dashed line) Henslow's sparrows flew from a nest in order to forage, expressed as percentage of total observations for each bird. (Modified from Robins, 1971, copyright 1971 by the Ecological Society of America.)

Individual variation in resource exploitation patterns is often high, which has some interesting consequences. Members of a food-limited population who collectively exploit a broad resource base lower competition in the process. This leads to a higher population density than could otherwise be reached. These factors may facilitate the establishment and retention of populations under marginally favorable conditions. Somewhat larger populations have a greater probability of survival, simply because of the greater number of individuals supported. However, if each individual confined itself to a few resources, a catastrophe that decreased the abundance of all resources would have a more severe effect than it would have on a group of individuals each exploiting a large number of resources. In the former instance a single resource or a few resources might be inadequate to permit the survival of any individuals, whereas in the latter, individuals might survive by feeding on several of the resources that the specialists treated as discrete.

Assuming that morphological differences improve the effectiveness with which individuals exploit different resources, much of the individual variation found in resource exploitation patterns is predictable. But the absence of morphological differences does not preclude differences in exploitation. For example, temporally variable conditions may select for behavioral foraging differences, but not morphological differences.

Synthesis

Given the diversity of food items that exist, it is not surprising that a wide range of feeding patterns occurs. However, these patterns represent variations on only a few themes. This is clearly seen in the case of discontinuous variation in foraging among animals with determinate growth patterns. The variation is almost inevitably associated with sexual dimorphism. The basis for this variation is a seemingly unrelated set of selective pressures, although the sexual differences could not develop unless ecological conditions permitted them.

Species living under changeable conditions may enjoy the highest probabilities of exploiting new opportunities, and it is possible that as a consequence they have the greatest evolutionary potential. This may occur in spite of (or because of) the fact that adaptations to their native areas may not attain the precision of those in species experiencing less variable regimes. Heavy reliance on learned behavioral patterns will improve the capability of animals to respond to variable conditions. Although learned patterns are well developed in only a few major taxonomic lines, they may be a major factor in determining the success of these groups.

It also appears that ways of dealing with environmental fluctuations are severely limited by phylogeny. For example, with a few minor exceptions, birds cannot wait out ebbs in resource availability by becoming dormant, and few mammals are capable of making long and rapid migrations out of temporarily unsatisfactory areas. Bats are the great exception to this rule, since many of them in temperate areas both hibernate and migrate. Nonetheless, they have not become a superior adaptive type able to exclude birds or other mammals (Morse, 1975a).

At the same time, it appears that the potential for variability in foraging exceeds the ability to retain it. Novel foraging patterns regularly occur in areas of low species diversity; either they arise there relatively more often than they do in areas of high diversity, or they are allowed to express themselves there for a longer period of time.

3 Food Selection

EACH OF THE MANY FOODS potentially available to an animal has a different nutritional value, a different pattern of spacing and abundance, and a different cost of capture and processing. In addition, each food exposes the animal to a different level of competition with other consumers and to a different risk of predation. Because the animal has only limited amounts of time and energy, its choices among different potential foods may critically affect its survival and reproductive success.

It is necessary to establish that efficiency in food gathering really matters. Some animals (particularly small endotherms) spend most of their waking hours foraging, at least during some periods in their lives. This preoccupation is illustrated in Gibb's study (1960) of tits (*Parus* spp.) and goldcrests (*Regulus regulus*) in England. During the winter these small, insectivorous, forest-dwelling birds spend most of the day foraging, and there is an inverse correlation between body size and the amount of time spent foraging (fig. 3.1). Goldcrests (the smallest species at about 6 g) forage almost continuously at certain periods in midwinter, leaving almost no time for other activities (such as grooming) that are necessary over a long period (see Kendeigh, 1970). Although Gibb demonstrated that some food is always present, it may become so difficult to obtain in the time available that individuals cannot maintain a neutral energy balance and are thus effectively food-limited. Winter is the season of coldest temperatures (leading to high energy costs), shortest day length (these

51

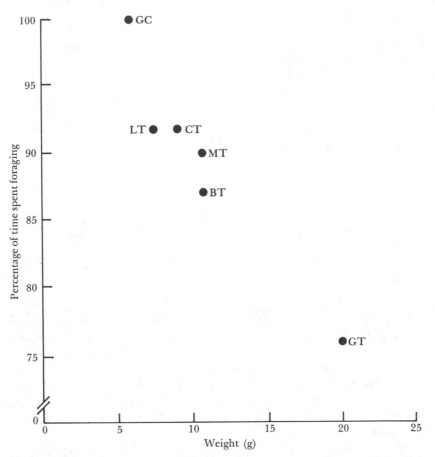

Figure 3.1 Percentage of daylight hours spent foraging by small English birds during cold weather in midwinter. GC = goldcrest (*Regulus regulus*), LT = long-tailed tit (*Aegithalos caudatus*), CT = coal tit (*Parus ater*), MT = marsh tit (*Parus palustris*), BT = blue tit (*Parus caeruleus*), GT = great tit (*Parus major*). (Modified from Gibb, 1954, with permission of the *Ibis*, journal of the British Ornithologists' Union.)

birds are diurnal), and the lowest rate of renewal of insect food supplies. Goldcrests are challenged most severely because of their high ratio of surface area to volume. At the northern edge of their winter range in Fennoscandia they suffer high mortality, with up to 90 percent of the residents dying during a winter in Finland (Österlof, 1966) and 92 percent in Norway (Hogstad, 1967). Inability to find adequate food in the time available appears to account directly for this catastrophic mortality.

Unfortunately, there have been only a few attempts to manipulate the food supplies of winter birds or other animals under natural conditions in order to test for food limitation, and the results of these studies have been equivocal (Krebs, 1971; Haartman, 1973). Several factors may confound simple experiments of this sort, and because it is known that food densities and environmental conditions may both differ markedly over time at a single location, more elaborate tests are needed to resolve the question. But the difficulty of demonstrating food-dependent population limitation is no reason to doubt its importance (see Diamond, 1978). Arguments on the importance of foraging efficiency need not rest solely on the question whether food directly limits population sizes. Even if food is not effectively limiting, selection should favor efficient exploitation, because the most efficient foragers will have the most time for other activities.

Rules for choosing foods

Optimal foraging theory attempts to predict the foraging patterns of animals on the basis of net energy gain per unit of time. Schoener (1969a) recognized three extreme types of foragers depending on the relative importance of searching and pursuit in their foraging repertoires: (1) those that expend neither time nor energy in food search alone and that pursue their prey (sit-and-wait predators such as *Anolis* lizards), (2) those that expend time and energy searching for prey but none in pursuit (foliage-gleaning insectivorous birds), and (3) those that expend time and energy in both search and pursuit (soaring hawks). Schoener then (1969b) recognized energy maximization and time minimization as alternative ways to maximize foraging efficiency. Energy maximizers seek to obtain the greatest amount of energy possible within a given period of time. Time minimizers attempt to minimize the amount of time required to obtain a given amount of food and thus have more time for other activities, such as predator avoidance or parental care. Nevertheless, the foraging patterns of animals using either alternative may be similar under many conditions. Schoener's categories suggest a variety of basic solutions to the problems of gathering and processing food, and each solution requires different patterns of behavior.

The role of several variables directly associated with foraging has been explored using optimal foraging theory. These variables include the type, distribution, and abundance of food; pursuit time; handling time; and feeding time. Emphasis has been placed on finding the smallest sets of variables that will give good fits to observations. Optimal foraging theory has been reviewed in detail by Pyke, Pulliam, and Charnov (1977). I will not discuss the subject exhaustively, but will outline its main aspects. Pyke and coworkers recognize four distinct categories of optimal foraging theory: optimal diet, optimal patch choice, optimal allocation of time to different patches, and optimal rate of movement.

Optimal diet

Most of the work on optimal foraging theory deals with optimal diet. Optimal diet is determined in the following way (nine authors, as summarized in Pyke et al., 1977) :

1. Each food is ranked by its ratio of food value to handling time.
2. Search time is incorporated, permitting calculation of the net rate of food gain for a diet. Search time is generally an inverse function of encounter rate.
3. The optimal diet is determined by adding food items in rank order until an addition to the diet ceases to exceed the net rate of food intake for a diet without this addition (fig. 3.2).

Three testable predictions result:

1. Whether a food is eaten depends not on its own abundance but on the abundance of higher-ranking items.
2. As a high-ranked food increases in abundance, lower-ranking foods will be forsaken in inverse order of ranking; the forager will specialize more and more.
3. Any given food will be either used or rejected; partial preferences will not occur.

Although it has often been observed that an animal's range of diet is apt to expand during food shortages (Ivlev, 1961; Beukema, 1968), few explicit tests of optimal foraging theory have yet appeared. Experiments conducted in the laboratory support this theory in part. Werner and Hall's studies (1974) on diet choice in bluegill sunfish (*Lepomis macrochirus*) are the most realistic of these. They tested the choices of sunfish exposed to three or four sizes of *Daphnia magna* in different densities and demonstrated that the fish became more selective as food abundance was increased, progressively dropping low-profit prey sizes from their diet (fig. 3.3). However, these studies were done in relatively small pools that probably provided only a rough equivalent to natural environmental conditions and lacked the normal range of competitors and predators. O'Brien, Slade, and Vinyard (1976) later pointed out that these fish might be responding to perceptual restraints. Under low-density conditions the apparent size of large prey would usually be small, because the closest individuals

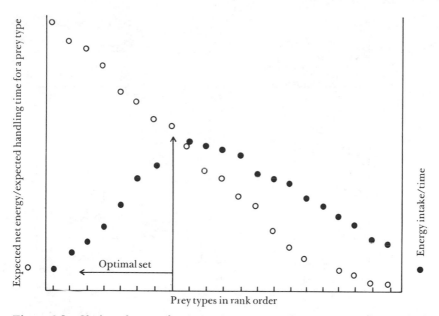

Figure 3.2 Choice of an optimal set of prey types. Prey types are first ranked by the ratio of expected net energy to expected handling time (1), and then the cumulative energy intake per unit of time (2) is calculated by adding prey types in rank order until (2) exceeds (1). (Modified from Charnov, 1976a, copyright 1976 by The University of Chicago.)

would on average be farther from the fish than at high prey density.

Krebs et al. (1977) tested the ability of hand-reared great tits to respond to changing concentrations of mealworms (*Tenebrio mollitor*) of two sizes (eight and four segments) provided for them on a conveyor belt. They reported a good fit to predictions 1 and 2, but not to prediction 3. The birds displayed partial preferences for the four-segment items, rather than rejecting them when their value dropped below the critical level. Laboratory experiments on foraging by great tits (Smith and Dawkins, 1971; Krebs, MacRoberts, and Cullen, 1972) demonstrated their ability to learn quickly to exploit a wide range of unnatural feeding situations, and thus it would be surprising if these birds did not conform to the predictions. But under natural circumstances the tits are exposed to a variety of potential competitors (Hartley, 1953; Gibb, 1954) and must be constantly on the watch for predators (Morse, 1973b; Geer, 1978). Although inadequately studied, their insect food supply is probably patchy and subject to continual change over time. Thus, the experiments reveal an ability on the part of these birds to forage in a way that will optimize their rate of food

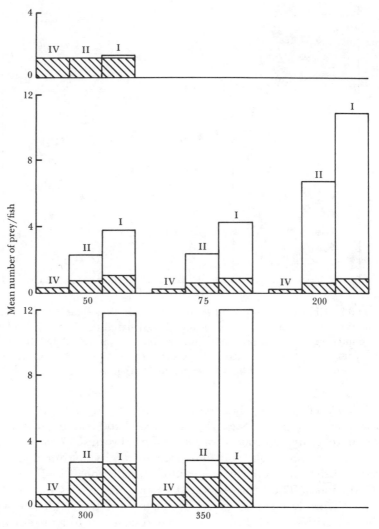

Figure 3.3 The mean number of each size class of daphnia eaten per blue-gill sunfish. The histogram in the top panel is the mean of eight experiments run at low density. The three histograms in the middle panel represent (from left to right) experiments at 50, 75, and 200 prey per size class. The lower panel similarly depicts the experiments at 300 and 350 prey per size class. The hatched areas superimposed on the histograms represent the number of prey expected in the stomachs of the fish if the items were eaten as encountered. Roman numerals refer to size classes of daphnia: Mean length of I = 3.6 mm, II = 2.5 mm, IV = 1.4 mm. (Modified from Werner and Hall, 1974, copyright 1974 by the Ecological Society of America.)

intake under extremely simple conditions but reveal little about the importance of such behavior under natural conditions, as noted by Krebs and coworkers.

Goss-Custard (1977a) demonstrated that in the field the sandpiper (*Tringa totanus*) prefers large polychaete worms to small ones; this appears to maximize the energy intake. However, Goss-Custard (1977b) also found that when amphipods (*Corophium volutator*) were present the birds preferred them to the polychaetes, despite the fact that the amphipods provide a smaller energetic reward than the polychaetes and thus should receive low priority. Goss-Custard (1977a) suggested that amphipods are easier to detect than polychaetes, but this explanation demonstrates the difficulty of testing simple models in complex field situations.

Two other recent field studies have found departures from predicted optimal diets. Kushlan (1978) found that great egrets (*Casmerodius albus*) regularly rob prey (mostly fish) from other wading birds, although this behavior appears to lower their capture rate by 10 percent. Morse (1979) reported that crab spiders (*Misumena calycina*) would obtain 7 percent more food if they specialized on capturing bumble bees visiting the flowers that they hunt, instead of attempting to capture any prey item that approaches them. In both cases prey availability changed rapidly over time, suggesting that the observed foraging patterns are compromises. Such patterns might optimize intake under poor rather than ideal conditions, as proposed by Thompson, Vertinsky, and Krebs (1974) on the basis of simulation studies of the foraging of bird flocks; if so, the pattern could be viewed as a long-term optimization (see Pyke et al., 1977). It is of interest that both Kushlan and Morse reported efficiencies approximately 10 percent below the predicted figure. It will be of interest to see if future studies produce similar figures.

To date the most impressive field study that generally supports predictions of the optimal diet model is that of Zach (1979). In both observational and experimental work, he found that northwestern crows (*Corvus caurinus*) were highly selective in their choice of the shelled prey that they would drop from the air in order to break. Crows dropped only large whelks (*Thais lamellosa*) in the area where Zach worked in British Columbia. Large whelks provided more calories than smaller ones and were more likely to break when dropped. The crows did not drop mussels, although they were eaten if found opened on the beach. Neither small whelks nor mussels in the study area were apparently large enough to provide an energetic gain when subjected to the costly dropping technique. Crows dropped their prey exclusively on rocks, the substrate most likely to break the shells.

Once they selected a whelk, they usually continued to drop it until its contents could be extracted. Zach found that the probability of a shell breaking remained constant from one drop to the next; therefore, abandoning a whelk after unsuccessful drops would merely necessitate additional searching for a substitute with the same qualities. Crows tended to fly up to the minimum height required for a shell to break when dropped, and they also typically released the shell only after they had passed the point of maximum ascent. Zach suggested that they could not clearly watch where the shell fell unless they assumed this position. If they dropped a shell from greater heights they would expend considerably more energy and also be farther away from the shell, perhaps losing a higher percentage of shells in this way.

Crows and their relatives are highly intelligent, and learned behavior probably plays an important role in their foraging patterns (Zach, 1979). In contrast, shell dropping in gulls appears to be much less flexible, and gulls show little discrimination in the substrate on which they drop their prey (Tinbergen, 1953a; Zach, 1978). How many animals are capable of foraging patterns as complex as those exhibited by Zach's crows? Based on this study one might predict that optimal foraging patterns involving complex decision making would be confined to animals possessing substantial learning abilities; such patterns should be widespread among mammals and certain groups of birds but poorly developed in many other groups of predatory animals. The existing laboratory tests do not answer this question, because they involve simple choices and a narrow taxonomic range of subjects.

It should be noted that shell-dropping behavior is only one aspect of the northwestern crow's broad spectrum of resource exploitation, which includes other invertebrates, berries, and the eggs and young of other species of birds. At times crows do not drop whelks even when they are readily available; at other times they are heavily exploited. Presumably this reflects changes in the profitabilities of the different items, but the observations required to demonstrate this would be difficult to obtain (Zach, 1979).

Ranking individual items may be inappropriate if a discrete unit (one food item) is not selected in each foraging movement (Stenseth and Hansson, 1979). This factor may not have been a problem in most studies considered up to this point, because the predators used (insectivores, for example) typically take one prey item at a time. However, if items are taken in the aggregate, as herbivores often do (for example, when they take a mouthful of grass), the aggregate net energy gained over a time period may be the relevant measure, rather than the instantaneous value of an item. Foods taken in the aggregate might be included in an optimal set if they were very abundant, thus providing greater energetic rewards than rarer, but individually more desirable, items. The actual rank may thus be in fact a function of abundance.

A few attempts have been made to use optimal foraging theory to predict differences in the dietary choices of coexisting species. Gill and Wolf (1975a) were able to predict whether three species of African sunbirds (*Nectarinia* spp.) would select open or closed flowers of a mistletoe (*Pharagmanthera dshallensis*). Although closed flowers offer larger rewards than open ones, they must be opened by the birds. In addition, closed flowers are less clumped than open ones, which should further decrease their rate of use. Thus, sunbirds should not specialize on closed flowers below a certain density, although if encountered they should always be used. The three species of sunbirds changed from specializing on clumps with large numbers of closed flowers to visiting flowers nonpreferentially at different relative frequencies of unopened flowers. These levels of change correlated well with the different energetic costs incurred by the three species in exploiting the two types of flowers. The problem of finding closed flowers appears to prevent any of these sunbirds from specializing on them.

Optimal patch choice

Optimal patch choice differs from optimal diet if foragers have the ability to recognize patches and to accept them or reject them prior to entering them. Patches are concentrated food sources separated by areas of lower density. Smith and Dawkins (1971) and Smith and Sweatman (1974) provided tits in the laboratory with patches that differed in density of mealworms. The feeding sites were located close to one another, such that according to optimal foraging theory the birds should completely exploit the richest site before moving on. The tits spent the most time in the richest areas, but they continued to exploit all the sites at low frequencies. Thus, they exhibited a graded response rather than an all-or-nothing response (fig. 3.4). Smith and Sweatman (1974) suggested that this type of foraging behavior may permit them to sample the environment continually in order to track changing patterns of resource abundance. This behavior is consistent with the foraging patterns of these birds in the winter, when they are constantly on the move.

Constant movement may also serve to patrol an area so that it is not encroached by other individuals. If an individual has exclusive use of an area or resource (for instance, a territory or a food), it could be of benefit to follow a long-term policy of optimization, perhaps even maximizing the condition of the food supply when it is at its

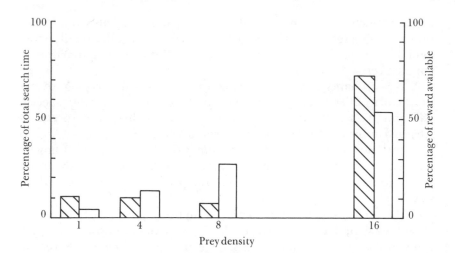

Figure 3.4 Distribution of hunting effort by great tits in the laboratory for patches of varying densities of mealworms. Hatched bars indicate the observed distribution of searching; white bars indicate the proportional distribution of food in the four food areas. (Modified from Smith and Dawkins, 1971.)

lowest point, rather than at less critical times (see Thompson, Vertinsky, and Krebs, 1974).

Not all food items in a patch are likely to be equal in value or ease of harvest. In that sense the results of Smith and coworkers' studies on tits may not reflect the conditions normally experienced by these birds. Tits would probably never totally exploit a patch in their natural environment. Their graded response to food items in different patches might have resulted from the different profitability of food items within the patches (including differing abilities to find those items), rather than any effort on the part of the birds to save some of the prey.

OPTIMAL ALLOCATION OF TIME TO PATCHES

When it becomes unprofitable to continue to exploit a patch, an individual should move on to another one. The time of abandonment should depend on the relative richness of the patch being used and the difficulty of getting to the next patch. Since the forager depletes the prey while exploiting a patch, capture rate should be a decreasing function of the time spent in the patch and thus might be used to decide when to leave it; an individual forager should leave a patch when its rate of uptake falls below a certain level. Because of the relative ease of finding another good patch in rich environments, the "giving-up" time should come sooner there than in poor environments (Krebs, Ryan, and Charnov, 1974; Charnov, 1976b).

In laboratory studies Krebs et al. (1974) found that black-capped chickadees (*Parus atricapillus*) did not spend a constant amount of time or gather a constant number of prey in a patch before leaving. However, the chickadees did exhibit a constant interval between their last catch and the point of leaving a patch (constant giving-up time) within an environment. This giving-up time was inversely related to the average capture rate for the environment. Using a simplified laboratory environment, Cowie (1977) verified a quantitative prediction about the time at which great tits would give up their foraging at one site to search for another.

Under more realistic conditions (large field arenas), Zach and Falls (1976) demonstrated that ovenbirds (*Seiurus aurocapillus*) took food from low-quality patches at a greater frequency than predicted by optimal foraging models. They hypothesized that the birds may have concentrated their activities on the most conspicuous prey, including those in low-quality patches, and in a sense were not foraging in a suboptimal way at all. However, Zach and Falls' results illustrate the difficulty of producing realistic models for animals that normally feed on small, cryptic insects. To correct this difficulty it would be necessary to include a factor that incorporated the predators' perceptual capabilities, a problem also raised by Elner and Hughes (1978) in their studies on dietary choices of green crabs (*Carcinus maenas*).

Some species may not even have flexible giving-up times. Krebs (1978), for instance, suggested that *Nemeritis canescens*, a wasp which parasitizes moth larvae, is unable to change its threshold in response to differences in the quality of hunting areas.

OPTIMAL PATTERNS OF MOVEMENT

The searching pattern of a hunting predator typically changes once it has found a prey item. In a uniform, unbounded environment the pattern shifts from meandering to one that tends to keep the animal in the vicinity of the food resource; this often means changing from straight or alternate left-right movements to a preponderance of either right or left turns (see Pyke, 1978). With adequate knowledge of the resources available in an area it should be possible to predict the optimal foraging pattern of individuals in response to patches of resources of varying characteristics.

Cody (1971) suggested that the optimal foraging pattern is one in which path crossing is minimized and reported that finch flocks move in this manner. He assumed that after hitting a boundary the birds would move backward (reflection). If the boundaries are physical (a canyon wall, for example) rather than biological (another group of finches), this pattern might be closely approximated. But, if the boundary is biologically determined, the result might be affected by whether the opposing individuals were present. Pyke (1978) argued that the assumption of a reflecting boundary is not realistic for many animals. He found that broad-tailed hummingbirds (*Selasphorus platycercus*) make more right and left turns on reaching a boundary than would be predicted by a reflecting boundary hypothesis. It seems likely that Pyke's hummingbirds were not *foraging* in an optimal fashion, at least not on a short-term basis. On reaching the boundaries of biologically determined space, many species show a tendency to move right or left rather than backward—for example, yellow warbler (*Dendroica petechia*) (Morse, 1966) and mixed-species foraging flocks (Morse, 1976b) —a behavior commonly associated with boundary patrol.

The utility of employing patterns of movement such as these should be a function of whether food resources are constant, declining as a result of exploitation, or replenishing themselves before a second visit is made. In cases where they are not self-replenishing, the predicted pattern of movement is of questionable use after individuals have exploited such an area once. Although they might obtain the most accessible resources initially and then continue to deplete the remaining resources to lower levels as time went on, the benefits to be gained from such a pattern would grow progressively smaller with each new visit. This type of pattern might characterize predators of seed crops once the seeds have ripened, or dormant insects over the course of a winter. On the other hand, where resources quickly replenish themselves, as insects do in warm weather when reproduction is occurring, resource levels may remain more constant. Cody (1971) suggested that his finches responded in such a way to the progressive ripening of seeds in the desert.

An optimal revisiting rate should exist if resources are self-replenishing. The rate of visitation would be a function of the rates of immigration or of reproduction by the prey, the rate of movement of the foragers, and the size of the area exploited. Rate of movement by the predators should be adjusted to the speeds at which the prey can be most effectively found (Pyke, 1978). In a study designed to test this factor, Ware (1975) determined that the rate of movement of planktivorous fishes varied as a function of the density of the prey items. However, other studies have reported widely varying rates of movement while foraging, suggesting that the rate observed may be a function of other demands as well. For instance, the foraging rates of four species of spruce-woods warblers (*Dendroica* spp.) differ

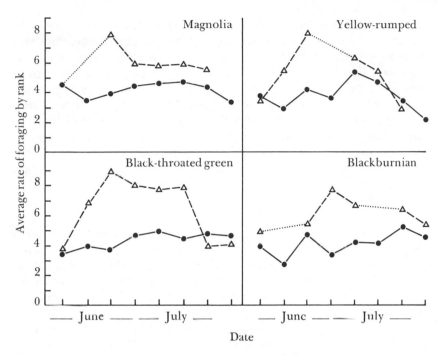

Figure 3.5 Mean rates of foraging by male (solid line) and female (dashed line) wood warblers over the period of the breeding season. Dotted line indicates data not available. A foraging rate of 1 = one to two foraging motions per minute; a rate of 8 = twenty-one to twenty-five motions per minute. (Modified from Morse, 1968b, copyright 1968 by the Ecological Society of America.)

drastically depending on the stage of the breeding cycle, with differences between males and females (Morse, 1968b) (fig. 3.5). It seems unlikely that these birds forage at an optimal rate at all stages of their nesting cycle, given the known patterns of insect prey abundance over the breeding season.

OPTIMAL FORAGING PERIOD

Many animals concentrate their feeding activities at certain times of the day. During the summer, insectivorous birds often forage most frequently during the cooler parts of the day (Morse, 1971a), which could be a response to lower availability of insects during the warmest part of the day or to increased costs of capture when the insects are highly active. Predatory reef fish concentrate their activities during twilight periods when their prey are most vulnerable because of the prevailing light conditions (Hobson, 1972; Major, 1977). The flowers of many species produce nectar only during restricted periods, and their visitors respond accordingly (Frankie, 1976). These observations suggest that animals concentrate their foraging activities if their prey vary in availability or in the costs or dangers of securing them. However, some animals forage incessantly, as is the case with gold-crests confronted with cold weather conditions during the short winter days in England (Gibb, 1954). Although this part of optimal foraging theory has not been formally developed, Schoener (1971) has suggested that foragers should expand their feeding periods when food is scarce, when energy demands are particularly large, or when food can best be converted into offspring.

LONG-TERM APPROACHES

Katz (1974) and Craig, DeAngelis, and Dixon (1979) have proposed optimal foraging models related to long-range objectives. They employ dynamic optimization techniques, which permit them to consider actions taken at one time that benefit an animal at a later time. For example, extra fat laid on by means of feeding techniques deemed suboptimal at the time might benefit an animal in the long run. The objective in Katz's model is to minimize total feeding time over a year. Applied to Ward's data (1965a, 1965b, 1971) on queleas (*Quelea quelea*), the qualitative correspondence between predicted and observed foraging patterns is reasonably good. Craig and coworkers found that their long-term model predicted some, but not all, of the shifts in foraging patterns observed in loggerhead shrikes (*Lanius ludovicianus*) over a year. They also constructed a short-term (daily) dynamic model in an attempt to explain why some of the long-term predictions were not followed by the shrikes. They discovered fluctuations in daily availability patterns (for example, insect prey are inactive and hard to find when temperatures are low early in the morning), which may complicate the long-term predictions. The long-term models tend to be more complex than the short-term static models with which I have been mainly concerned, and the required data base is more difficult to gather. The relative applicability of the two techniques to natural populations is unclear at this time. Both techniques require predictable foraging environments for the predators; if a high level of unpredictability exists, these deterministic approaches may be inherently unworkable (Oaten, 1977).

CRITIQUE

Until very recently most work on optimal foraging has been theoretical (MacArthur and Pianka, 1966; Emlen, 1966; Schoener, 1969a, 1971; Rapport, 1971; Pulliam, 1974; Estabrook and Dunham,

1976). Relatively few explicit tests of the theory have been published, and some of these have been performed in simplified microcosms (Werner and Hall, 1974) or highly artificial laboratory settings (Cowie, 1977; Krebs et al., 1977). The simple models work much better in the laboratory than under field conditions. Evidence from natural situations is still sparse, and experimental field studies are few. Thus it is not yet possible to give a critical assessment of the potential contribution of optimal foraging theory. Animals might not conform to the simple rules of optimal foraging for several reasons:

1. Interference from competitors may prevent access to high-ranked foods.
2. Danger from predators may result in modifications of foraging patterns.
3. Nutrient requirements of foragers may dictate that they include items of low rank.
4. Shifts of resource characteristics in space and time may occur rapidly enough to preclude conformity to the simple model.

Some disagreement exists over the possible importance of these factors. MacArthur and Pianka (1966) assumed that competitors have a major impact on optimal foraging patterns, whereas Schoener (1971) and Pyke et al. (1977) argued that competitors are generally unimportant. Evidence from the field suggests that these factors regularly complicate the attainment of optimal patterns.

A large body of literature suggests that many communities are highly influenced by competitive and predatory activities (Hairston, Smith, and Slobodkin, 1960; Paine, 1966; Connell, 1975). It is becoming increasingly obvious that many animals have a wide range of behavioral patterns that facilitate flexible responses to such pressures (Miller, 1967; Morse, 1974). Optimal foraging theory has not as yet been fully integrated with the results of these other areas of ecology, in spite of their potential importance in determining an animal's foraging style. Nor, with one exception (Pulliam, 1975), have dietary constraints been entered into the models. In general the literature suggests that foraging patterns comprise a variety of factors (competition, predator avoidance, nutrient demands, and so forth). More complex models that incorporate these constraints on foraging can be built, but they will necessarily be less general than their predecessors. Hughes (1979) has made a start in this direction by producing a model that incorporates variable handling and recognition times. Optimality should be more closely approached in mutualistic plant-pollinator systems than in predator-prey systems, even if a plant's strategy over evolutionary time is to obtain pollination services for as low a cost in nectar and pollen attractants as possible. On a short-term basis the goal of a plant is to *attract* pollinators, whereas that of a prey individual is to *avoid* being captured by a predator. For this reason it is not easy to extrapolate from one type of system to the other.

Although optimal foraging theory does lend itself to the generation of testable hypotheses (Krebs, 1978), some studies have attempted to explain away results not entirely in accord with the original hypotheses presented for testing. As Maynard Smith (1978) and Lewontin (1979) point out, it is not legitimate to explain discrepancies between theory and observation by introducing a new hypothesis and then claiming that the now-modified theory has been confirmed.

Foraging behavior

Search patterns, capture proficiency, and the techniques used to process captured food items all affect the range of food items an animal can use.

SEARCH TECHNIQUES

The relative importance of search, capture, and processing differs markedly with the characteristics of the prey. Predators that concentrate on small and cryptic (but otherwise almost helpless) prey tend to be particularly good at searching but relatively inept at chasing and handling more formidable items. Small birds that feed on foliage-inhabiting insects are the classic example.

Cryptic prey and specific search images. Cryptic prey create a major perceptual problem for their predators. L. Tinbergen (1960) found that until cryptic, foliage-inhabiting insects reached a certain density, great tits brought them to their young less frequently than predicted by chance. When a prey species reached a certain density (which depended on its size, palatability, and conspicuousness), it was supplied to the young more frequently than predicted from its density alone. However, when the resource reached an extremely high density, fewer individuals were brought to the young tits than predicted by chance. Tinbergen proposed that the tits were forming "specific search images" through a learning process; that is, they were developing perceptual mechanisms that efficiently distinguished prey from background by detecting key visual cues. He suggested that individuals stopped relying on the search image at high densities, choosing rarer items in order to ensure variety in the diet.

Tinbergen (1960) made the simplifying assumptions that prey are distributed randomly and that tits hunt randomly. Any deviation in the number of prey taken from that expected from its density would represent a change in the recognition rate of prey. But the data of Tinbergen (1960) and Royama (1970) show sizable runs of particular prey items, even rare ones, suggesting that the predator's searching or the prey's distribution, or both, are not random.

Tinbergen's data (1960) included only items brought to nestling tits. Thus the argument assumes that adults and nestlings eat similar distributions of food items. However, adults of several birds, including great tits (Tinbergen, 1949; Royama, 1966a, 1970) and oystercatchers, *Haematopus ostralegus* (Lind, 1965) and *H. bachmani* (Hartwick, 1976), feed their nestlings significantly larger objects than they themselves eat. Tinbergen (1949) was aware of this problem but apparently did not consider it critical to his argument. By concentrating on the diet of nestlings, Tinbergen was able to gather a large body of data; similar information for adults is more difficult to obtain.

Royama (1970) pointed out a number of potential objections to Tinbergen's interpretation of how cryptic prey are found. He noted that great tits sometimes shift their attention abruptly from a prey species while its abundance is still increasing rapidly and before it reaches the density predicted by Tinbergen's model. Generally these shifts coincide with the appearance of other new prey (large lepidopteran larvae) that offer larger rewards than those previously chosen. Royama (1970) also found that a single encounter with a prey item often led to several consecutive captures of that species, as did single presentations of certain foods to carrion crows (*Corvus corone*) in another study (Croze, 1970). Tinbergen's lower critical level (associated with a critical lower frequency of encounters with a particular type of prey) is inappropriate under these circumstances (Royama, 1970).

Profitability. As an alternative to Tinbergen's hypothesis, Royama (1970) introduced the concept of profitability, a modification of Holling's (1959) type 2 functional response curve (fig. 3.9). Here the energetic gain of foraging for a given food object increases with its density up to a certain point, above which prey are so abundant that an increase in their numbers cannot further increase their profitability. For example, equal numbers of prey might be taken per unit of time, whether five hundred or a thousand individuals were present per cubic meter; in fact, great tits sometimes experience larval prey densities of this order in oak woodlands. Under such superabundance almost all time devoted to feeding activities would be spent processing prey (transporting them to young, preparing them for young). Thus Tinbergen's argument of avoidance of monotony is not necessary to explain why the use of particular food items does not exceed a certain level. A new food should be selected when its profitability exceeds that of the previously preferred food. Because great tits feeding nestlings on abundant prey spend much of their nonsearching time carrying food to their nests, and because they carry only one item at a time, substitution of a substantially larger item should take place even when that item is much less abundant than the first item. In fact Royama found that rare, profitable prey were taken to nests in series more often than were less profitable prey of similar density, and this could explain the shifts in food choices that he noted.

By bringing to the nest the largest prey that their young can process, parents minimize the amount of food delivered, assuming they can carry only one item at a time. Small items would be most profitably consumed by the adult on the spot, and Royama observed that adults regularly fed on them when foraging. Long-billed marsh wrens (*Telmatodytes palustris*) (Verner, 1965) and blue-gray gnatcatchers (*Polioptila caerulea*) (Root, 1967) also take relatively large prey to their young. After fledging, young tits and oystercatchers following their parents are given food more similar in size to that eaten by the parents than to the food they were given earlier in the nest (Royama, 1970; Lind, 1965; Hartwick, 1976).

Whenever food must be carried to a central place, the size of the item will be an important consideration. Similar problems should be experienced by communally roosting birds and other animals (Hamilton and Watt, 1970), by thrushes that break open their food on a single anvil (see Cain and Sheppard, 1954), and by shrikes that use thorn trees to secure their prey while ripping it apart (see Cade, 1967).

Since carrying more than one item simultaneously would permit the transport of greater amounts of food, why do some birds carry only one item at a time? Perhaps these birds are limited to one item by morphology (bill size), by the difficulty of retaining a first item while a second is being captured, or by the need to process items before they are fed to the young. For example, oystercatchers bring items to their young one at a time; they would have to drop the first item processed (a bivalve, for example) to gather a second one. Since they invest substantial time and energy procuring each item (Norton-Griffiths, 1967, 1969), the loss of an item to another individual might be serious. In fact, Baker (1974) has reported that gulls (*Larus* spp.) in New Zealand commonly rob oystercatchers that are opening bivalves.

Other species such as swifts (*Apus apus*) (Lack and Owen, 1955), puffins (*Fratercula arctica*) (Lockley, 1953; Cockhill, 1973), some terns (*Sterna* spp.) (Hays, Dunn, and Poole, 1973), and herons (*Ardea cinerea*) (Owen, 1955) do gather several items at once. These birds either swallow and regurgitate food to the young (swifts and herons) or hold it in the beak (puffins and terns).

Royama found little evidence to suggest that tits choose food items for the sake of variety per se, as Tinbergen had argued in order to explain why the most abundant prey are often not taken as frequently as predicted. Royama argued that a diet selected for its profitability meets all dietary requirements; since foragers must pay attention to various types of prey in different places in order to maximize profitability, their diet will automatically attain adequate variety. The data are not sufficient to determine whether many species are automatically assured of adequate variety. The unexpectedly high frequency with which the tits fed spiders to their one- to five-day-old nestlings does not seem to support this argument. The mixed deciduous forest in which Royama worked offered a greater variety of potential prey than did the pine plantation in which Tinbergen worked. Individuals that regularly forage in optimal areas may be relatively likely to obtain balanced diets automatically.

Patches versus specific prey. Royama suggested that the tits exploited productive patches of prey, which they located by searching for specific sites or backgrounds rather than for the prey itself. Starlings (*Sturnus vulgaris*) and oystercatchers have also been observed to concentrate their foraging activities on profitable patches of prey, mainly cranefly larvae (*Tipula paludosa*) (Tinbergen, 1976) and cockles (*Cerastoderma edule*) (O'Connor and Brown, 1977), respectively.

Alcock (1973) explored the relative importance of searching for specific prey and searching for productive patches (fig. 3.6). In the laboratory, red-winged blackbirds (*Agelaius phoeniceus*) used cues associated with both specific foods and the locations at which the foods were presented, and they quickly shifted their foraging patterns when confronted with new configurations.

A few other studies demonstrate the ability of searchers to pick out specific prey. With practice, domestic chicks (*Gallus domesticus*) show an increasing ability to select cryptic grains of rice in laboratory experiments (Dawkins, 1971), and woodpigeons (*Columba palumbus*) search for the types of grains they have previously encountered in abundance (Murton, 1971a). By selectively feeding tits with one color morph of a caterpillar, den Boer (1971) influenced their color choice when two morphs were subsequently offered simultaneously; no such choice had been noted previously.

Choice of foraging area may be similarly influenced by previous success. Great tits and blue tits show a strong tendency to forage in the

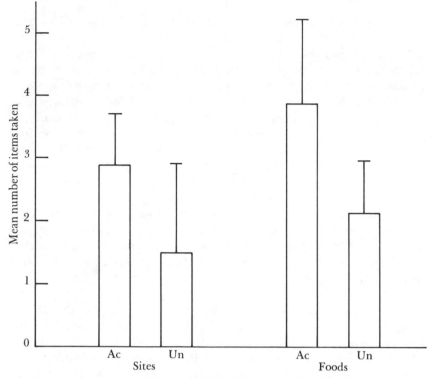

Figure 3.6 Foraging patterns (±1 SD) of red-winged blackbirds at previously accustomed and unaccustomed sites *(left)* and foods *(right)*. Ac = accustomed; Un = unaccustomed. (Modified from Alcock, 1973.)

same area or type of area in which they have previously found a high density of prey (Smith and Dawkins, 1971; Krebs, MacRoberts, and Cullen, 1972; Smith and Sweatman, 1974). In the laboratory, great tits choose sites with large prey over those with small prey (Smith and Sweatman, 1974). Croze (1970) and Murton (1971a) provided additional anecdotal evidence that crows and woodpigeons favor areas similar to those where they have previously found food.

Thus it seems clear that animals use search images to find specific food items and to choose areas in which to feed. Obviously, both techniques might be used simultaneously, but this possibility has seldom been investigated.

If individuals are able to recognize areas that offer particularly good foraging, they should develop ways to continue exploiting them as long as they remain profitable. Smith (1974a, b) demonstrated that European blackbirds (*Turdus merula*) and song thrushes (*T. philomelos*) show a strong tendency to concentrate their efforts near the area of

initial capture, an effective technique for clumped prey. This may
be accomplished by turning more frequently, reducing the length
of moves, or changing the pattern of successive turns (fig. 3.7). Most
of the difference in the thrushes results from altering the sequence of
turns. Using bits of pastry dough, which were readily accepted as
artificial earthworms, Smith induced the thrushes to forage in ways
very similar to those seen under natural circumstances. Area-restricted
searching has been demonstrated in several other birds, insects, and
fish (references in Tinbergen, Impekoven, and Frank, 1967; Smith,
1974a; Thomas, 1974).

Oddity and conspicuousness. Although predators are often assumed to select odd and conspicuous prey, most evidence about the
importance of these two factors in food capture is anecdotal (see Salt,
1967). For these purposes, *odd* means differing from the majority in

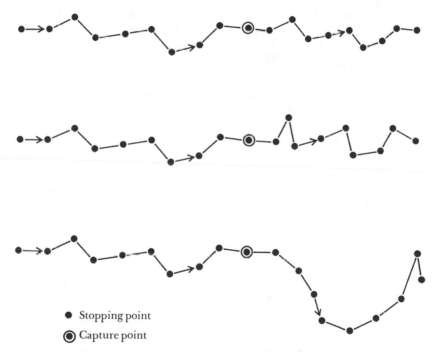

● Stopping point

◉ Capture point

Figure 3.7 Three ways in which the ten moves following prey capture by
a thrush may cover three-quarters of the beeline distance covered by the ten
moves preceding capture of a prey item. The top track achieves the reduction
by the thrush making shorter moves, the middle by making larger turns, and
the bottom by altering the sequential patterning of its turns. (From Smith,
1974b.)

any way, and *conspicuous* means not matching one's background. If an animal is part of a group that itself forms a substantial part of the background, that individual, if differing from all or most of the others, may be considered both odd and conspicuous (Curio, 1976). These terms can refer to behavioral as well as to morphological characteristics. The distinction between odd and conspicuous depends on the way a predator evaluates the background (Curio, 1976).

Using hawks and mice in the laboratory, Mueller (1968, 1971, 1974b) attempted to distinguish the relative importance of oddity and conspicuousness and to determine whether they are functions of previous experience. Although he referred to the results of previous experience as the development of a "search image," it is unclear whether Mueller used the term in the sense of "learning to see," as have most recent workers (Curio, 1976).

Mueller's predators were American kestrels (*Falco sparverius*) and broad-winged hawks (*Buteo platypterus*), and his prey were white mice (*Mus musculus*) dyed gray or undyed, against gray or white substrates. The hawks were trained to hunt from a perch in the laboratory, and the mice were placed on pedestals about them. He concluded that a searching image is the most important factor influencing prey selection, and that oddity (contrast with other individuals) is considerably more important than conspicuousness (contrast with the background). These results are consistent with Royama's hypothesis (1970) about predation of demonstrably cryptic forms, that profitability should be maximized by feeding on familiar prey. Rabinowitch (1969) and R. G. B. Brown (1969) have similarly emphasized the importance of past experience on the subsequent choice of noncryptic seeds by zebra finches (*Taeniopygia castanotis*) and pigeons (*Columba livia*). Choices shift quickly in response to changing rates of success. Mueller's finding that odd-looking or odd-acting individuals are particularly vulnerable to predators accords with other descriptions in the literature (Errington, 1946; Tinbergen, 1946; Rudebeck, 1950–51). In natural populations odd prey might include young, old, or sick individuals.

It is not clear how far Mueller's experiments on captive hawks and mice can be generalized. Kaufman (1973a) has suggested that Mueller's design may not really be a test for oddity, because it does not eliminate the possibility that the other mice might be forming the background. Although this possibility seems unlikely because mice covered only a small fraction of the test site (Mueller, 1973), Kaufman's comment points out the difficulty of designing experiments that completely separate these alternatives.

By conducting experiments in the field or in large enclosures where prey might hide, Kaufman (1973b, 1974a) found that conspicuous prey (laboratory-reared white mice) were attacked more frequently by owls and shrikes than were inconspicuous prey (laboratory-reared agouti mice). He attempted to control for oddity in these experiments by presenting prey in pairs, one conspicuous and one inconspicuous. White mice were taken more frequently than agouti mice when released simultaneously in the presence of wild, free-ranging loggerhead shrikes (*Lanius ludovicianus*) if cover was dense, which supports the argument for conspicuousness. However, in sparse cover the predation pattern was reversed (fig. 3.8), which suggests that the shrikes were using a search image (Kaufman, 1973b). Other studies using owls (*Tyto alba* and *Otus asio*) in large field enclosures gave similar results (Kaufman, 1974a). Kaufman did not indicate whether his owls were wild-caught or hand-reared, so the effect of past experience cannot be

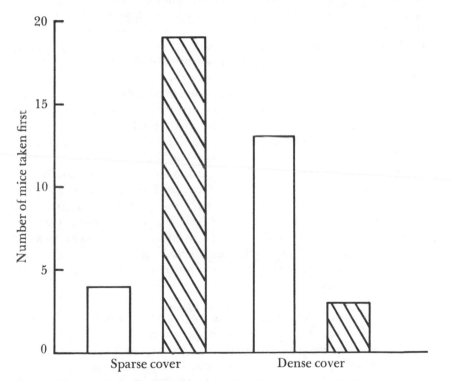

Figure 3.8 Capture of white and agouti mice in sparse and dense cover by loggerhead shrikes. White bars = white mice; hatched bars = agouti mice. (Modified from Kaufman, 1973b.)

firmly rejected as a possible basis for his results. Also, the two strains of mice may have differed in behavior as well as the conspicuousness of their coat colors.

In some experiments conducted in field enclosures with different-colored substrates, using old-field mice (*Peromyscus polionotus*) of varying natural colors as prey, Kaufman (1974b) found that owls tended to select conspicuous mice. His conclusions support Dice's studies (1947) of owl predation on deer mice (*P. maniculatus*) of varying colors against contrasting backgrounds. This suggests that predation may exert powerful selection on the visual characteristics of prey.

Many aberrant individuals are substandard in one way or another. Krischik (1977) has investigated this problem in the laboratory, using as prey juvenile bluegill sunfish that were either normal or infected with a microsporidean parasite that caused abnormal coloration and morphological changes affecting their swimming behavior. Largemouth bass (*Micropterus salmoides*) were used as predators. Krischik found that both oddity (color and behavior) and conspicuousness contributed to the vulnerability of these fish, although the effects of oddity were usually stronger. She pointed out an additional factor that may be of considerable importance to such choice behavior in the field. Normal individuals were inevitably socially dominant over aberrant individuals of similar size and forced them to marginal sites. Under natural conditions, substandard individuals might thus regularly occupy sites where they would be more vulnerable to predators, and this may be a major factor in differential predation.

There seem to be no clear overall indications of the relative importance of oddity and conspicuousness as cues used by predators. However, although odd individuals are necessarily in the minority, conspicuous individuals are not. Oddity will merge into conspicuousness only if animals are strongly clumped. Search images might even be based on oddity or conspicuousness; a predator's reaction to specific types of prey may originally have developed in response to cues that reflected these traits.

Functional response. In many instances predators increase their attacks on prey as the number of prey increases (Solomon, 1949). This phenomenon is referred to as a functional response. Holling (1959) described three types of functional responses (fig. 3.9): type 1, which produces density-independent mortality up to a satiation point; type 2, which produces an inversely density-dependent mortality over all ranges of prey abundance; and type 3, which produces direct density-dependent mortality up to a point at which the predator is

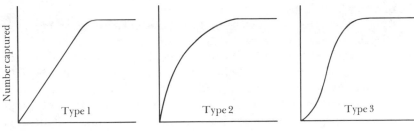

Figure 3.9 Three types of functional response curves. (Modified from Holling, 1959.)

satiated or runs out of time. Type 1 responses appear to be restricted to certain filter feeders and hence are not directly related to this discussion of predator responses. Most responses that have been studied are of type 2. Royama's profitability model (1970) is based on this pattern of response. The type 2 curve is usually considered characteristic of invertebrate predators, because type 2 responses have primarily been obtained from them (Holling, 1959). The type 2 curve assumes a situation in which predators feed on only one species of prey, a condition not always typical of the insect parasitoids used in these studies. The idea behind the type 2 response is that as prey become more abundant, predators should initially improve their feeding ability on them. However, the proportion of the time involved in killing, processing, and so on, soon takes up a sizable portion of the time available, producing the descending function plotted in fig. 3.9.

Type 3 curves are typically ascribed to vertebrate predators, which usually take more than one prey species. The sigmoid shape of the type 3 curve is related to the fact that the predator in question does not switch immediately to the prey species under investigation when that prey species is at low density, instead continuing to concentrate on another species of prey. Vertebrate predators should demonstrate a type 2 response if confronted with only one prey species, whereas invertebrates that are not specialists on a single prey species should and do show a type 3 response if additional prey species are present (Murdoch and Oaten, 1975).

Switching. Murdoch and his collaborators explored the response curves of predators exposed to alternate prey (Murdoch, 1969; Murdoch and Marks, 1973; Murdoch and Oaten, 1975; Murdoch, Avery, and Smyth, 1975). They set up a series of switching models that extend Holling's functional response models to multispecies conditions. Switching, as defined by Murdoch (1969), is a state in which the num-

ber of attacks made by a predator on a prey species is disproportionately high when the prey species is abundant relative to other prey,
and disproportionately low when the prey species is scarce relative to
other prey species (fig. 3.10). As he noted, the concept is not a new
one, being implicit in Elton's book (1927) on animal ecology. Switching presumably is attributable to conditioning, and the curve generated
resembles the one ascribed by Tinbergen (1960) to predators developing a search image. In the nonswitching alternative (null hypothesis),
attacks will be proportional to the densities of different prey present.

The switching model differs from Royama's profitability model in
that the latter requires that different prey species occur in separate

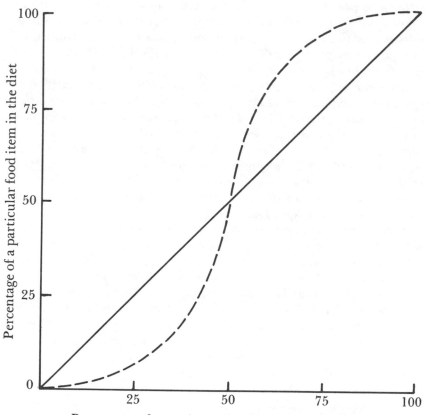

Figure 3.10 Switching. The diagonal line is the line of no preference. (Modi-
fied from Murdoch and Oaten, 1975, with permission from *Advances in
Ecological Research* 9:63, 1975. Copyright by Academic Press Inc. [London]
Ltd.)

patches, thus demanding separate searching efforts (Murdoch and Oaten, 1975). The prey in Royama's model are cryptic to varying degrees, but this is not crucial to the differences between the models. Alternative prey are readily available in the switching models, so that search time is effectively eliminated.

Murdoch and coworkers found support only for the nonswitching hypothesis (number of attacks proportional to density of different prey), using snails as predators on mussels and barnacles (Murdoch, 1969) and ladybird beetles as predators on two species of aphids (Murdoch and Marks, 1973). Consequently, they predicted that switching will occur only if predators show searching and selection behavior, available prey are distinctly different, and predators have to hunt differently to obtain each species. On the other hand, if preferences are strong, even intensive training may fail to modify choice behavior. Santos' experiments (1976) on a predatory mite (*Zetzellia mali*) and three herbivorous mites frequenting apple trees are consistent with Murdoch's predictions of which species will and will not switch. Santos could not train his mites to show switching behavior. They appear to encounter their prey by random search, using tactile cues.

The best evidence for switching comes from studies using notonectid bugs (back-swimmers) as predators on mayfly nymphs or isopods (Lawton, Beddington, and Bonser, 1974) and guppies (*Poecilia reticulata*) as predators on *Drosophila* or tubificid worms (Murdoch et al., 1975). The predation on mayfly nymphs by back-swimmers fits the switching curve almost perfectly. Although direct field evidence for switching is scarce, Murdoch and Marks (1973) concluded that indirect support comes from its similarity to apostatic selection (direct frequency-dependent selection) by predators on different morphs of a prey species. Popham (1941), Clarke (1969), and Allen (1976) gave examples of apostatic selection by insects and birds, and Allen predicted that it will turn out to be common in nature.

Unfamiliar objects. Food must not only be discovered; it must also be palatable. Many animals (and plants) have defenses that make them unpalatable or even dangerous to predators. Often such prey are also conspicuous in some way, in which case they are said to have warning or aposematic characteristics. Not surprisingly, many predators are extremely reluctant to handle unfamiliar food items (Coppinger, 1969, 1970; Croze, 1970). Wild-caught and hand-reared blue jays (*Cyanocitta cristata*), common grackles (*Quiscalus quiscula*), and red-winged blackbirds studied by Coppinger often avoided novel stimuli. In some cases Coppinger used tropical butterfles with color patterns unlike those found within the range of these jays, suggesting that specific innate responses did not account for the avoidance. The response was not confined to distasteful prey. This reluctance may explain why capture rates often start out well below those predicted by the search image hypothesis when a new item is present at low density.

Some objects do seem to elicit avoidance. For example, young reed warblers (*Acrocephalus scirpaceus*) avoid wasps (Davies and Green, 1976); being stung could kill one of these young birds or at least prevent it from feeding for a period of time. Other species of birds apparently must learn to avoid noxious insects (Mostler, 1935). S. M. Smith (1975, 1977) has demonstrated that some neotropical birds (flycatchers and motmots) instinctively avoid the red-yellow ring patterns characteristic of coral snakes.

Food choice may be socially conditioned. Rats (*Rattus norvegicus*) show a reluctance to feed on novel food objects. If such objects are eaten and a rat becomes ill, that information is apparently communicated to others. Young rats themselves may obtain their preferences by following their parents, probably a common pattern among animals that care for their young (see Lore and Flannelly, 1977).

However, some animals exhibit exploratory behavior toward novel objects (Berlyne, 1966). This behavior is common in young animals that are just beginning to forage for themselves (Davies and Green, 1976) but diminishes as they become more experienced in separating profitable from unprofitable food items. The tendency is retained to some degree in other species, especially ones that exploit wide ranges of foods, such as crows (see Croze, 1970). Individuals often exhibit more exploratory behavior when satiated than when hungry (Chapman and Levy, 1957). This suggests that having met its immediate needs, an individual may monitor other resources of possible use in the future. This is consistent with the partial preferences found in studies of optimal diet, which suggest long-term optimization.

In apostatic selection the rare species or morph, by definition the odd one, enjoys an advantage; it is preyed on less frequently than predicted by chance. This runs counter to the idea that odd and conspicuous individuals are especially likely to be taken by predators. The contradiction may be resolved by reference to other qualities of the prey. Apostatic selection has usually been sought in animals that employ crypsis, whereas studies of predatory risk on the odd and conspicuous have generally dealt with active prey; the latter probably enjoy very limited degrees of crypsis, if any. Unfamiliar prey are more likely to be odd or rare than are other potential food items. If they are avoided because of their unfamiliarity, attack frequencies might resemble those associated with apostatic selection based solely on relative prey densities. Studies are needed that incorporate novelty, oddity, and conspicuousness into a single design.

The models discussed in this section differ from optimal foraging models in predicting that as the density of a prey item increases, the response to it will change. Optimal foraging theory predicts that items rare in a diet will not be increased solely as a consequence of their increase in numbers. On this point the two bodies of theory are in direct conflict. According to Murdoch's arguments, switching is most likely to occur when a predator does not perceive great differences in the value of alternative prey but shows a tendency to specialize on one or the other alternative. Studies of switching have concentrated on situations in which prey differ in value only modestly. Tests of optimal foraging theory have used alternative prey differing conspicuously in their value, in part to simplify interpretation of the results (J. D. Allan, personal communication). Optimal foraging theory is concerned primarily with the way natural selection acts on the foraging behavior of individual predators; switching theory is more concerned with the role of predators in stabilizing prey populations. Accommodation between the two bodies of theory seems to be both possible and desirable, in that each is incomplete as presented.

NUTRIENT CONTENT

Adequate quantities of food do not necessarily provide an adequate diet. Little is known about nutritional balance in the wild (Kear, 1972), although considerable work with captive and domesticated animals has made clear the nature of the problem (Mitchell, 1964). The apparent good condition of most animals in natural populations suggests either that energetically profitable strategies automatically provide all essential nutrients (Royama, 1970), or that animals subtly discriminate in favor of critical nutrients. The problem of obtaining adequate variety probably has a phylogenetic dimension, such that entire taxa do not extend into areas where their nutritional needs are not regularly met with ease, or their densities remain low where certain items are severely limited. Natural selection should be able to modify dietary requirements so that essential nutrients can be included in a way that permits maximally efficient foraging.

The evidence suggests that active discrimination regularly occurs in favor of foods with particular nutrient qualities. Red grouse (*Lagopus lagopus*) prefer heather leaves high in nitrogen and phosphorus, the two most likely limiting nutrients (fig. 3.11). Discrimination in favor of such food is particularly intense just before egg laying (Moss, 1972; Moss, Miller, and Allen, 1972). Although Moss and coworkers

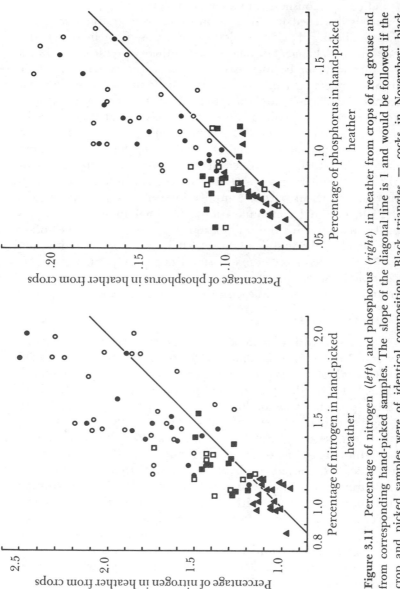

Figure 3.11 Percentage of nitrogen (*left*) and phosphorus (*right*) in heather from crops of red grouse and from corresponding hand-picked samples. The slope of the diagonal line is 1 and would be followed if the crop and picked samples were of identical composition. Black triangles = cocks in November; black squares = cocks and white squares = hens, December to March; black circles = cocks and white circles = hens, April to early May. (Modified from Moss 1972.)

did not discover how the grouse selected these leaves, they established that it was not based solely on the size and woodiness of the food. Data from Finnish red grouse support the conclusions of Moss and coworkers (Pulliainen and Salo, 1973). Pulliainen (1970) made similar observations for the rock ptarmigan (*L. mutus*). Royama's observation (1970) that great tits frequently feed spiders to their newborn young may also reflect special (but unknown) nutrient requirements. Arnold (1964) listed several other examples, mainly taken from work on grazing range mammals, showing preferences for leaves over stems, and for green material over dry. In some cases the chosen food is particularly high in nitrogen, sugar, and phosphorus and low in fiber. Price (1978) found that a Coke's hartebeest (*Alcephalus buselaphus cokei*) increased its protein intake through selective grazing.

The problem of overall food quality (energy content and nutrient balance) is probably important wherever an animal's diet is dominated by abundant, relatively low-quality foods. Pigeons feeding on *Brassica* greens (brussel sprouts, rape, turnips) had a rate of digestion only one-third that of their ingestion rate; because of this slow gut clearance they could not forage for substantial periods and were unable to maintain a constant weight on brassicas alone during the English midwinter (Kenward and Sibley, 1977). Some laboratory work suggests that certain items of low nutritive value are actively refused. Rats will refuse diets that have serious amino acid imbalance, vitamin deficiency, or mineral deficiency (Rodgers, 1967; Rogers, Tannous, and Harper, 1967; Rozin, 1967). Such refusals are usually associated with systematic sampling of other food items (Rozin and Rodgers, 1967). Recent studies of primate food choice (see Clutton-Brock, 1977) show parallels to the animals described here.

Dietary needs may reflect a species' evolutionary history, as Janis (1976) has demonstrated by comparing the demands of rumen and cecal digestion in artiodactyls and perissodactyls, respectively. However, considerable variation may exist in the dietary needs even of closely related species. The bay-breasted warbler (*Dendroica castanea*), which typically exploits outbreaks of lepidopteran larvae, can thrive on an almost pure diet of moth larvae, but this does not adequately sustain congeneric species from similar habitats, who are not adapted to tremendous increases of a single food source (Morse, 1971a).

In some cases specific nutrient requirements appear to be obtained only by resorting to foods or other materials of negligible caloric value. At certain times Asiatic elephants (*Elaphas maximus*) in Ceylon consume large amounts of grass having very low caloric value (insufficient to sustain them as a staple), suggesting that they obtain some essential nutrient in this way (McKay, 1973). Reichman (1977) found that in addition to seeds rich in energy, several desert heteromyid rodents select some that are very poor in energy.

Striking differences often exist between the diets of adults foraging for themselves, adults producing or caring for young, and growing young. The diet of many normally herbivorous birds does not contain adequate protein for young or for females forming eggs, and these individuals typically make up this deficit with animal protein (Ward, 1955a; Snow, 1971; Morton, 1973; Morse, 1975a). Only a very few species of frugivorous birds are able to raise their young totally on fruit, and even in these cases the developmental period is greatly extended, increasing the probability of nest predation (Morton, 1973). Similar examples are widespread. Pregnant or lactating female mammals experience high protein demands as a consequence of their condition and may shift their diets as a result. Many hymenopterans are carnivores as larvae and nectar feeders as adults.

We are only beginning to appreciate how animals obtain balanced diets in the field. Considerable evidence contradicts the hypothesis that animals typically obtain a balanced diet simply by eating the most energetically profitable foods, thus throwing doubt on simple models of optimal foraging (see Pulliam, 1975). Special feeding techniques required to obtain all necessary nutrients further complicate the problem of obtaining adequate energy in the time available.

Capture techniques

Consumers such as tits, snails, and ladybird beetles do not experience problems in capturing their food once they have found and selected it, but other predators experience more difficulty in capturing food items than in finding and selecting them. A wide range of capture techniques minimizes this difficulty. At one extreme are predators who sit and wait for prey or stalk it, and thus capture it completely by surprise; at the other are those who run down their prey. Stalking and ambush are most commonly associated with lone predators and with habitats containing adequate cover. Chasing usually occurs in open areas, where rapid, sustained pursuit is possible and prey derive little protection from the cover. Many chasers hunt in social groups (Kleiman and Eisenberg, 1973).

Sit-and-wait techniques are usually used by small predators, partly because their prey are typically more abundant than those of larger predators and partly because sit-and-wait techniques depend on concealment, which becomes progressively more difficult with increasing size. Although the large boid snakes have overcome this problem (see Bellairs, 1969), they are ectotherms, with demands only a fraction that of similarly sized endotherms. Because of their high energy demands, one would expect sit-and-wait techniques to appear in large birds and mammals only where an extremely profitable prey base existed. Although bobcats (*Lynx rufus*) and lions often use ambush techniques, they regularly move between sites, and lions employ ambush only in areas where large prey concentrate (Schaller, 1972).

Relatively few predators are capable of capturing prey in a long-distance chase. Even the fastest must get within a critical distance from their prey, typically by stalking, because they can maintain their great speeds only over a relatively short distance (Schaller, 1972; Eaton, 1974). Most prey of the lion can outrun it, so the lion's success depends on capturing the prey before it reaches its full speed (Schaller, 1972). Tigers (*Panthera tigris*) will not continue an attack unless they succeed in making contact with the prey almost instantly (Schaller, 1967).

Only social mammalian predators do not appear to be seriously constrained by this limitation. Several predators of large mobile prey hunt in social groups—for example, wolf (*Canis lupus*) (Mech, 1966), wild dog (*Lycaon pictus*) (van Lawick and van Lawick-Goodall, 1971), and spotted hyena (*Crocuta crocuta*) (Kruuk, 1972b). Cooperation permits capture of otherwise inaccessible prey by minimizing the conflict between killing ability and speed. A group can kill far larger prey than can a lone predator, and can also run down the prey cooperatively, which lessens the premium on cursorial adaptation. However, social groups of predators can only be sustained where prey are abundant.

Predators specializing on large, active prey often exhibit a particularly strong bias toward young, old, and ill animals—for example, wolf (Mech, 1966), spotted hyena (Kruuk, 1972b), lion (Schaller, 1972; Rudnai, 1974), and grizzly bear (*Ursus arctos*) (Cole, 1972). Many predators of large prey even scavenge when the opportunity presents itself. Of all the large East African carnivores, only cheetahs avoid carrion (Schaller, 1972). Goshawks (*Accipiter gentilis*) tend to capture underweight or injured woodpigeons, which are less alert and also less able to outdistance the predators (Kenward, 1978) (fig. 3.12).

Errington (1946) suggested from his studies on muskrats (*Ondatra zibethica*) that individuals (often young or old) forced out of populations were preyed on with high frequency. Animals unfamiliar with their surroundings were more conspicuous and consequently more vulnerable than residents. Although this conclusion is frequently stated as a general truth in the literature, few quantitative data exist to back it up.

Where substandard prey are actively chosen, the predator appears to be opting for a high probability of a modest reward over a low prob-

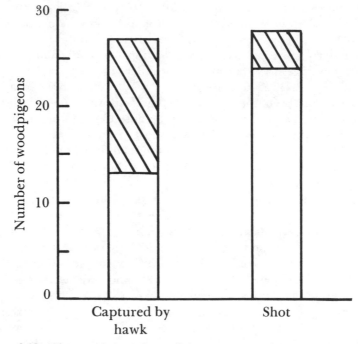

Figure 3.12 Characteristics of woodpigeons captured by goshawk and of those shot by investigator. White areas of histogram = healthy woodpigeons; hatched areas = diseased woodpigeons. (Modified from Kenward, 1978.)

ability of a large reward. Prime individuals should be taken unless they provide more food than the predator or predatory group can use, require special expenditures or risks, or are invulnerable. Actively seeking less than maximal rewards represents a basic difference between predators for which capture involves the major expenditure of time and energy, and those for which searching does.

Predators ought not to attack potentially dangerous prey unless the benefit is large. Schaller (1972) has compiled a large number of such observations on lions, showing that hunting potentially dangerous prey carries a high probability of injury. In some areas casualties suffered while attacking dangerous prey are a major source of death, especially where lions feed heavily on buffalo (*Syncerus caffer*) and large antelopes. Hornocker (1969) suggested that the mountain lion (*Felis concolor*) is highly adapted to making quick kills without being hurt, and this probably also holds for tigers (Schaller, 1967).

Certain prey species unable to retaliate actively may nevertheless present hazards to their predators. Ospreys (*Pandion haliaetus*) occasionally drown as a result of capturing a fish so large that they cannot fly into the air with it and, at the same time, are unable to extricate it from their claws (Bent, 1937). Similarly, large food objects may become lodged in a predator's esophagus if swallowed whole, a source of mortality in many species (Rogers, 1957; Schoener, 1971). The spiny defenses of animals such as pufferfish and sticklebacks (Moodie, McPhail, and Hagen, 1973) are also dangerous to their predators.

FOOD MANIPULATION

Ease of manipulation is a major consideration in food choice, especially when considerable processing time is required, and it probably becomes the dominant factor when costs of search and capture are low (Smigel and Rosenzweig, 1974). This problem has been explored primarily in the laboratory by providing foods of different sizes and shell thicknesses to seed-eating birds (Kear, 1962; Hespenheide, 1966; Myton and Ficken, 1967; Willson, 1971; Willson and Harmeson, 1973) or small mammals (Rosenzweig and Sterner, 1970; Smigel and Rosenzweig, 1974; Mares and Williams, 1977). Usually these animals select the seeds that they can husk most readily; their choices differ with bill or body size and show only weak correlations with known energy content (Willson and Harmeson, 1973).

Willson (1971) noted that large finches take a wider range of food items than small ones. This is consistent with a strategy of taking the most readily available items, given that large individuals are physically capable of processing a wider range of foods than small individuals (Schoener, 1971). If processing time is low, as it would be for small seeds eaten by large birds, such a trend would be predicted if the reward were at least modest in size and the small items taken were common relative to the large ones. This trend is not always seen, however. Baker (1977) presented evidence suggesting that large shorebirds tend to be more specialized feeders than small ones. The choice taken probably reflects the food-size distribution normally encountered (see chapter 2). This problem is by no means confined to seed eaters. Craig (1978) found that it took loggerhead shrikes as much as 90 minutes (±65 minutes) to capture, subdue, and manipulate mice before they were able to begin feeding on them; insects provide the most common foods of these shrikes, and for these the handling time is negligible.

Processing time may vary with the consumer's satiation. Zach and Falls (1978) found that ovenbirds (*Seiurus aurocapillus*) required longer to process large insects the longer they had been feeding on them. Ware (1974) and Werner (1974) have reported similar results from experiments on trout (*Salmo gairdneri*) and sunfish (*Lepomis* spp.), respectively.

A more subtle problem may arise where some items contain rewards and similar items do not. Ligon and Martin (1974) found that piñon jays (*Gymnorhinus cyanocephalus*) could distinguish between piñon containing a store of endosperm and those that did not (fig. 3.13). Superficially, aborted seeds resemble viable ones. A large proportion of a seed crop may be aborted, so this type of discrimination is of major importance. Since many nuts are stored and used at times when other foods are scarce, discrimination should be accurate even at times of temporary plenty. In this case, visual, tactile, and auditory cues all seemed to be of importance. Jays quickly learned to reject artificially weighted seeds that contained no endosperm. Many plants produce seeds in husks that often have aborted or insect-damaged endosperm, so the problem of discrimination is probably widespread. Chickadees

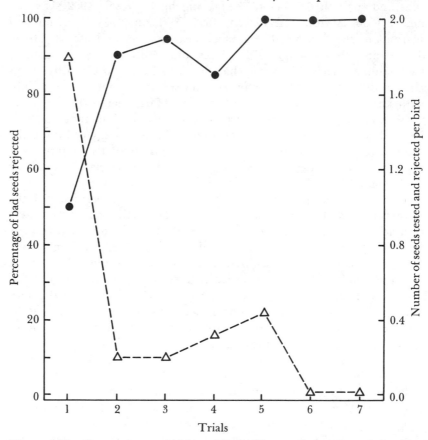

Figure 3.13 Change in recognition of bad piñon seeds by nine adult piñon jays. Solid line = percentage of bad seeds rejected; dashed line = number of seeds tested and rejected per bird. (Modified from Ligon and Martin, 1974.)

(*Parus* spp.) readily reject aborted seeds (Morse, 1970b) as well as ones that have previously been attacked by insect larvae (Morse, unpublished). Sloan and Simmons (1973) demonstrated that chipping sparrows (*Spizella pusilla*) can recognize and reject budworm larvae and pupae parasitized by hymenopterans. Lenteren and Bakker (1975) provided a similar example in which the cynipid wasp *Pseudeucoila bochei* apparently learned to avoid depositing its eggs in *Drosophila* larvae that had already been parasitized.

Synthesis

Simple models based on energy maximization are not very successful in predicting the foraging patterns of animals under natural circumstances, apparently because these models do not incorporate the constraints imposed by competitors, predators, the spatial and temporal heterogeneity of resources, and the problem of nutritional balance. Heterogeneity alone imposes several different limitations on the forager's capabilities. Variation in their prey forces predators to learn new food sources, and responses to shifts in supply cannot be instantaneous. The exploratory behavior of foragers noted by many workers probably represents a short-term cost that is repaid only on a long-term basis. Prey should consequently be under considerable selective pressure to maximize the amount of learning required by the predator; unpredictability in space and time and polymorphism appear to be responses to this pressure. Such responses should in turn favor predators that respond effectively to whole groups of prey (for example, seeking out and recognizing all caterpillars). This then favors prey species that are as different from one another as possible. The ability of predators to respond to these perceptual challenges could to some degree determine the morphological diversity of prey, the overall diversity of species, and the relative abundance of individuals at different trophic levels.

For a food to be of high rank it must be readily available. A food that predators have not learned how to find efficiently cannot qualify as a high-quality item, even if it would rank high in terms of its energy content. This complicates the evaluation of optimal foraging theories because it is difficult to determine whether a prey item is ignored because it is not perceived or because it has a low rank.

A similar operational difficulty occurs in the attempt to determine whether a predator is simply not foraging optimally or is instead practicing long-term optimization. Pyke et al. (1977) assumed that long-term optimization occurs among animals having exclusive access to a food source, as in a territory; it might pay to exploit the territory in a way that brought the greatest cumulative long-term harvest of the greatest rate of harvest at some critical time in the future (see Morse, 1976a; Thompson, Vertinsky, and Krebs, 1974) .

Work on optimal foraging theory has not yet been well integrated with related studies on the niche relationships of foraging animals, although this may represent one of its greatest potential contributions to ecology. Notable exceptions include studies by Gill and Wolf (1975a) on African sunbirds and by Werner (1977) on sunfishes and bass. Optimal foraging theory allows predictions based on energetic considerations to be tested against the actual resource exploitation patterns of groups of coexisting species with similar resource exploitation patterns (guilds) . Here it is used as a tool, rather than as an end in its own right. Conformity to prediction would support the argument that energy was the main factor responsible for the patterns observed, and nonconformity would serve to measure the importance of other factors. Combining the concepts of foraging efficiency and niche partitioning may help to reveal the factors that determine boundaries of coexistence in a community (niche partitioning, more formally treated as species-packing theory) .

4 *Habitat Selection*

ALL ANIMALS to some extent control where they will feed, breed, and carry out other important activities, and many do so with great specificity. Indeed, field identification guides sometimes distinguish between similar species on the basis of habitat. For example, in the eastern United States the least flycatcher (*Empidonax minimus*) typically nests in gardens and orchards, whereas the virtually identical Acadian flycatcher (*E. virescens*) haunts the middle stratum of moist, deciduous woodlands.

The choice of site will depend on the time and energy available to search for that site, the probability that additional search will improve the choice, and the degree to which continued search for an even better site will change the searcher's fitness (see Levins, 1968). Excessive search for a better site after a satisfactory site has already been found may substantially increase the probability of prereproductive death or waste resources that might otherwise be allocated to reproduction. The importance of these factors will differ with the attributes of the searching animal (such as its mobility, energy resources, and adult longevity), the variety and spatial characteristics of suitable habitats (large or small, homogeneous or patchy), and the occupancy of sites by other individuals. Habitat selection is in several respects similar to food selection.

Although the terms *habitat selection* and *habitat utilization* are roughly equivalent, they make a subtle distinction. I use the term habitat selection to signify the active choice of an area by an individual,

89

and habitat utilization to signify that an individual occupies an area. Any case of habitat selection is also a case of habitat utilization, but the converse is not true.

Distributional patterns need not reflect active choices. In some species that are planktonic in the larval stage and sessile in the adult stage, the dispersal of young is for the most part random. In the most extreme case, the habitat supporting such a population may simply reflect differential survival in favorable and unfavorable environments. A random settlement pattern may entail considerable mortality, but it is suited to the occupation of environments that are extremely patchy relative to the exploratory capabilities of their inhabitants.

Many planktonic larvae do, however, have highly selective settling behavior, and their ability to remain waterborne for varying periods is well known (D. P. Wilson, 1970; Meadows and Campbell, 1972a, b). Meadows and Campbell suggested that the local distribution of such marine animals is largely determined by habitat selection and the appropriate behavior accompanying such choices. Moore (1975) adopted a more conservative view than Meadows and Campbell, emphasizing the unpredictability of conditions at spawning times, the dependence on physical factors for successful dispersal, and the uncertainty that space will be adequate to support another individual. Here I consider mainly cases in which animals actively select, concentrating on vertebrates whose dispersal abilities usually permit them ready access to favorable sites. Where appropriate I compare host-plant, host-animal, nest-site, food, and mate selection with habitat selection.

Mechanisms

A decrease in the amount of time spent exploring usually precedes selection of a site. Sale (1969) formalized this by suggesting that site selection is regulated by a negative feedback mechanism sensitive to characteristics of the immediate environment. Manini (*Acanthurus triostegus*), a surgeonfish, show a high intensity of exploration in an inadequate environment (uncovered water), but exploration decreases in an adequate environment (shallow, covered water) (fig. 4.1). Sale's hypothesis is a slight modification of Reese's explanation (1963) for the way hermit crabs select their shells. Reese suggested that shell selection (a special case of habitat selection) occurs through a reduction of shell exploration by crabs in response to sensory cues obtained from a satisfactory shell. Critical factors appear to include the overall configuration of the shell, the size of the aperture, and the weight and volume of the shell. These factors affect several aspects of fitness in hermit crabs, including protection from predators (Vance, 1972), clutch size (Childress, 1972; Fotheringham, 1976), and egg production

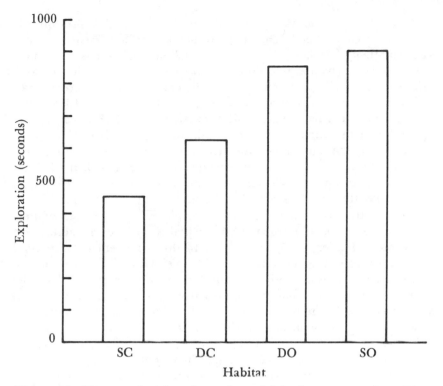

Figure 4.1 Mean exploration time of manini in four test environments. SC = shallow, covered; DC = deep, covered; DO = deep, open; SO = shallow, open. (Modified from Sale, 1969.)

(Bach, Hazlett, and Rittschof, 1976). Sale found that fish previously conditioned to tanks without covers switched, when given the opportunity, to sites containing cover. Apparently the manini retained an innate preference for cover.

Thorpe (1963) suggested that selection behavior is goal directed in that the organism has a cognitive map (*Sollwert*), not necessarily the product of learning, which it continually compares with the stimulus experienced. (Thorpe actually used this concept to explain patterns of nest building by birds and insects.) Such a map could provide a mechanism responsible for the habitat selection observed by Sale, Reese, and others. However, no independent evidence exists for a *Sollwert*, and in its present form this model cannot answer such questions as why exploratory behavior varies in different unsatisfactory habitats, as reported by Sale. Neither has Thorpe placed this concept within the framework of natural selection.

Sometimes selection has a simple basis. Studies of seed-size choice

(Kear, 1962; Willson, 1971) demonstrate that even naive birds quickly learn to select the food items that can be used most efficiently, a function of their morphological attributes. Choice of food is a problem akin to that of habitat selection. The basis for discrimination in seed-size choice is readily apparent, even if the neural correlates are unclear. Habitat selection might occur in response to food alone, if food differs predictably from habitat to habitat (Verner, 1975). However, if supplies fluctuate, animals may be unable to respond directly to food, even if it is the most important attribute of the area to be selected. Other simple factors such as the openness or denseness of the habitat in relation to an animal's ability to move about in it (see Hildén, 1965) may then provide cues that facilitate habitat selection.

Frequently it is assumed that the principle of heterogeneous summation (Seitz, 1940; Tinbergen, 1951) explains whether a habitat will be chosen (Hildén, 1965). According to this argument the threshold for release, resulting in selection, would depend on the internal motivation of an organism. Motivation, in turn, would be modified by the external stimuli, both positive and negative, perceived by an organism. Not every habitat must thus possess all the features characteristic of the ideal environment, the overriding factor being that the combined effect exceeds the threshold of the settling action. Questions have arisen as to whether heterogeneous summation models are appropriate for this or other situations (see Hinde, 1970; Baerends, Bril, and Bult, 1965). For instance, Curio (1969) has suggested that releasing stimuli (in a different context) may operate as a multiplicative interaction rather than an additive one. In reality, it seems highly unlikely that any two factors would have equal weight, or that the interaction between any two sets of variables would take an identical form. It would seem to be important that some negative factors act in an absolute way under extreme conditions. Obviously, tree-nesting birds will find a habitat containing no trees unacceptable, regardless of the other merits. It is commonly assumed that selection occurs in response to cues that can be quickly ascertained, but little is known about the properties of such cues, about the number of cues actually employed in any given case, or about the ways in which they are combined to reach a decision.

Ultimate and proximate factors

The adequacy of an area may depend on factors totally different from the cues that govern the settling response. Proximate factors (the psychological factors of Lack, 1937) may in themselves be irrelevant to the success of an organism, although they must correlate strongly with the relevant ones (ultimate factors). Ultimate factors include food, protection from predators and climate, breeding sites, and conditions imposed by the structure and functional activity of the animal involved (for example, the effect of tall, thick grass on a short-legged hopping bird, such as a sparrow).

Ultimate factors are often of little help in assessing the suitability of an area. This problem is especially acute for animals that live in seasonal habitats. For instance, the number of foliage-dwelling insects found by birds early in the spring may provide erroneous information about the food supply when young have to be fed later in the year. When black-throated green warblers (*Dendroica virens*) establish territories in spruce forests during early June, food intensity in white spruces (*Picea glauca*) is often two to three times that of red spruces (*P. rubens*) (Morse, 1976a) (fig. 4.2). However, these warblers establish higher population densities in red spruces than in white spruces. During times of maximum resource demands in late June and July, when nestlings and fledglings must be fed, no significant difference occurs in the abundance or type of food in the two species of trees. The birds apparently feed more easily on red spruce foliage than on that of white spruce. Several lines of evidence suggest that food may be scarce during the breeding season (Morse, 1967a, b, 1971b). If the warblers used food at the time of settling as a cue, they would establish denser populations in white spruces than in red spruces. The birds exhibited considerable constancy in territory size, even in a year in which their populations crashed, which could be a response to proximate rather than ultimate factors (Morse 1976a). The proximate factor has not been established, but subtle differences in foliage could create differences in the ease of traversing it and finding prey in it. White spruce foliage lies perpendicular to branches; red spruce foliage lies flat. The former will force the warblers to hop higher in order to progress and will wet their plumage more thoroughly during inclement weather. White spruce boughs are also thicker and shorter than red spruce boughs, making prey in white spruces less accessible to black-throated green warblers, which must hop over the top of the foliage (Morse, 1976a). The birds thus appear to respond to cues that are easily monitored and highly correlated with the ultimately important factors, rather than to the existing conditions.

Some animals appear to assess their food supply directly. Pitelka, Tomich, and Treichel (1955) believe that jaegers, hawks, and owls can directly estimate the abundance of lemmings, their major food supply during years of plenty. A direct, accurate assessment of critical resources may be particularly important to species such as these, which feed on food that is highly unpredictable. Their reproductive behavior is very flexible and is geared to fluctuating food supplies. When lemmings are scarce these species may completely forego breeding; when lemmings

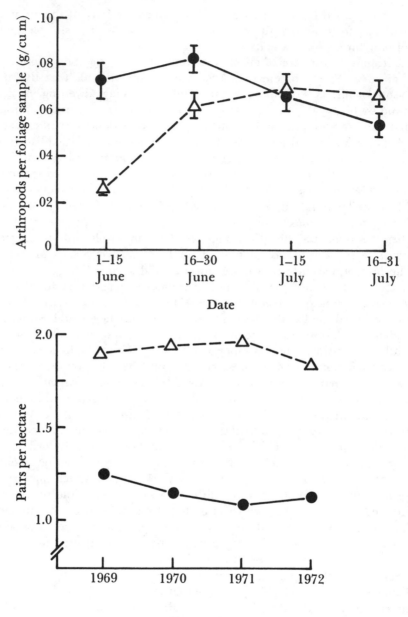

Figure 4.2 Standing crop of insects (±1 SE) (*top*) and density of black-throated green warblers (*bottom*) in red spruce (dashed line) and white spruce (solid line) forest. (Modified from Morse, 1976a, copyright 1976 by the Ecological Society of America.)

are abundant they have large broods, which considerably exceed those of closely related species not dependent on such variable food supplies.

Studies of habitat selection have emphasized responses to positive stimuli, but negative stimuli may be as important. Hildén (1965) suggested that the presence of a territorial male may prevent other males from settling, although the display of a territorial individual is often believed to attract other males to the general vicinity (Svärdson, 1949). The conspicuous presence of predators surely acts in a similar way. Stein and Magnuson (1976) showed in laboratory studies that crayfish (*Orconectes propinquus*) selected substrates providing a maximum amount of protection (pebbles rather than sand) when placed with smallmouth bass (*Micropterus dolomieue*) large enough to eat them. Control crayfish showed no preference for one substrate over the other (fig. 4.3). These results are consistent with distributions of crayfish in the field.

Where conditions are predictable but fluctuating, proximate cues may be important. Where conditions are extremely constant, forecasting is unnecessary, and animals might respond directly to ultimate factors, but this prediction cannot be tested because the relevant studies have not been conducted on animals living in extremely constant environments.

Learned versus innate behavior

The relative importance of innate and learned elements in habitat selection is unknown. Basing his review primarily on field work with birds, Hildén (1965) placed heavier emphasis on inherited factors than did Klopfer and Hailman (1965), although he did invoke learning to account for certain rapid changes in habitat use. Klopfer (1963) found suggestions that early learning influenced choice of pine or oak foliage by chipping sparrows; wild-caught adults preferred pine needles to oak leaves, but young reared in the presence of oak showed a decreased preference for pine. Klopfer and Hailman (1965) predicted that habitat selection in slow-developing (altricial) birds such as chipping sparrows is less rigid than in fast-developing (precocial) birds. But Partridge (1974) found that hand-reared naive blue and coal tits (*Parus caeruleus* and *P. ater*) express foliage preferences that match those of normally foraging birds in the field (for oak and pine, respectively; see fig. 4.4). It appears that habitat preferences have a large inherited component under natural circumstances, but that these preferences are also subject to some modification by learning.

Wecker (1963, 1964), working on prairie deer mice (*Peromyscus maniculatus bairdii*), provided evidence that early experience affects subsequent habitat selection, but he could not induce these animals

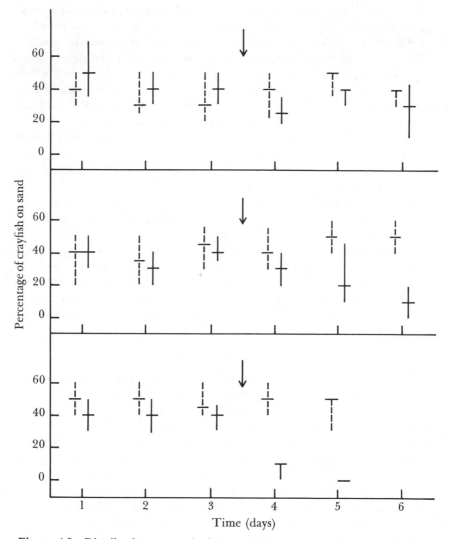

Figure 4.3 Distribution on sand of large adult (*top*), small adult (*middle*), and juvenile (*bottom*) crayfish in aquaria with equal portions of sand and either gravel (for juveniles) or pebbles (for adults). Dashed line = control; solid line = treatment. Median and interquartile ranges are based on forty (adults) and sixty (juvenile) observations in four and six tanks, respectively. Each observation was based on ten animals per tank in controls and one to ten animals per tank in treatments. Arrows indicate when smallmouth bass were placed in treatments. (Redrawn from Stein and Magnuson, 1976, copyright 1976 by the Ecological Society of America.)

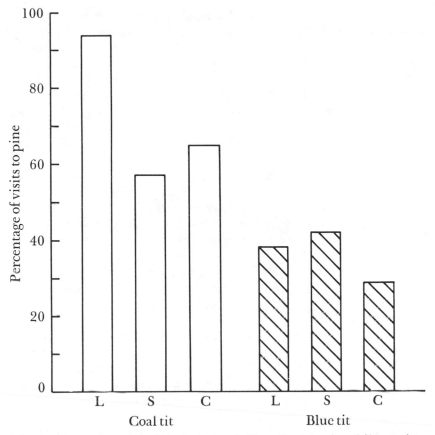

Figure 4.4 Preferences of coal tits and blue tits for pine foliage when presented with a choice between pine and oak foliage. L = leaves; S = branches stripped of leaves; C = complete branches. (Modified from Partridge, 1974.)

to choose woodland over their normal grassland habitat. Grant (1971) had little success altering habitat choice in meadow voles (*Microtus pennsylvanicus*). Some meadow voles occupy forested areas, but Grant concluded that these individuals disperse there when their favored grassland habitats are saturated. Subsequent investigations of habitat selection in rodents have produced similarly mixed results (Randall, 1978).

Neither Löhrl (1959), studying collared flycatchers (*Ficedula albicollis*), nor Catchpole (1974), studying reed warblers (*Acrocephalus scirpaceus*), found a high correlation between the habitat in which an individual nested and the one in which it was reared. If experience in the period immediately after fledging tends to determine where an individual will settle, a patchy distribution of habitats could mask a real relationship between the kind of habitat containing the nest in which the individual was reared and the habitat in which it will nest. That relationship could only be revealed by more careful studies that document the movement of recently fledged young. Catchpole's studies on reed warblers were conducted in just such a heterogeneous area. Since his birds experienced high rates of success in several habitats, selection for discrimination would not be high in this case although it may be in others.

Wiens' work (1970, 1972) on red-legged (*Rana aurora*) and cascade tadpoles (*R. cascadae*) suggests that both early experience and innate preference may be of importance in these species' choice of habitat. Laboratory-reared red-legged tadpoles prefer striped backgrounds, which correspond to the submerged sticks, cattails, grasses, and filamentous algae that characterize their natural habitats. Cascade tadpoles prefer backgrounds composed of squares, which correspond to the gravelly substrates over which they typically live (fig. 4.5). Wiens found that if he reared tadpoles of these two species on inappropriate substrates (accomplished by switching the substrates) they showed no preference for the substrate on which they would normally be reared.

Evidence such as that gathered by Wecker and Wiens suggests that innate factors may sometimes provide the coarse tuning and learned factors the fine tuning for habitat selection. Because choices may be based on information acquired at different stages of development, these are likely to be critical periods.

The question how migratory animals first choose their wintering grounds is closely related to this topic. In many species (deer, geese, swans, cranes) the young travel with the adults, sometimes even in family groups (Hochbaum, 1955). But adults and young of some species clearly move separately. For example, young of the herring gull (*Larus argentatus*) often winter far to the south of the adults (Kadlec and Drury, 1968). The effect of early experience on the breeding grounds may prompt juveniles to settle in similar areas in a new region. Observations of migratory species on their wintering grounds and on migration lanes generally support this explanation (Kadlec and Drury, 1968). However, conclusive observations do not exist, because the available data do not include species in which it is known that adults and immatures migrate completely separately. In many species, once individuals choose a wintering area they tend to return to it in subsequent years (Nisbet and Medway, 1972; Diamond and Smith, 1973; Ely, 1973).

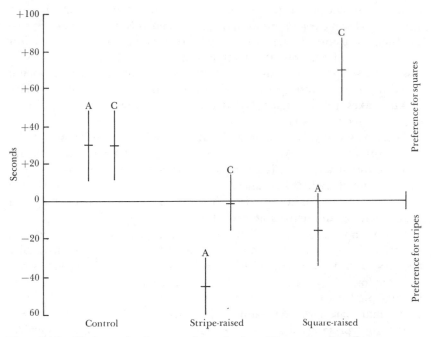

Figure 4.5 Choices of substrates by tadpoles. Time spent by *Rana aurora* (*A*) and *R. cascadae* (*C*) tadpoles in an artificial environment with equal areas of striped and square-patterned substrate. Tests ran 180 seconds. A point plotted on the horizontal line = 90 seconds on striped substrate and 90 seconds on square-patterned substrate; +20 = 110 seconds on stripes and 70 seconds on squares, and so forth. Vertical lines include 95 percent confidence interval on either side of mean. Controls were raised on a featureless substrate. (Modified from Wiens, 1972.)

Site tenacity

Experience gained by using a site once may compensate for the higher success rate that might eventually be attained at another site. This advantage is responsible for site tenacity, the tendency to remain in an area, once established. Richdale and Warham (1973) noted that several Buller's mollymawks (*Diomedea bulleri*) returned to the same nest sites year after year over periods of up to twenty-five years. In fact, site tenacity, particularly in the case of males, may be considerably greater than mate fidelity in procellariiform birds (albatross, shearwaters, storm petrels) . So-called mate fidelity in some species apparently results from the fact that both individuals return to the same site with extremely high predictability—for example,

Madeiran storm petrel (*Oceanodroma castro*) (Allan, 1962). Laysan albatross (*Diomedea immutabilis*) have been observed attempting to use their accustomed sites after they were turned into a construction area (Rice and Kenyon, 1962). It is not known whether nesting success in these species increases site tenacity, although this is probable (Wilbur, 1969). Nesting success definitely increases nest-site tenacity in kittiwakes (*Rissa tridactyla*) (Coulson, 1966), which are cliff-nesting gulls (see table 4.1).

These cases involve stable habitats. Looking at gulls and terns (Laridae) as a group, McNicholl (1975) argued that nest-site tenacity in these long-lived birds is strongly developed in stable habitats, and that it is much reduced in unstable habitats. Only a species with low nest-site tenacity could colonize unstable areas such as marshes and prairie lakes with fluctuating water levels.

Among short-lived passerine birds the evidence for tenacity is weaker, although several aspects of tenacity resemble those in sea birds. Site tenacity does not increase with age in most passerines (Darley, Scott, and Taylor, 1977), although such an increase does occur in the hole-nesting pied flycatcher (*Ficedula hypoleuca*) (Haartman, 1949). Passerine males are more tenacious than females, and successful breeders return to a previous site more frequently than unsuccesssful breeders (Darley et al., 1977).

Although an individual usually settles in an area used by other members of its species, it may choose an atypical site if its normal habitat is saturated. It may subsequently remain in the new habitat even if other sites become available. If the new site is part of a vacant niche, to which the species is preadapted, the probability of successfully colonizing this habitat is enhanced if the initial settlers display strong site tenacity.

In practice, individuals who move into unusual sites usually shift

Table 4.1. Proportion of kittiwake gulls that retained or changed mates related to hatching success in previous year. (From Coulson, 1966.)

Gulls	Retained Same Mate	Changed Mate	Percentage Changed Mate	Percentage Changed Mate, Excluding Deaths
Whose eggs hatched in previous season	233	92	28	17
Whose eggs failed to hatch in previous season	21	36	63	52

back to more typical habitats when they become available (Hildén, 1965). American redstarts *(Setophaga ruticilla)* usually nest in mature deciduous forests, but a high proportion of first-year birds use other habitats including immature deciduous forests and coniferous forests (Ficken and Ficken, 1967; Morse, 1973a) (fig. 4.6). The proportions of old and first-year birds in different habitats strongly suggest that a shift occurs, although differential survival could be part of the explanation. Krebs (1971) has documented shifts between habitats by individually marked English great tits *(Parus major)* when nearby sites become available in normally preferred habitats. The preferred habitat in this case is deciduous forest and the alternate habitat is adjacent hedgerow. These moves are made more frequently by first-year birds that have just established territories for the first time than by individuals that have already bred successfully (Krebs, 1971). This observation suggests a site tenacity adaptively tempered by past experience. Past success should serve as an indicator of a site's suitability. Since the characteristics of a new site (best foraging area, nest location, sing-

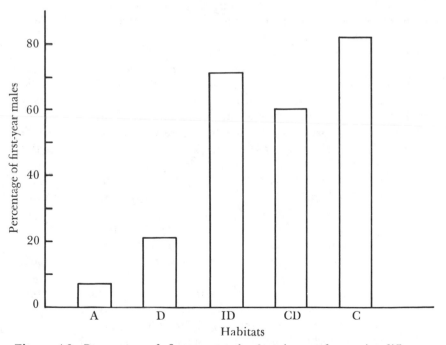

Figure 4.6 Percentage of first-year male American redstarts in different habitats. A = alder; D = deciduous; ID = immature deciduous; CD = mixed coniferous-deciduous; C = coniferous. (Modified from Ficken and Ficken, 1967.)

ing perch, places to hide from predators) must be learned, initial reproductive success on a new site might be lower than on an old one, even if the new site has greater ultimate potential. If the number of subsequent broods to be produced is small, as is the case with song-birds such as the redstart and the great tit, a relatively poor showing in the first year following a move might not be recouped in the individual's lifetime. Thus unless the two habitats differ greatly, site tenacity may be favored, especially if initial investment is high (for example, when a nest hole or burrow is constructed subsequent to establishment). Also, tenacity should increase with time. The risk of decreasing fitness by changing sites is lower in a first breeder, because its future remaining reproductive output is greater than that of an older bird, and it has not yet learned the characteristics of an area.

When established individuals persist in sites that change radically during their tenure, a new habitat may become colonized by their offspring (see Hildén, 1965). Several habitats disturbed by man have been colonized rapidly in this way; for example, the mistle thrush (*Turdus viscivorus*) has expanded its range from coniferous forests into parks (Peitzmeier, 1947). Offspring could perpetuate an initial error either by returning to this area or by imprinting on the novel habitat and choosing a similar area elsewhere. Presumably their chances of success are low in these new habitats, and some evidence suggests that young may be repelled by conditions to which adults return faithfully—for example, terns (*Sterna* spp.) (Austin, 1949). But even occasional successes would maintain the saturation of most habitats, including the new habitats created by human activities.

Drost (1955, 1958) sent about one thousand herring gull chicks to be reared in several zoos in the interior of Germany and found that, at reproductive age, several of these birds returned to their rearing sites or to nearby locations. However, some subsequently appeared at distant inland sites similar to those at which they were reared and others appeared at colonies on the coast. These results suggest that several different factors may be responsible for the settling patterns of individuals, even within a single population. If the ages of the chicks differed when they were removed, this could have affected the results.

New permanent populations are sometimes established by invasion. For example, red crossbills (*Loxia curvirostra*) have established breeding colonies in several areas following large winter incursions. Isolated populations far south of their usual breeding grounds (England, Philippines, Indochina) probably arose in this way. Southerly breeding of northern winter birds is not unusual after such incursions, but new populations are not often established (Newton, 1970). However, low success rates over long stretches of time suffice to saturate habitats (crossbill populations in England were established in pine plantations, where previously no satisfactory habitat had existed).

Thus in the few cases investigated, events occurring early in life apparently exert major effects upon subsequent choice of habitat. Experience obtained from using a site once probably often compensates for higher success rates than might be attained at other sites, particularly when the period of use is limited. Since it has been shown in a few cases that site tenacity or imprinting to habitat can provide the mechanism for habitat expansion, it is of interest to ask what the relative importance of this factor might be in colonization. The matter has theoretical importance because it has been suggested, and in certain cases demonstrated, that habitat shifts represent the first stages of ecological release, rather than changes of foraging behavior per se. A systematic analysis (of more evidence than is now available) is required to assess this matter adequately.

Tradition

Tradition is the attachment of a lineage of animals to a site from one generation to another. It appears to be important in determining the sites occupied by group-forming animals. King (1955) demonstrated that adult female prairie dogs (*Cynomys ludovicianus*) often desert their social unit (coterie), leaving their young behind and moving to the outside of the colony. This behavior should have an extremely conservative effect on the areas used by young female prairie dogs, although the effect on the adults is unclear. This system bears a remarkable similarity to the swarming of honeybees (*Apis mellifera*), which often results in the old queen leaving her hive to a daughter (see Michener, 1974). Herds of red deer (*Cervus elaphus*) are typically led by mature females who presumably know the best areas for feeding, calving, and so forth (Darling, 1937). Again, the result will be that active selection by young individuals plays a relatively small role in patterns of habitat exploitation.

In several species the young of one sex, usually the female in mammals and the male in birds, often remain with the group permanently (Schaller, 1963; Dittus, 1977; Brown, 1978). In some species the juveniles of the dispersing sex subsequently form bachelor groups—for example, toque monkey (*Macaca sinica*) (Dittus, 1977)—and in others they are usually solitary—for example, gorilla (*Gorilla gorilla*) (Schaller, 1963). These individuals have to make independent choices of the habitats they will exploit, which may differ from the prime habitat being used by the established groups. Young male mammals subsequently attempt to infiltrate or take over other groups. If they succeed, they will presumably be influenced by the preferences shown by this group. If they split off the main group with some members of the group they have infiltrated, forming a satellite group, this group may have to make new habitat choices, unless they establish themselves on a portion of the parent group's territory. Several factors thus exert an effect on the frequency with which social animals select sites, either ones similar to those in which they were reared or novel ones. It would be of interest to study whether patterns of habitat selection differ between the dispersing and nondispersing sexes.

Social stimulation

Social stimulation is the attraction of individuals to the presence of other individuals, usually but not necessarily members of the same species. Although the overall importance of social stimulation to settlement is unclear, it may decrease the frequency of independent habitat choice. For example, the presence or absence of other individuals may determine whether an area is used by some social species. Social species frequently occupy certain sites to the exclusion of other seemingly appropriate locations. Gull colonies may even switch to previously unoccupied sites from time to time, suggesting that discrimination of one site at a particular time to the exclusion of others does not result from subtle cues that make seemingly attractive areas inappropriate (Klopfer and Hailman, 1965). Sociality, in turn, results from other advantages, such as improved acquisition of food or avoidance of predators.

Social stimulation may also occur at an interspecific level and account for the presence of individuals in habitats that they otherwise seldom occupy. Juncos (*Junco hyemalis*) often follow flocks of chickadees and titmice as they forage along the edges of forests in winter. However, the chickadees and titmice are typically forest-dwelling species and usually soon retreat into tall growth. The juncos sometimes follow them through the forest for considerable periods of time, although these areas are not typical of the edge and open-country habitats that they usually occupy at this season (Morse, 1970b) and probably provide them with relatively poor foraging (Fretwell, 1969b). Probably the juncos' tendency to follow chickadees and titmice, with which they overlap little in diet, results from a protective advantage enjoyed by consorting with species that typically forage in trees. The arboreal species often sound alarms in response to aerial or terrestrial predators long before the ground-foraging species would see them.

The presence of conspecific individuals may indicate that a site is satisfactory. Male birds displaying territorially often attract later arrivals (Persson, 1971), despite the fact that they will be rebuffed from the residents' own sites.

The presence of conspecifics may stimulate settling in several species with planktonic larvae, including oysters. The larvae have chemoreceptors that respond to metabolic products of the individuals already present (Crisp and Meadows, 1963; Crisp, 1967).

Other factors

Little effort has been made to compare the mechanisms of habitat selection among closely related species. Grant and Ulmer's comparison (1974) of shell selection in two similar sympatric hermit crabs is of particular interest because they found marked differences between the species. *Pagurus acadianus* selects periwinkle (*Littorina littorea*) shells even when they are in low supply, and *P. pubescens* selects either periwinkle or dog whelk (*Thais lapillus*) shells roughly in proportion to their abundance (fig. 4.7). Where the two hermit crabs overlap in subtidal areas of northeastern North America, these preferences should affect their relative abundances. Although Grant and Ulmer believed that segregation does not result from interspecific aggression, the differences in shell preference between the two species are consistent with

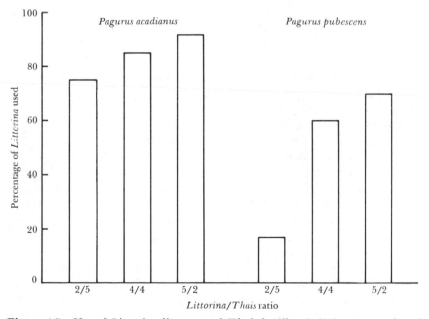

Figure 4.7 Use of *Littorina littorea* and *Thais lapillus* shells by two species of hermit crabs, *Pagurus acadianus* and *P. pubescens*, when the shells were provided in varying ratios. (Modified from Grant and Ulmer, 1974.)

acadianus displacing *pubescens* for shell supplies. Interspecific fights for shells have been recorded among other species of hermit crabs (Hazlett, 1970), which increases the plausibility of such a mechanism. Aggressive interactions over resources or space are widespread between species in many groups (see chapter 11; Morse, 1974).

The preferences of an individual may change over time independently of environmental fluctuations. This tendency may regularly accompany indeterminate growth patterns because in a heterogeneous environment increasing body size will be accompanied by changes in the sites that can be most effectively exploited. The manini studies of Sale (1968, 1969) demonstrate continually changing preferences for progressively increasing water depth and cover as these fish increase in size, a shift closely reflecting their changing food preferences and vulnerability to predators. Crayfish of different sizes vary in choice of substrate, with the small ones choosing areas that permit rapid burrowing. Individuals that are too big to be caught by bass typically choose coarser substrate (Stein and Magnuson, 1976).

Different members of a population, even those of the same age class, often occupy distinctly different habitats. These differences could result simply from the varying pressures of intraspecific and interspecific competition (Svärdson, 1949), which when heavy would prevent many individuals from occupying preferred habitats. Although this factor often seems largely responsible for distributional patterns (see Morse, 1974), Catchpole (1974) has demonstrated that active differences in habitat choice exist within populations of reed warblers (*Acrocephalus scirpaceus*), with some individuals preferring reed beds and others upland habitats. Similarly, individual differences are widespread among larval invertebrates (Meadows and Campbell, 1972b).

A population's distribution may be affected by selective predation, with differential survival between habitats or sites. The restriction could be the immediate result either of active selection for habitats that give protection from predators, or of direct differential predation. Rosenzweig (1973) hypothesized that the absence of short-legged pocket mice from the open areas frequented by long-legged kangaroo rats may result from habitat selection mediated by high predatory pressure in open areas. On the other hand, Stein and Magnuson's (1976) work on crayfish shows that they may vary their choices depending on the presence or absence of predators themselves.

Unfortunately, most studies of selective predation in different habitats deal only with strongly cryptic morphs. Examples include the well-known studies of industrial melanism in the pepper moth (*Biston betularia*) (Kettlewell, 1955; Bishop, 1972) and of shell color and pattern in land snails (*Cepaea* spp.) (Cain and Sheppard, 1954). Although other factors such as physiological differences of the various morphs may act at times (Bishop, 1972), the evidence for selective predation on conspicuous morphs remains strong.

The occupation of new habitats

Habitats totally new to a species cannot be colonized through site tenacity, imprinting to habitat, tradition, social stimulation, or other factors but may be entered through a series of steps. Individuals may occupy a habitat sharing some characteristics with that of their parents, but having novel characteristics as well. Subsequently, individuals from this new population may colonize another habitat that shares some characteristics with their parents' habitat, but not those of the ancestral population. The oystercatcher (*Haematopus ostralegus*) has colonized farmlands in the British Isles, probably by entering the interior along riverbeds (Heppleston, 1972). Although this habitat is strikingly different from the seashore, it does contain sandy or gravelly areas comparable to those on which oystercatchers usually nest. As populations increased, the birds may have started feeding away from the water. The agricultural areas opening up in the valleys provided an obvious opportunity. Saturation of streamside nesting areas might force some individuals to leave the river valleys entirely, especially if feeding was easy in farmlands lying considerable distances from the valleys themselves. Nesting areas in the farmland bear some resemblance to the ancestral seashore sites in that both are quite open. The populations just described reside in northeastern Scotland. Oystercatchers in some other parts of Britain now appear to be entering the initial stages of this process. Birds in northeastern England now show an increasing tendency to nest in river valleys, where they have not nested in the recent past (Heppleston, 1972).

Another example is the apparently recent colonization of the boreal spruce forest by mourning doves (*Zenaida macroura*) (Morse, 1975b). This is one of the few areas in North America that they had not previously occupied successfully (Aldrich and Duvall, 1958). In the late 1960s a few birds frequented large openings produced by hurricanes. Mourning doves typically inhabit open, brushy sites and often nest in conifers, so these areas readily met their needs. In 1974 a pair of these birds occupied a site in the midst of a largely undisturbed forest adjacent to the hurricane-damaged area. They foraged primarily in the spruce forest itself and successfully raised at least one young bird, a feat repeated by a pair the following year (D. H. Morse, unpublished). Presumably these birds came from the small population in the hurricane-damaged area, since no other dove populations exist in the immediate vicinity.

It may seem difficult to employ this model for the occupation of urban habitats, but at least some of the species that occupy highly urban situations occur in suburban areas as well (see Weber, 1972), implying a stepwise colonization. Certain species living about office buildings and other sites providing little in the way of natural conditions occupy cliffs and similar situations in the wild. The common street pigeon or rock dove (*Columba livia*) lives on cliffs over many parts of its natural range, and a major predator of the pigeons, the peregrine falcon (*Falco peregrinus*), is typically a cliff nester (Bent, 1938; Hickey, 1969). Until recently the peregrine falcon nested on skyscrapers in several major urban areas.

Competition

Preferred sites may all be taken where sizable numbers of conspecifics are present, necessitating a contest for already filled sites and occupation of secondary areas. The relegation of first-year great tits and American redstarts to atypical and presumably suboptimal habitat can be accounted for in this way. Where population sizes fluctuate markedly, as in several microtine rodents, occupation of secondary habitat may be commonplace at times (Grant, 1971). In the presence of larger dominant individuals, small bluegill sunfish (*Lepomis macrochirus*) preferentially choose temperature regimes known from previous control runs not to be favored, rather than enter areas of preferred temperature frequented by larger dominant bluegills (Beitinger and Magnuson, 1975).

Other species may act in a similar way, their importance depending on their abundance, social dominance, and similarity of preferences (Svärdson, 1949; Morse, 1974). The commonness of ecological release or niche expansion (MacArthur and Wilson, 1967; Morse, 1974) illustrates the importance of other species in this regard. In experiments with meadow voles and red-backed voles (*Clethrionomys gapperi*) in large outdoor enclosures, tendencies to occupy secondary habitats depend on whether the other species occupies the unaccustomed area (Grant, 1969; Morris and Grant, 1972).

If competition is regularly experienced over a long period of time, it may be the selective factor promoting genetic differences in preferences among species, automatically reducing the intensity of competition. Past competitive interactions could well account for differences in habitat selection between kangaroo rats (*Dipodomys merriami*) and pocket mice (*Perognathus penicillatus*); the species prefer different heights of cover (Rosenzweig, 1973) and appear to have different predator avoidance capabilities. States (1976) believed that past competition may be responsible for differences in habitat choice among chipmunks (*Eutamias* spp.) in the western United States. Such considerations led Klopfer and Hailman (1965) to predict that individuals living with many competitors should exhibit more rigid habitat selection than others. Unless individuals are predictably successful in interactions with other species, however, release will still occur. Flexibility in habitat selection may only be discovered if a population is sampled at times when the densities of their competitors differ.

Nest-site selection

The choice of a site in which to lay eggs or give birth to young is a factor of great importance to the reproductive success of many animals. It is not surprising to find that selection mechanisms for nest sites are often quite precise, given the impossibility of moving the eggs or young for a substantial period of time. Not only must the site be suitable for the young at all times, but the parents themselves may have particular requirements if they care for their offspring. In marsh-nesting laughing gulls (*Larus atricilla*), the site must have a landing place (Klopfer and Hailman, 1965). Seabirds requiring considerable distances in which to become airborne must nest either adjacent to places where the adults have a runway or where they can jump off cliffs—for example, gannet (*Sula bassana*) (Nelson, 1966). Birds such as loons or grebes, which can scarcely walk on land, must choose locations at the water's edge (Palmer, 1962). Colonial seabirds, which typically have little or no nest concealment, must choose sites where predators are rare or absent (Kruuk, 1964), and this restraint is probably responsible for their usual confinement to isolated islands.

Nest-site selection in small terrestrial birds may also include some rather unusual factors. Individuals nesting in sedges or rushes should not attach their nests partly to a dead stalk and partly to a live, growing one. This mistake is occasionally made by red-winged blackbirds (*Agelaius phoeniceus*), causing nests to be tipped on their sides, emptying their contents into the water or onto damp earth below (Allen, 1914).

Commensal or mutualistic relationships with other species may determine nest placement. N. G. Smith (1968) found that oropendulas (*Zarhynchus wagleri*) and caciques (*Cacicus cela*) frequently nest in close association with wasps (*Protopolybia* and *Stelopolybia*) or stingless bees (*Trigona*). These wasps and bees effectively prevent the approach of botflies (*Philornis*) that would otherwise parasitize the young birds heavily, causing considerable mortality. Similarly, several species of small birds sometimes nest within the huge platformlike nests of ospreys, which are exclusively fish eaters and typically drive off other raptors (Bent, 1937).

Synthesis

In a broad sense the selective basis for habitat selection is apparent; it will result in improved reproductive success. For example, individual great tits nesting in broadleaf woodlands fledge more young than those in adjacent pinewoods (Tinbergen, 1960). However, Krebs (1971) has pointed out that great tits nesting in pinewoods may represent a different subunit of the population than those of the broadleaf woodland and that success rates of these individuals might be low even in broadleaf if they were able to enter it. Few studies incorporate individual variation into considerations of habitat selection. For vertebrates it is often difficult to design experiments that will readily circumvent this difficulty and at the same time measure a relevant aspect of fitness. Studies on shell choice, suitability, and use in hermit crabs (Bach et al., 1976) are of special interest in this regard, for they can readily test such characteristics as vulnerability to predation and clutch size. On the other hand, the compact set of requirements satisfied by an appropriate shell may bear little resemblance to the large, all-purpose activity spaces required by highly mobile vertebrates.

As a type of choice behavior, habitat selection bears definite similarities to food selection, and many of the same strictures may apply. In fact, habitat selection might appropriately be considered a logical extension of patch selection, with the major difference from foraging theory lying in the time constraints introduced by foraging in patches. The choice of habitat is typically of longer term, frequently a permanent one. The nature of the food supply may in some cases be the most important determinant of habitat choice. Often food acts as an ultimate rather than a proximate factor. This would seem to be the case with the spruce-forest warblers, which prefer to establish territories in places that at the time of choice provide poorer food supplies than do their secondary habitat choices.

The overall perceptual basis for habitat selection remains an open question, as do the roles of innate and learned components. Early learning may be difficult to separate from innate patterns of habitat selection, which complicates analysis. It is easy to envision both innate and learned factors playing a role, the coarse and fine tuning of the system, but it is not yet possible to say how frequent such a pattern may really be.

Habitat selection turns out to be related in several instances to the social system of the population in question, with patterns of land tenure frequently substituting for active selection. Social systems themselves are now believed to be determined in part by the environmental conditions experienced (Crook, 1965; E. O. Wilson, 1975). The relative conservatism in habitat selection by animals with and without tradition should provide evidence about the importance of continually experienced environmental factors. So too should differences in the mechanisms of habitat selection between the sexes of species with tradition, where one normally disperses and the other does not.

This is a very good time to study the mechanisms of habitat selection, in that many natural habitats are undergoing changes as a result of man's activities. These changes provide animals with potential opportunities and challenges, ones that in many instances they may never have previously experienced. Under this regime the numbers of many species decline precipitously, sometimes to extinction. Yet other species prosper, often occupying habitats which only a short time in the past were either rare or nonexistent. Some of the successful species exploit conditions that, at least to our eyes, are quite different from those in which they evolved. It would be instructive to compare mechanisms of habitat selection in some of these highly adaptable species, such as rock doves, house sparrows (*Passer domesticus*), or house mice (*Mus musculus*), under both pristine and modified conditions.

5 *Avoiding Predation*

DEFENSE AGAINST PREDATION reduces the efficiency with which most animals can feed (Stein and Magnuson, 1976). A few species elude their predators by outgrowing them (Paine, 1976), but this appears to be unusual. One of the most invulnerable of mammals, the African elephant (*Loxodonta africana*), loses its young to lions on occasion (Sikes, 1971); that this is a serious problem may be inferred from the well-developed protective behavior of the parents (Sikes, 1971; Schaller, 1972). In fact, protecting the young from predation is a major part of parental care.

The compromise that must be made between efficient resource exploitation and the avoidance of predators will depend on the abundance and type of predators, the kind of resources exploited by the potential prey, and various attributes of the prey themselves (fig. 5.1). For example, it seems unlikely that a small, highly efficient, fleet-footed ruminant will evolve. Rumination, an effective way to break down plant material containing large amounts of cellulose, requires a large and complex digestive system; there should be a minimum size at which this system can operate at high efficiency. A small ruminant might be so overburdened by its digestive system that it could not easily escape predators (see Janis, 1976). The smallest antelopes in fact feed selectively on high-energy foods, and this appears to have affected the social organization of several small ungulates (Jarman, 1974).

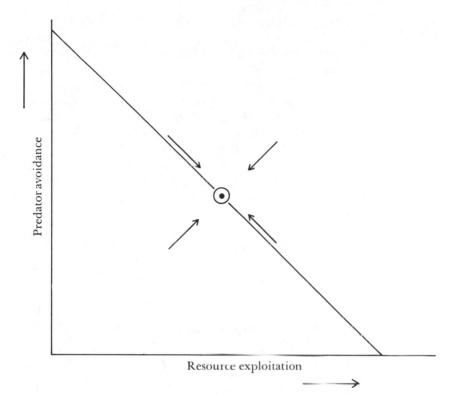

Figure 5.1 Interaction between efficiency of resource exploitation and success at predator avoidance. Resource exploitation is measured as net energy gain per unit of time, predator avoidance as the reciprocal of predation rate. The diagonal line represents the maximal development of the two traits at steady state. Neither can remain above this line; below the line, efficiency of both can increase. The circle represents a stable equilibrium point for a case in which both traits are of equal importance.

Responses to predators

For convenience, I shall separate responses to predators into several categories: ignoring a predator, moving quietly out of contact, freezing, running away, stopping other activities and keeping the predator under surveillance, mobbing, other evasive responses, confronting or attacking a would-be predator, and other defensive postures.

IGNORING A PREDATOR

No animal can ignore a predator unless it is unlikely to attack (see Smythe, 1970b; Young, 1971). In some instances a predator may incidentally inform vulnerable prey that they will not be attacked;

this probably happens with food-engorged lions (*Panthera leo*) (Schaller, 1972) and wild dogs (*Lycaon pictus*) (Malcolm and van Lawick, 1975). However, it is difficult to know whether an animal has actually perceived a predator.

Individuals cannot always know that predators are or are not present. Considerable evidence suggests that hungry individuals often pay less attention to predators than do those who are well fed. This phenomenon can be seen in many birds immediately after periods of very severe winter weather (Morse, 1970b; Griscom, 1941; Batten, 1971), or after cold periods during the spring migration of small insectivorous birds, when their accustomed insect prey are temporarily unavailable (Zumeta and Holmes, 1978). In extreme cases this tameness may be no more than an expression of physiological insensitivity, but the end result is that the affected individual devotes all of its efforts to searching for food, at a high potential cost. It could, of course, be advantageous to ignore predators when the nourishment at hand is absolutely required for survival.

Moving out of contact

Prey may hide and sneak away, such that the predator remains unaware of their presence. Since no overt response is given to the predator, it may again be difficult to determine whether the prey is responding to the predator. Prey living in heavy cover probably have the best opportunities to elude predators without drawing attention to themselves. Still, there is danger in this strategy because the prey loses contact with the predator and thus increases the probability of later being taken by surprise (Smythe, 1970b), or of encountering another predator in the area into which it has moved to avoid the first (Young, 1971).

Social groups using this strategy may emit calls of such low intensity that they are not perceived by the predator. For example, turkeys (*Meleagris gallopavo*) give soft singing notes (Schleidt, 1961).

Freezing

If the potential prey is not conspicuous, freezing (remaining motionless) may be a feasible alternative as long as the predator's search path is unlikely to cover the prey's location (Smythe, 1970b). Because the prey has the predator under surveillance, it can flee if the predator approaches to within a critical distance. Freezing is a practical option when the distance between the predator and prey is too short to allow moving out of contact, and it is often associated with the development of colors and patterns that are cryptic against the prey's typical background (see Cott, 1940). This kind of defense, as well as the previous one, will be effective only where the predator hunts in areas much larger than those frequented by the prey. Otherwise, the prey will spend an inordinate amount of time hiding. This is probably not often a serious constraint because most predators cover areas that are large relative to the activity range of their prey in order to obtain adequate food (Schleidt, 1961; Schoener, 1968b). Difficulties do arise if the predator has long-range sensory abilities (for example, visual predators in open areas).

Animals may freeze in response to calls given by conspecifics or by members of other species. This behavior regularly occurs in flocks of birds, when one individual first spots a predator. Chickadees and titmice (*Parus* spp.) freeze immediately on hearing a call from another individual if they are located in leafless trees at considerable distance from cover (Morse, 1970b).

RUNNING AWAY

Only by stalking or ambushing can predators approach cursorial prey, whose top speed is almost always well above that of the predator's (Schaller, 1972). For example, most prey are not vulnerable to a tiger (*Panthera tigris*) unless it is initially within 20 to 30 m (Schaller, 1967), while the area of danger around a cheetah (*Acinonyx jubatus*) or a pack of wild dogs may be 100 m or more (Schaller, 1972). Flight distance may vary within a species depending on the individual's reproductive condition or other variables (Rowe-Rowe, 1974).

If a predator suddenly confronts prey at close range, the prey may have no alternative but to run away or to stand its ground and fight. The hunting strategies of roaming predators must maximize the frequency with which this dilemma confronts prey, because in this situation the prey cannot attain flight speed before attack. Even so, if the prey is highly maneuverable, it may elude the predator during this critical period (Howland, 1974).

The cursorial condition limits the other modes of defense that might be developed in prey species. To take an extreme example, it is incompatible with heavy armor. Not surprisingly, the sensory capabilities of cursorial prey are acute, and they typically confine their activities as much as possible to locations where ambush is unlikely. Wildebeest (*Connochaetes taurinus*) show extreme reluctance to ford streams where they may be ambushed by lions; if it is necessary, they stampede through them, sometimes drowning in large numbers as a result (Schaller, 1972). If potential prey are social they can minimize the chance of surprise through group alertness (see chapter 12).

The three alternatives discussed so far can be used by prey to cope with predators at different distances from them. Running is usually a response to a predator that has already detected the prey. Moving into burrows or other inaccessible areas (usually known in advance, or even constructed by the users) is an analogous strategy of predator avoidance. A given individual may employ one technique or another depending on the circumstances. For example, titmice surprised by a hawk in the midst of the leafless canopy of a deciduous forest in winter will freeze in position, but if they see this predator well before it approaches closely they will dive into dense cover, usually in the understory (Morse, 1970b). Often birds that have initially frozen in a seemingly vulnerable position move to a lower height and more protected location subsequent to their initial response. This suggests that the protection obtained by freezing is not as great as that obtained by reaching adequate cover, in which hawks are unable to attack on the wing. However, flight at the initial point of discovery would presumably result in even higher mortality if the predator were too close. The relatively great facility with which some bird-eating hawks (*Accipiter* spp.) pursue prey on foot through dense cover (Bent, 1937) is probably an adaptation to the mixed defense employed by these small birds.

Keeping the predator under surveillance

Prey may be able to spot predators at considerable distances in open areas; in such places they may be within view of predators much of the time. Under such circumstances habitual fleeing is not feasible, if only for energetic reasons. Many workers believe that cooperative surveillance of predators is a major advantage of participation in social groups (see chapter 12).

Some prey respond to different behavioral characteristics of the predator. Resting lions, for instance, do not typically elicit an evasive response from grazing mammals as long as they remain visible (Schaller, 1972). Vocalizations and visual displays of the prey species may keep others apprised of the predator's location. At such times prey simply keep a modest flight distance between themselves and the lions, which seldom attack (Schaller, 1972).

Mobbing

Some prey mob their predators; they approach, display, and sometimes even attack a threatening predator. Often many individuals, even of several different species, will mob a predator simultaneously. Mobbing has been investigated most thoroughly in birds (Hinde, 1954a, b; Curio 1969, 1975, 1978), although it occurs in mammals (Smythe, 1970b; Schaller, 1972) and fish (Eibl-Eibesfeldt, 1964). In birds mobbing is usually accompanied by loud vocalizations, which attract other birds. Such displays alert prey to the presence of a predator, and in some instances drive the predator from the area. Mobbing also tells the predator that prey are aware of it and are able to escape (see Smythe, 1970b). Small birds mob most frequently where dense vegetation provides cover and the predators involved are ambush hunters. However, open-country species such as antelope may also mob, keeping a critical distance from the predator (Schaller, 1972).

Sometimes mobbing animals are captured by predators (Bent, 1937; Smith, 1969; Myers, 1978). Small falcons of the neotropical forest may even provoke mobbing as a hunting technique (Smith, 1969). Species that mob are well able to interpret the behavior of the predator and respond quickly. Chaffinches (*Fringilla coelobs*) and many other species will mob a perched hawk or owl, but they respond very differently to the same bird in flight (Hinde, 1954a, b). The time and energy devoted to mobbing and the mortality sometimes associated with it indicate its great significance.

There is controversy over whether the greatest benefit of mobbing accrues to the first mobbers or to others (see Harvey and Greenwood, 1978). The first mobbers might profit from any of the advantages suggested above, but they could also be the most vulnerable. Since joiners are presumably as alert as the initiator, the advantage of initiating remains obscure. In family groups, initiating might be favored by kin selection (chapter 7); the benefits would tend to fall on relatives of the initiator, increasing the initiator's inclusive fitness (Hamilton, 1964) in spite of the direct cost it might suffer.

EVASIVE RESPONSES

Many small animals have elaborate alarm responses consisting of vocalizations and evasive maneuvers. The predator is made aware that potential prey have seen it, even if it had not previously seen the prey, which makes this sort of evasion different from moving out of contact. Evasive responses are widespread among social birds, although by no means confined to them.

It is believed that some of the calls given repetitively by small diurnal birds are difficult for a predator to locate (Marler, 1955, 1957). In the only tests of this hypothesis, barn owls (*Tyto alba*), pygmy owls (*Glaucidium* spp.), and goshawks (*Accipiter gentilis*) were able to locate all vocalizations presented to them (Shalter and Schleidt, 1977; Shalter, 1978a) (fig. 5.2). These calls thus must have a different basis than previously attributed to them (Shalter and Schleidt, 1977), although they nevertheless play a role in escaping predators. However, the predators did exhibit lower levels of response to these notes than

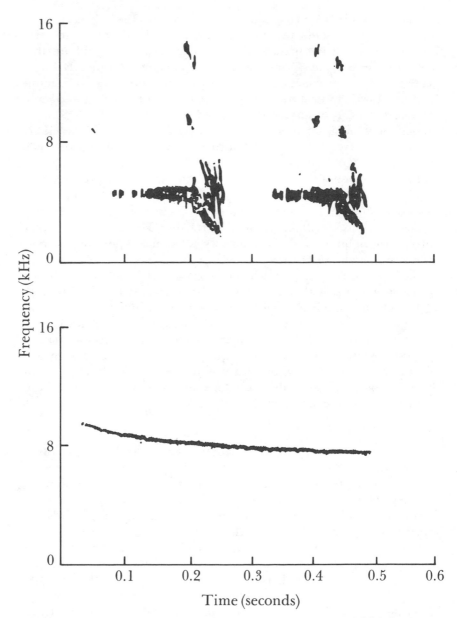

Figure 5.2 Sound spectrograms of mobbing call (*top*) and high-frequency alarm call (*bottom*). The alarm call is generally believed difficult for a predator to localize (however, see text), whereas the mobbing call presumably presents no such difficulties. The mobbing call is from a chaffinch (*Fringilla coelobs*); the alarm call, a European blackbird (*Turdus merula*). (Redrawn from Shalter, 1978a.)

to the vocalizations of mobbing birds; this was particularly true of individuals who had hunted under natural conditions. Shalter and Schleidt (1977) and Shalter (1978a) suggested that the calls tell the predator it has been discovered and will stand little chance of catching prey. They attribute the strikingly different physical characteristics of these notes and other calls to Darwin's (1872) principle of antithesis, which postulates that structurally distinct signals will minimize the possibility of error. Harvey and Greenwood (1978) criticize Shalter and Schleidt for using owls in these studies, pointing out that owls are nocturnal predators. But it is well known by ornithologists that owls frequently hunt on dark days when small birds are still active.

Rooke and Knight (1977) have challenged Marler's hypothesis on theoretical grounds, arguing that alarm calls are designed to be audible to predators. Shalter and Schleidt's experiments are consistent with Rooke and Knight's arguments.

Some small birds, such as chickadees and titmice, give two different vocalizations in response to flying predators under different contexts; one is given when a predator appears suddenly at extremely close range and the other is given when a predator is first detected at a greater range (Morse, 1970b). In either case evasive behavior follows, and individuals become silent or dive into the nearest dense vegetation. The exact response depends on how close the individual is to cover at the moment of alarm. If it cannot quickly reach cover under such circumstances, flying will only call attention to itself (Morse, 1970b). Schleidt (1961) noted that turkeys use four distinct calls in such contexts.

A "confusion chorus" (Miller, 1921), in which several birds utter calls, often follows these initial notes and responses. It is believed that vocalizations coming from several points in space make it impossible for a predator to concentrate on one individual. These calls are generally continued while the predator remains in the vicinity. A different "all-clear" call causes group members to resume their previous activities (Morse, 1970a). This pattern of behavior is used even by birds that experience low frequencies of attack.

Social animals give several other kinds of displays that warn group members of danger. In general, visual signals predominate where they can be easily seen (open country). For example, many plains-dwelling species, such as antelopes or pronghorns, have characteristic flash marks. It is essential to the theory of natural selection that the individual profit from giving these displays, either directly or through kin selection. Zahavi (in Dawkins, 1976) suggested that the visual displays given by ungulates decrease their individual vulnerability by drawing attention to less agile animals. Although little precise information exists about the relatedness of the members of most natural populations, a few studies indicate close relationships in relatively permanent groups (Brown, 1970; Ridpath, 1972; Schaller, 1972; Zahavi, 1974).

The distracting displays given by parents seem likely to increase their vulnerability but are easily accounted for by the effects on their offspring. There is good evidence that a giver of warning calls may increase its vulnerability to predators. Sherman (1977) found that Belding ground squirrels (*Spermophilus beldingi*) characteristically call when a predator (weasel, badger, dog, coyote, marten) enters their colony. However, the tendency to call is not randomly distributed among the group members. Old females with surviving offspring give the most calls, males the fewest (fig. 5.3). Because males are the dispersing sex, they are less closely related to other members of the group than females. Sherman documented regular predation during this study and found that callers were captured several times.

Distraction displays by parents may lure predators away from immobile or vulnerable young. Parents often pretend to have a broken wing or some other injury (fig. 5.4). The distractor retains a critical flight distance; otherwise the predator would stand an excellent chance of gaining an even larger reward than the young would provide. These displays are widespread in birds (Armstrong, 1947; Simmons, 1952), but also occur in some mammals (Rush, 1932; Kruuk, 1972b).

Smythe (1970b) attempted to explain the stiff-legged displays of cursorial mammals (stotting) in the presence of predators. He suggested that these displays advertise the performer's knowledge of the predator's presence and its ability to escape should the predator attack. This "pursuit-invitation" model should work for both the predator and the prey. It would be to the prey's advantage to be interrupted for as short a time as possible from its other activities, and it would be to the predator's advantage not to waste time and energy on a pursuit with low probability of reward. Smythe noted that some species giving these displays are solitary. Here, he argued, their main function must be to tell the predator that the prospective prey is aware of its presence. The antelopelike caviid rodent *Dolichotis patagonum* of the Argentine pampas appears to signal directly to the potential predator rather than to a conspecific; individuals observed by Smythe lived in pairs, and when approached both began displaying simultaneously.

Pitcher (1979) argued that stotting may be maladaptive as a pursuit-invitation display where social predators such as lions (*Panthera leo*) or African wild dogs (*Lycaon pictus*) are involved. Stotting Thom-

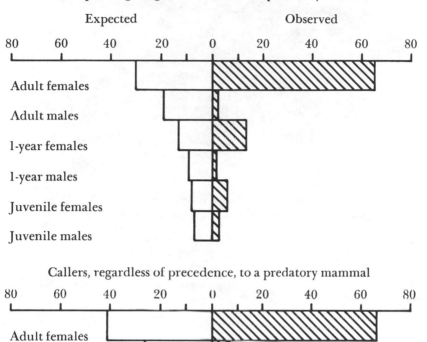

Figure 5.3 Expected (white bars) and observed (hatched bars) frequencies of alarm calls by various sex and age classes of Belding ground squirrels. Expected values were computed by assuming that animals call randomly, in direct proportion to the number of times they are present when a predatory mammal appears. (Modified from Sherman, 1977, copyright 1977 by the American Association for the Advancement of Science.)

son's gazelles (*Gazella thomsonii*) are frequently chased by predators and sometimes caught. Pitcher suggested that the bounding leaps permit an individual to spot concealed members of the hunting group.

Charnov and Krebs (1975) further developed the idea that the

Figure 5.4 Distraction display of kentish plover (*Charadrius alexandrinus*) about its nest. (Redrawn by Jaquin Schulz from Simmons, 1952.)

performer of warning calls in a group may lower its vulnerability relative to others. This advantage is obtained by perceiving the danger first and acting first. The response warns other individuals who then act, increasing their own survival probabilities, but to a degree less than that of the sender (fig. 5.5). It is assumed that the predator's best strategy is to redirect its attention to individuals who have had less time in which to respond. The hypothesis does not depend on group or kin selection and thus could account for these responses in temporary seasonal groups, such as wintering flocks of migratory sparrows, which are unlikely to be composed of closely related individuals. Owens and Goss-Custard (1976) independently derived a similar explanation for the flocking tendencies of shorebirds. They proposed a number of tests that might distinguish between the models invoking kin selection and individual selection.

Chance and Russell (1959) and Humphries and Driver (1967, 1970) argued that erratic movements (protean behavior) lower the vulnerability of prey (fig. 5.6). Protean behavior prevents a predator from predicting in detail the next position or action of the prey. A priori, protean displays would seem a good strategy for an individual suddenly confronted by a sit-and-wait predator and without the opportunity to plot a strategy. Some examples of possible protean behavior include the irregular flight patterns of flocked birds such as starlings (*Sturnus vulgaris*) attacked by hawks, distress calls given by cornered prey, and the deceptive flight of several butterflies such as bright-colored morphos (*Morpho* spp.). Human beings and birds find the latter almost impossible to capture, apparently because of their

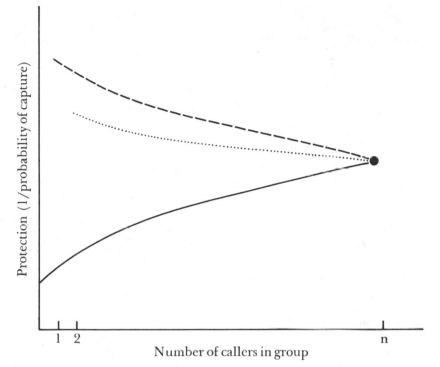

Figure 5.5 Protection to various members of a flock accruing from the Charnov Krebs hypothesis. Dashed line = first caller; dotted line = second caller; solid line = noncallers. Callers subsequent to the second caller will fall between the dotted and solid lines until all members of a flock become callers.

rapidly and irregularly flashing color patterns (Young, 1971). Several moths and insects show erratic flight patterns upon experiencing ultrasonic pulses, such as those produced by insectivorous bats (Roeder, 1962, 1975). Bats may improve their chances with irregularly flying prey by covering the prey with their wings as it is encountered (see Webster and Griffin, 1962), thus gaining a second attempt if they fail to seize it initially. Humphries and Driver argued that protean displays are particularly resistant to learned countermeasures, which makes them unique as antipredatory devices.

In more predictable situations, winged predators may single out an individual that responds differently from the others. Hawks attacking house sparrows or starlings in flocks most frequently captured those individuals that failed to execute movements made by the other individuals (Tinbergen, 1946; Rudebeck, 1951). The tendency of

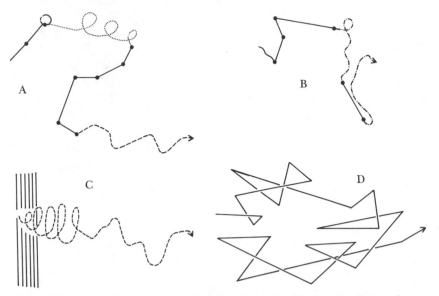

Figure 5.6 Examples of erratic displays. *A,* psychodid fly (*Psychoda phalae-noides*) disturbed by a tactile stimulus. *B,* escape reaction of a duck flea (*Ceratophyllus garei*). *C,* typical reaction of a chironomid midge disturbed from its resting place on a tree trunk. Solid line = hopping; dotted line = spinning; dashed line = flying; alternate dashes and dots = running. *D,* stickleback (*Gasterosteus* sp.) chased by a merganser duckling. (Redrawn from Humphries and Driver, 1970.)

predators to choose odd prey (see Mueller, 1971; chapter 3) could make predator avoidance strategies relatively conservative. Even a potentially superior escape technique might initially be exposed to a high intensity of predation, simply because it was different. Unfortunately, this effect would be difficult to test.

These sophisticated patterns of predator avoidance are not confined to birds and mammals. A number of relatively motile intertidal and bottom-dwelling invertebrates respond to the presence of predatory starfishes or gastropods with escape behaviors entailing high degrees of discrimination (Phillips, 1976; Hoffman and Weldon, 1978). Phillips (1976) has reported that two species of limpets (*Acmaea* spp.) evade species of starfish that normally prey on them, but not those that do not. They also do not evade species of starfish that would accept them as prey but that normally occur in different habitats (fig. 5.7). Although the discriminatory cues are chemical in nature, they appear to reach a level of specificity comparable to those seen in birds and mammals.

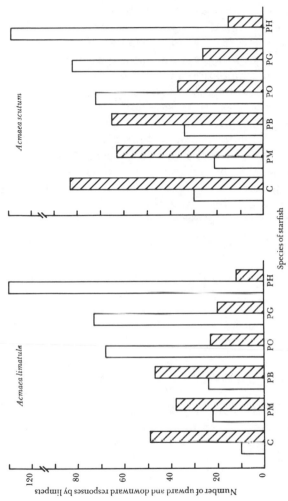

Figure 5.7 Responses of 200 intertidal limpets, *Acmaea limatula* and *A. scutum*, to several species of large starfish. Only those limpets moving more than 1 cm up or down are included. The criterion for a positive response to a starfish predator was increased upward movement by the population. White bars = upward movement; hatched bars = downward movement. Each starfish species was represented by approximately the same total wet weight (1,000 ± 50 g). C = control; PM = *Patiria miniata*, an omnivorous scavenger; PB = *Pisaster brevispinus*, predaceous but strictly subtidal when it occurs on rocky shores; PO = *Pi. ochraceus*; PG = *Pi. giganteus*; and PH = *Pycnopodia helianthoides*; all three are molluscan predators that at least occasionally inhabit the intertidal zone. (Modified from Phillips, 1976).

CONFRONTING A PREDATOR

Prey species that confront their predators are usually large relative to the predator or have specialized means of defense. Groups employ aggressive defense more often than individuals. Large herbivores such as elephants, rhinoceroses, buffalo, and gaur (*Bos taurus*) regularly defend against and even kill the largest species of predators (lions, tigers) that attack them or their young (Schaller, 1967, 1972). One might predict that many of the unlucky predators would be young individuals because discrimination in food selection is often learned; in some cases this is known to be true (Schaller, 1967). Predators' prey preferences may have a considerable effect on their own survivorship patterns. Some prides of lions take buffalos much more frequently than do others, and prey-inflicted injuries make up an appreciable component of their overall mortality (Schaller, 1972). Prey preference in lion prides may have a traditional basis (Schaller, 1972), a conclusion also supported by Rudnai's observations (1974) of long-term changes in cropping patterns by lions.

Prey that are slow relative to their predators are most likely to develop an active defense strategy, in many cases as a group, as exemplified by the circular defensive formation employed by the adult males of musk oxen (*Ovibos moschatus*) herds against wolves (*Canis lupus*) (Tener, 1965; Wilkinson, 1974). Large primates sometimes attack prospective predators outright (Eisenberg, Muckenhirn, and Rudran, 1972). Even flocks of small birds may turn on their predators, although this does not often happen. Gersdorf (1966) has several times observed a flock of starlings turn on a pursuing sparrow-hawk (*Accipiter nisus*); the starlings have actually driven the hawks into the water or onto the ground at high speed, sometimes with fatal results.

Species capable of retaliation against would-be predators often employ warning coloration or behavior to inform the predator of their unsuitability as prey. Other animals may avoid attack by mimicking these dangerous or unpalatable species (see Wickler, 1968). Although this deception is often largely morphological, it must usually be accompanied by appropriate behavior, particularly if the morphological similarity is relatively imperfect. For example, several hole-nesting titmice hiss if disturbed while incubating, and Sibley (1955) hypothesized that they are auditory mimics of snakes. While hissing, the birds sway with their mouths open, which presumably adds a visual component to the mimicry. The thought of a titmouse mimicking a snake seems ludicrous, but the overall effect may be quite compelling in the dim light of the cavities in which they nest.

Certain animals apparently do not defend themselves when left with no other alternative. Estes and Goddard (1967) observed that a wildebeest attacked by wild dogs often looks "less the victim than the witness of its own execution." However, this is uncharacteristic of the vast majority of prey.

OTHER DEFENSES

Large species often experience fewer predatory pressures than small ones. Certain top carnivores (tiger, lion) and large herbivores (elephant, rhinoceros), which tend to be vulnerable only when young, enjoy very low risks of predation. The feeding strategies of such animals may thus differ from those of animals at high risk. Consequently, it may be profitable to compare large and small animals, or mainland and island populations of the same or similar species, which often experience markedly different predatory pressures (Curio, 1969; Carlquist, 1974).

The formidable integument of some animals (armadillos, pangolins, porcupines, turtles) provides them with considerable protection against most predators. This kind of defense requires protective behavior (withdrawing or otherwise covering vulnerable parts), forcing the animal to cease other activities. But it also places large demands (for time and energy) on the predator who would break through such defenses. In spite of their advantages, armorlike defenses are not widespread among extant vertebrates. Armored endotherms are almost always omnivores, herbivores, or myrmecophages (for example, armadillos and pangolins) (see Matthews, 1971); they do not have to cope with highly mobile food items, and their foods are usually abundant. Most species with elaborate antipredatory mechanisms of this sort have necessarily compromised their foraging efficiency or their foraging options. The mobility of a heavily armored herbivore will necessarily be lower than that of one without armor. Consequently, the theoretical maximum rate of resource exploitation would be lower than that of an unarmored form, if the animal must move around to obtain its food. Armored species are concentrated in areas of relatively high productivity in which travel and search time is low, particularly if competition with unarmored species is likely to occur.

Among animals that have just entered a new adaptive zone (a fundamentally new way of life; Simpson, 1953), there is often a high frequency of heavily armored forms. This can be seen several times in the fossil record of the vertebrates (see Romer, 1966), followed in most cases by a progressive diminution of armor. This recurring pattern could be attributed to the difficulties of using new conditions efficiently, with the result that considerable amounts of time are required for their exploitation, exposing the animals to high risk of predation. In addition, armor may provide a solution if new opportunities take individuals to sites in which their mobility is inadequate for escape.

The presence of predators may also slow down the foraging rates of animals, decreasing their overall efficiency and perhaps decreasing the range within which they are able to forage. Stein and Magnuson (1976) found in laboratory studies that crayfish (*Orconectes propinquus*) exposed to smallmouth bass predators (*Micropterus dolomieui*) modified their activity patterns, and foraged less than when the predators were absent (fig. 5.8). This effect was more pronounced in small, more vulnerable crayfish than in larger ones. Large crayfish also remained in more exposed locations than small ones, which agrees with data from the field. Murdoch and Sih (1978) found that juvenile back-swimming bugs (*Notonecta hoffmanni*) fed up to ten times as rapidly when not in the company of the cannibalistic adults as when with them. Solitary woodpigeons (*Columba palumbus*) spend more time peering about (presumably searching for predators) than do flock members; this slows their foraging rate (Murton, Isaacson, and Westwood, 1971), as do the excited responses of members of titmouse flocks temporarily separated from the group (Morse, 1970b). Kenward (1978) noted that in winter woodpigeons tend to avoid filling their crops completely until just before roosting time. He suggested that they might be keeping themselves as light as possible in order to improve their ability to escape from large flying predators such as goshawks (*Accipiter gentilis*). These examples illustrate how predators may limit the foraging rates of their prey, causing them to depart from the simple energy maximization models discussed in chapter 3.

Alternatively, the development of poison or distastefulness might serve as a way to offset an animal's basic inefficiency at resource exploitation. This has happened many times among invertebrates (Cott, 1940; Edmunds, 1974), but seldom among vertebrates, which in general have life-styles in which they actively avoid their predators (see Marshall, 1965; Noble, 1931; Bellairs, 1969; Dorst, 1974; Matthews, 1971). The exceptions are amphibians and snakes, which have explored this pathway with some success, but which do not represent the mainstream of vertebrate evolution and generally exhibit little flexibility in defense.

Certain species of birds associate with other species that give them protection as a result of their activity in repelling predators. Small passerine birds that sometimes nest in the bulky superstructure of osprey (*Pandion haliaetus*) nests derive such an advantage (Bent, 1937). It has also been hypothesized that birds nesting in the presence

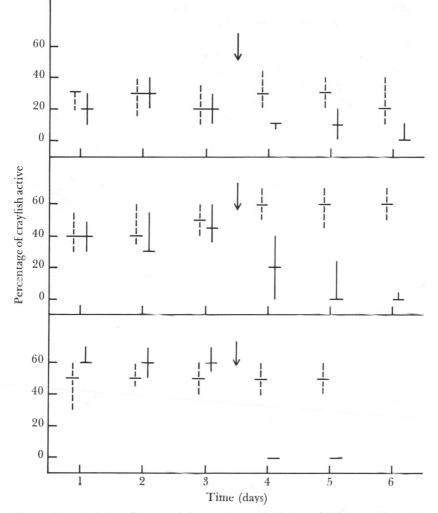

Figure 5.8 Activity of large adult *(top)*, small adult *(middle)*, and juvenile *(bottom)* crayfish in aquaria with equal portions of sand and either gravel (for juveniles) or pebble (for adults) substrates. Dashed lines = control; solid lines = treatment. Arrows indicate when smallmouth bass were placed in treatments. Experimental design as in fig. 4.3. (Redrawn from Stein and Magnuson, 1976, copyright 1976 by the Ecological Society of America.)

of certain loud and alert shorebirds benefit from their alarms and from their attacks on potential nest predators. Göransson et al. (1975) tested this idea by placing chicken eggs in areas within the ranges of lapwings (*Vanellus vanellus*) and curlews (*Numenius arquata*) and

in undefended areas. They found a striking difference in egg predation in the two areas (approximately threefold), which resulted
largely from the depredations of crows and gulls. Efforts were made to
avoid attracting predators to the nest sites, but obviously the results
would have been somewhat different had normal parents been about
the eggs. Nevertheless, the striking difference combined with the observed behavior of the protecting species suggests strongly that the
observed effect is a real one. It is not known whether the species obtaining benefits is responding to these protecting species, although
one might predict that the latter are a factor associated with nest-site
selection.

Andersson and Wiklund (1978) repeated and extended Göransson
and coworkers' experiments, testing the effect of the fieldfare (*Turdus
pilaris*), an Old World thrush, as a deterrent to nest predators (crows,
jays, magpies). They set eggs in the vicinity of colonies and of isolated
nests of fieldfares, as well as at sites away from them. As predicted, egg
predation declined as fieldfare numbers increased (fig. 5.9).

Sometimes animals obtain protection from predators only at a considerable cost. Certain birds often nest in gull colonies, despite the
fact that gulls are to varying degrees predatory on their eggs and
young—for example, sandwich terns (*Sterna sandvicensis*) nesting in
black-headed gull (*Larus ridibundus*) colonies (Fuchs, 1977). The
black-headed gulls incidentally provide protection to the terns from
nest predators such as herring gulls (*L. argentatus*) and carrion crows
(*Corvus corone*), although they take sandwich tern eggs when they
can. Black-headed gulls are more effective in driving the large herring
gulls from the colony area than are sandwich terns. Herring gulls
readily take young as well as eggs, whereas the black-headed gulls
largely confine their activities to eggs. Because the flight responses of
black-headed gulls and sandwich terns are similar, the black-headed
gulls have only limited opportunities to eat sandwich tern eggs
while the terns are off their nest. In Fuchs' study in Scotland, black-
headed gulls took about 10 percent of the sandwich terns' eggs.

Sandwich terns may also nest with common and arctic terns (*S.
hirundo* and *S. paradisaea*), species which are also very good at driving off gulls, but which do not rob their eggs. It would seem that
these other terns are preferable associates for the sandwich tern because they do not rob eggs. But, they nest about four weeks later than
do the black-headed gulls and sandwich terns. Fuchs suggested that
the advantage of rearing young early in the year, even at the cost of
some predation, outweighs the advantage of breeding in synchrony
with the other terns later in the season.

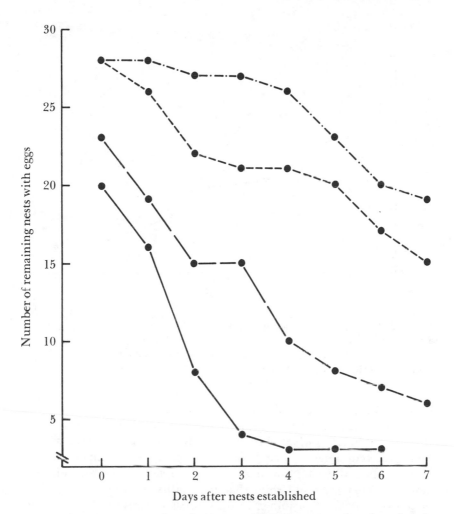

Figure 5.9 The pattern of predation at artificial nests as a function of their proximity to colonies or solitary pairs of fieldfares. Plots from top to bottom: nests inside fieldfare colonies, nests outside fieldfare colonies, nests near solitary fieldfare pairs, nests away from solitary fieldfare pairs. (Modified from Andersson and Wiklund, 1978.)

Energetic demands and the challenge of predators

The distribution of foraging animals may be measurably affected by predators. Hewson (1976) documented a tendency of mountain hares (*Lepus timidus*) to forage on pioneer heather (fig. 5.10) rather than on nearby taller heather that contains more nitrogen and

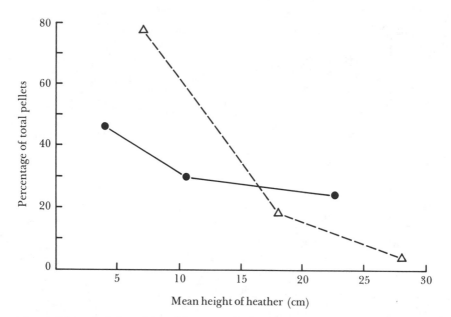

Figure 5.10 Activity of Scottish mountain hares in heather of varying height at two localities. Activity was measured by determining the relative frequency of fecal pellets. Solid line = Corndavon study area; dashed line = Lochnagar study area. (Modified from Hewson, 1976.)

phosphorus. He attributed this preference to the better visibility in the short vegetation, which permits them to watch for predators, especially red foxes (*Vulpes vulpes*). Closely related species with similar ecological requirements may even partition the environment on the basis of their differing abilities to avoid predators. Rosenzweig and Winakur (1969) hypothesized that the ability of kangaroo rats (*Dipodomys merriami*) to forage in open areas is a result of their acute senses and great agility. Sympatric pocket mice (*Perognathus penicillatus*), which are neither as good at detecting predators nor as able to evade them, confine their activities largely to areas with relatively heavy cover. Similarly, States (1976) suggested that the habitat partitioning of *Eutamias* chipmunks in the western United States may reflect their differing vulnerability to predation.

Seabirds that nest on the ground in areas free of terrestrial predators provide another interesting case. Nesting areas are often far removed from feeding grounds. Where safe nesting areas are scarce, individuals become so concentrated that feeding near the colony is impossible. The additional commuting time required may reduce feeding efficiency and thus limit the size of the population that can be supported. Starlings from huge roosts may move as much as 80 km from these sites in search of food (fig. 5.11). If a limitation of safe roosts is responsible for such great buildups, and if movements as long as these are required to obtain food, foraging restraints may again limit the ultimate size of both the roost and the local population (Hamilton et al., 1967; Hamilton and Gilbert, 1969).

Maiorana (1976) has discussed the life history patterns of species in which avoidance of predators takes a high priority, often to the point that foraging activity is seriously reduced. Reproductive output and growth rates tend to be low, presumably as a direct result of the effort spent on predator avoidance. Maiorana believed that this compromise is of particular importance in secretive species, such as plethodontid salamanders.

Habituation to predators

If possible, an animal should avoid responding to harmless objects that resemble predators. For example, Schleidt (1961) demonstrated experimentally that turkeys (*Meleagris gallopavo*) readily habituate to predatorlike objects, and songbirds will eventually stop mobbing a stationary owl (Hinde, 1954a, b). However, flocks of titmice in eastern North America almost always respond to mourning doves (*Zenaida macroura*) flying rapidly through the forest (Morse, 1970b). This long-tailed dove bears a superficial resemblance to the principal predator on small birds in these areas, the sharp-shinned hawk *Accipiter striatus*. When a dove appears suddenly, literally bursting through the trees, there is probably not enough time for the titmice to distinguish it from a hawk.

Shalter (1975, 1978b) has attempted to determine why prey species do not readily habituate to predatorlike objects in the field. Using both galliform birds and pied flycatchers (*Ficedula hypoleuca*), he found that habituation resulted from a position effect. When the position of a stimulus was continually changed the birds did not habituate, which accords well with the natural situation, in which two attacks are unlikely to come in the same place and from the same direction.

Failure to habituate has important implications for prey species, if responses to predators and predatorlike objects are seriously time-consuming. A prey species that must spend most of its time foraging, as often happens during winter or the breeding season, could be excluded from an area even if it was very rarely taken by the predator. Harrassment by the predator could have an effect on the size of the prey population similar to that which would be caused by actual predation, although the predator population would gain nothing.

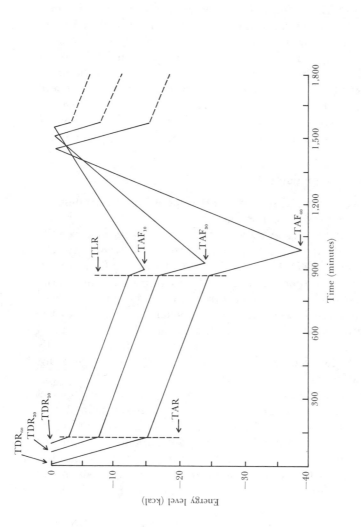

Figure 5.11 Relative energy flux in kilocalories (1 kcal = 4,186 joules) in roosting starlings. TDR_i = energy level at the time of departure for roost from the ith feeding range; TAR = time of arrival at roost; TLR = time of departure from roost; TAF_i = time of arrival at ith feeding range. Activity of the roost individuals is staggered, resulting in a total activity time of more than 24 hours (1,440 minutes). (Redrawn from Hamilton et al., 1967, copyright 1967 by the Ecological Society of America.)

Both prey and predator populations should be depressed as a result, an aspect of predator-prey relationships that has not been much studied.

Effect of predation on responses of prey

Although the titmice mentioned above live in an area that appears to have no flying predators, they almost invariably respond to the predatorlike silhouettes of mourning doves. In midwinter such responses, each of which takes five to twenty minutes out of the day's foraging, constitute a large expenditure for predator avoidance because at that time of year the titmice are often resource-limited (Gibb, 1960). The fact that titmice continue to respond to doves and other objects in places devoid of avian predators suggests that strong selection for such behavior, over long stretches of evolutionary time, produced a relatively unmodifiable genome. Curio (1969) investigated this resistance to change in Galapagos finches. The Galapagos finch (*Geospiza difficilis*) of Tower Island has no aerial predators, but it still responds to them, although somewhat less intensely than do other finches (fig. 5.12).

Similarly, congeneric titmice experiencing substantial predator pressure in Wytham Woods, Oxford, England, distinguish predators from inappropriate stimuli more efficiently than those in eastern North America (Morse, 1970b, 1973b). The English flocks respond inappropriately only in exceptional situations. The Wytham tits may be demonstrating a sophisticated time-minimizing response to the frequent presence of sparrowhawks (Morse, 1973b). Often the sparrowhawks first fly over a flock and then attack at a low height. Members of the flocks typically remain under cover longer following a fly-over than after an attack. The hawks seldom attack twice in rapid succession and are usually successful only when they surprise the tits. The failure to attack twice is itself probably a response to the alertness of the prey. The tits have made the most of this consequence, cutting to a minimum the amount of time devoted to antipredatory activities.

Large game such as deer, traditionally preyed on by wolves (*Canis lupus*) and mountain lions (*Felis concolor*), are probably attacked much less frequently now than in the past. Like the songbirds, they probably have innate defensive behaviors that are maladapted to their current situations (see Lack, 1965).

Synthesis

The interpretation of many antipredator responses has changed over the last few years. Traditionally these behaviors have been viewed as altruistic or as deceptive. Undoubtedly some displays are truly

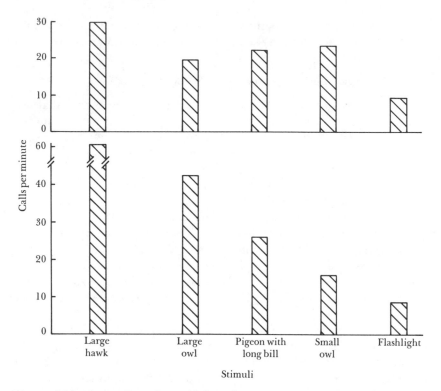

Figure 5.12 Strength and specificity of responses to predators by experimentally naive Galapagos finches from islands with extremes in predatory pressure. *Top = Geospiza difficilis* on Tower Island; *bottom = G. fuliginosa* on Indefatigable Island (highest predator pressure). (Modified from Curio, 1969.)

altruistic, especially those that directly benefit offspring unable to defend themselves; distraction displays of many ground-nesting birds (Simmons, 1952) are good examples. Protean displays (Humphries and Driver, 1967, 1970) are almost certainly deception intended to confuse a predator. But warning calls have recently been interpreted by some workers as purely selfish; the caller is taken to be minimizing its own danger at the expense of others, by bringing other individuals to the predator's attention when they respond to the caller (Charnov and Krebs, 1975). Certain displays may serve a time-saving function, signaling the predator that the prey knows of its presence and is able to escape should the predator attack (Smythe, 1970b). Predators that respond by failing to attack could save considerable time and energy themselves. But the relationship between predator and prey ought to remain dynamic, because the prey species would be under pressure to push this advantage as far as possible, to the point of deception. Similarly, the predator that saw through such deception would benefit considerably.

Unfortunately, the more novel ideas discussed here have not been tested adequately (indeed, neither have many of the more traditional ones). Sorely needed are tests of models in the field, and more detailed field work will itself surely contribute to the diversity of ideas being considered. It does seem that clear antipredator behavior involves many kinds of strategies and many kinds of selection pressures. Theory on the implications of natural selection is currently an area of intense interest and is providing many new models relevant to this subject.

6 *Behavioral Thermoregulation and Maintenance*

ALL ENDOTHERMS and many ectotherms operate most effectively at relatively high constant body temperatures (see Hamilton, 1973). Endotherms generate this temperature but may supplement it with heat from extrinsic sources. Ectotherms largely depend on environmental sources of heat, but they often actively capture this heat, rather than passively varying in temperature with the environment. Sometimes ambient temperatures exceed an animal's maximum preferred or tolerated temperature, and the animal must act to reduce the burden. Many different behavioral adaptations serve these ends.

Obtaining and retaining heat

Both endotherms and ectotherms expose themselves to objects warmer than the surrounding environment in order to reduce or even reverse their outflow of heat. Sunbathing lizards and snakes are familiar examples (Bogert, 1949, 1959; Cloudsley-Thompson, 1972), but sunbathing is practiced by a wide range of other animals as well, including a variety of arthropods and endotherms. Many animals display remarkable abilities to retain the heat they have generated internally or gathered externally; for example, they may use insulation or carefully avoid exposure to the wind.

ENDOTHERMS

On cold sunny mornings during the winter, chickadees and titmice often perch, exposing themselves to the sun (Morse, 1970b).

One might expect them to be foraging rapidly, since the gradient between body temperature and early morning ambient temperature is especially large. On cloudly mornings they do forage at a normal rate. Similarly, Morton (1967) found that on cold winter days white-crowned sparrows (*Zonotrichia leucophrys*) fed at lower rates when it was sunny than when it was cloudy; indeed, the sparrows regularly indulged in sunbathing. Yellow-bellied marmots (*Marmota flaviventris*), which live at high altitudes, orient themselves to the sun and take up flattened positions on rocks early in the morning (Travis and Armitage, 1972). These observations all suggest that sunning may significantly lower energy demands.

The stimulus for sunbathing may be a rapid increase in the temperature of the environment (Lanyon, 1958) or an increase in light intensity (Morton, 1967), which would normally correlate with temperature. Morton found that the foraging activity of his sparrows decreased as sunrise approached. Chickadees and titmice also reduce their foraging rates as the sun's rays begin to strike the treetops (Morse, 1970b).

Birds sometimes sunbathe under hot conditions as well (Hauser, 1957). Functions served by this behavior may include the production of vitamin D, the removal of ectoparasites, and simple comfort during molting (Hauser, 1957; Goodwin, 1968; Kennedy, 1968, 1969). In addition to the usual basking behavior, which does not involve exaggerated stances, Mueller (1972) noted exaggerated, spread-winged displays in hawks; these were given primarily in response to light, rather than heat. Mueller suggested that this behavior was not involved in thermoregulation, because it was often performed at the hottest time of day. Potter and Hauser (1974) interpreted it as relief from discomforts associated with molting.

Body color can affect the efficiency of behavioral thermoregulation. Hamilton and Heppner (1967) and Heppner (1970) have shown that black-dyed zebra finches (*Taenopygia castanotis*) act somewhat like black bodies and may obtain as much as 23 percent metabolic economy in artificial sunlight at 10°C, as determined by oxygen consumption rates of dyed and undyed (white) birds (fig. 6.1). Black-dyed birds absorb more radiant energy during the day and emit about the same amount as other individuals during the night.

Lustick (1969) extended Hamilton and Heppner's (1967) experiments, but to avoid dyeing the plumage he used male and female cowbirds (*Molothrus ater*) (black and grayish brown, respectively) and albino and dark gray zebra finches. His results were similar to those of Hamilton and Heppner, showing a 20 percent greater economy for sunbathing black birds than for white ones, although the white birds

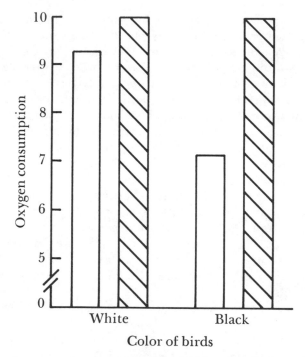

Figure 6.1 Average oxygen consumption of white zebra finches and the same white zebra finches dyed black, when exposed and not exposed to artificial sun (sunlamp). Test temperature was 10°C. Units are milliliters of oxygen per gram of body weight per minute. White bars = sun on, hatched bars = sun off. (Modified from Hamilton and Heppner, 1967, copyright 1967 by the American Association for the Advancement of Science.)

attained at least a 6 percent economy over birds that were not exposed to such a source of radiant energy. Although Hamilton and Heppner's experiments may have been affected by dyeing the birds and Lustick's by using different genetic stocks, the agreement between their results clearly implies that the effect is real.

These results suggest that light-colored animals should be favored in very sunny locations where excess heat loads may be a problem. Light-colored animals should be common in desert areas, although light coloration there may also be explained as camouflage (see Edmunds, 1974). In deserts with particularly high temperatures, even light-colored animals must find shelter during the hottest part of the day (Hamilton, 1973), and this may reduce the advantage of being light-colored. Deserts, particularly ones with high temperatures, have notoriously wide ranges of temperature; the advantage may in fact be to those who can most effectively exploit periods of intermediate temperatures, for example, cool mornings. Dark-colored animals that can heat up quickly could in principle use the surface of the desert for the longest period of time (Hamilton, 1973). Hamilton found color patterns of desert animals consistent with this interpretation. Most desert-dwelling tenebrionid beetles are black; only in relatively cool deserts, where maximum temperatures do not often exceed those tolerable, did he find white beetles as well. Although Hamilton mainly considered ectothermic animals, he noted that several desert homeo-therms have black feathers, fur, or skin—for example, wheatears (*Saxicola* spp.). He suggested that the effect is general among small animals that can readily avoid high environmental temperatures by some means such as burrowing or sheltering in the shade.

As would be expected, morphological and behavioral characteristics relating to thermoregulation are well integrated. The roadrunner (*Geococcyx californianus*) is a large, brown-and-white stripped, ground-dwelling relative of the cuckoo. Roadrunners often assume an unusual posture, in which their backs face the sun, their wings are drastically deflected with body feathers held erect, and black patches of skin in their axillary areas are exposed to solar radiation. Energy consumption may drop more than 40 percent when they perform this activity during cold weather (Ohmart and Lasiewski, 1970) (fig. 6.2). Even nestling roadrunners, whose skin is black, perform a sophisticated pattern of behavioral thermoregulation. When not being brooded in the early morning, they expose themselves to solar radiation, and during warmer periods they move into patches of shade provided by the vegetation surrounding the nest and by the contours of the nest itself. It is essential that the parents be free to hunt during the early part of the day, because the lizards that form the bulk of their diet are most active then, largely because of their own thermo-regulatory adaptations (Ohmart, 1973).

Bateman and Balda (1973) noted that nestling piñon jays (*Gymnorhinus cyanocephalus*) are pigmented black before feathers cover them, a condition they consider related to the early nesting period of this species. Nests of piñon jays are typically built with southerly exposures, and the females generally leave the nest to forage in the middle of the day, when solar radiation minimizes the loss of heat from the young. European red squirrels (*Sciurus vulgaris*) also build their winter nests on the exposed sides of tree trunks; their nest material (*Usnea* lichen) has excellent insulating properties, enabling them to maintain temperatures 20–30°C above ambient at high latitudes (Pulliainen, 1974).

Grebes are probably the only group to have been investigated systematically from this point of view (Storer, Siegfried, and Kinahan,

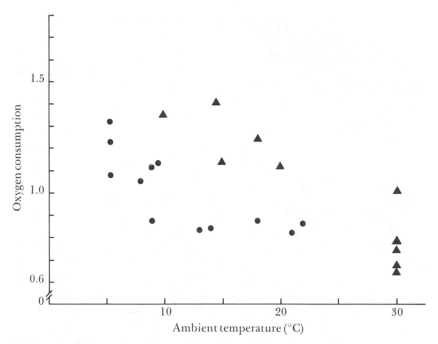

Figure 6.2 Relationship between oxygen consumption and ambient temperature in adult roadrunners. Oxygen consumption is in milliliters of oxygen per gram of body weight per hour. Triangles = birds in dark; circles = birds posturing in sun. (Modified from Ohmart and Lasiewski, 1970, copyright 1970 by the American Association for the Advancement of Science.)

1975). Some species raise the black-based white feathers of their back and face, exposing to the sun the heavily pigmented skin lying underneath. This behavior and pigmentation is characteristic of the small species, with only two exceptions, both of which are rather large birds confined to high Andean lakes. Large, lowland species do not show these adaptations (table 6.1). This is exactly as would be predicted from the more severe thermal restrictions of small animals because of surface-volume relationships.

   Black bodies not only gain heat but also emit heat more rapidly than bodies of other colors. So how can an animal realize a long-term gain simply by being dark? Kirchoff's law (that a good absorber is also a good emitter) refers only to a particular wavelength at a given temperature. Most energy emitted by animals falls in the 4–20 micron range (Hammel, 1956), a function of their modest body temperatures, whereas much of the solar energy absorbed is of shorter wavelengths (see Hamilton, 1973). Thus, in living animals Kirchoff's law applies only to wavelengths longer than those of most of the energy in

Table 6.1. Occurrence of sunbathing and associated pigmentation in grebes. (From Storer, Siegfried, and Kinahan, 1975.)

Species	Mean Weight (g)	Skin Pigment	Feathers with Black Base	Sunbathing Posture
Least grebe				
(*Tachybaptus dominicus*)	131	present	present	present
Alaotra grebe				
(*Tachybaptus rufolavatus*)	189	present	present	unknown
Red-throated dabchick				
(*Tachybaptus ruficollis*)	192	present	present	present
Madagascar grebe				
(*Tachybaptus pelzelnii*)	unknown	unknown	present	unknown
Australian dabchick				
(*Tachybaptus novaehollandiae*)	207	present	present	present
Rolland's grebe				
(*Rollandia rollandia*)	247	present	present	present
Hoary-headed grebe				
(*Poliocephalus poliocephalus*)	249	present	present	present
New Zealand dabchick				
(*Poliocephalus rufopectus*)	253	unknown	present	present
Silver grebe				
(*Podiceps occipitalis*)	323	present	present	present
Eared grebe				
(*Podiceps nigricollis*)	334	present	present	present
Pied-billed grebe				
(*Podilymbus podiceps*)	396	absent	absent	absent
Horned grebe				
(*Podiceps auritus*)	405	absent	absent	absent
Taczanowski's grebe[a]				
(*Podiceps taczanowskii*)	427	present	present	present
Atitlan grebe				
(*Podilymbus gigas*)	699	absent	absent	absent
Flightless grebe[a]				
(*Rollandia microptera*)	706	present	present	present
Red-necked grebe				
(*Podiceps grisgena*)	990	absent	absent	absent
Great-crested grebe				
(*Podiceps cristatus*)	1,042	absent	absent	absent
Western grebe				
(*Aechmophorus occidentalis*)	1,087	absent	absent	absent
Great grebe				
(*Podiceps major*)	1,478	absent	absent	absent

[a] Large species living at high altitude.

solar radiation. Dark individuals absorb and retain considerably more energy when in sunlight than light ones (see Norris, 1967; Heppner, 1970). Behavior may reduce losses even further. For instance, birds' feathers insulate much more effectively if they are fluffed out, forming dead air spaces, than if sleeked. This point should hold both for retaining internal temperatures and resisting external ones (see Øritsland, 1970). Thus, minimizing one's insulatory capabilities when exposed to a heat source and maximizing them when away from it should result in a net energy gain.

Lavigne and Øritsland (1974) point out that Hamilton and his coworkers overlook the possibility that light animals may be effective heat absorbers at certain wavelengths. They noted that on ultraviolet film polar bears (*Thalarctos maritimus*) appear black; by this means polar bears absorb some wavelengths of solar energy and at the same time remain cryptic to their vertebrate prey.

When temperatures become high, animals may need to reduce their heat loads. Respiratory cooling occurs in a wide variety of endotherms, regularly taking the form of panting, gular flutter, or sweating. Some species, such as storks (Kahl, 1963), New World vultures (Ligon, 1967), and seals (Gentry, 1973), employ evaporative cooling by releasing urine or diluted feces on themselves. Some ectotherms also use this technique (Bellairs, 1969).

Thermoregulatory factors underlie much of the behavior of seals on their rookeries, putting constraints on their space utilization and, consequently, their social systems. As highly modified aquatic mammals, seals are faced with a particularly stringent environmental challenge when they come ashore during the breeding season. They must reproduce on land where their highly modified limbs (flippers) and heavy layers of insulatory fat are severe impediments to movement. As a result these animals seldom move far from the water's edge, even where no terrestrial predators occur. At the edge of the water seals are exposed to direct solar radiation, which, combined with their heavy insulation, may have particularly severe heating effects. They may also be exposed to heat from the substrate and, occasionally, to temperatures lower than those in the adjacent water.

Pinnipeds have several behavioral patterns that actively facilitate temperature regulation under these circumstances, including panting, waving their flippers, changing posture, flipping sand over their bodies, moving into shade or water, and urinating on their pelage. At low temperatures flippers may be concealed, thus exposing minimal surface area. Some species, such as Stellar's sea lion (*Eumetopias jubatus*), huddle under cold conditions, although others, such as fur seals (*Arctocephalus fosteri*), do not (Gentry, 1973). At temperatures over 30°C both of these species retreat to the water. Males use this behavior only as a last resort because their reproductive success depends on defending prime areas within which females take up residence. By putting on large amounts of fat prior to the breeding season they have already reduced or eliminated the requirement of feeding during the reproductive period, but this layer may complicate problems of heat loss on warm days. When thermal problems become severe, males holding territories away from the shore are forced to intrude through other territories in their trips to and from the water, resulting in long periods of absence and high frequencies of aggression. This probably results in lower reproductive success.

Lustick, Battersby, and Kelty (1978) found that herring gulls (*Larus argentatus*), large white birds with gray backs and wings, could control their heat budgets in open nesting colonies solely by changing their posture and orientation to the sun. Although temperatures are modest during the incubation period of these birds (April and May), many colonies are located in the open where little or no protection is available from the direct sun. As a consequence, the birds may overheat unless they respond directly to these conditions. Lustick and coworkers noted that birds in the colony exposed to direct sunlight at temperatures between 5°C and 10°C panted if they were incubating but not if they were standing on their nests, which would result in heat loss through the feet. As noted earlier, reflectance from white plumage is much greater than that from darker plumage. Lustick and coworkers found that birds in the sun constantly oriented to it. Given the angle of the sun at that time of year (about 55°), the white surfaces of the birds received direct solar, reflected, and ground radiation, and the darker parts of the birds received diffuse and atmospheric radiation. Radiative heat gain was thus minimized by reducing the directly exposed surface area (producing a minimal silhouette) and by exposing the highly reflective white surfaces to the solar radiation. On cloudy days, in shade, and at high wind velocities, the gulls did not orient. Later in the season when the sun is at a higher angle and temperatures are higher, orientation would be of less importance in minimizing heat gain, although posture would still be of significance. However, by this time the majority of the birds are in the water, where any heat gain can easily be countered by loss through the feet.

REPTILES AND TERRESTRIAL AMPHIBIANS

As a group reptiles use an impressive variety of methods to exploit sources of radiative energy; these include movements between sun and shade or other hot and cold sites, and changing shape, posture, position or even color when exposed to a heat source (see Huey, 1974). For the most part reptiles make far more use of the sun in thermoregulation than do endotherms (Templeton, 1970). The sun-bathing abilities of certain lizards are spectacular; an Andean lizard (*Liolaemus multiformis*) can raise its body temperature to over 30°C by basking in the sun when the shade temperature is below 0°C (Pearson, 1954). In addition to these activities, reptiles also employ physiological functions (normal muscular activity, respiratory cooling, and reradiation) in temperature regulation (Templeton, 1970). Behavioral thermoregulation constrains the way reptiles can use other resources (see Avery, 1973). Huey (1974) and Huey and Webster (1975) found that tropical *Anolis* lizards living in open areas regularly bask in the sun, whereas individuals of the same species living in the adjacent forest do so much less frequently, tolerating a lower and more variable thermal regime (fig. 6.3). There are few exposed perches in the forest, and reaching them requires the expenditure of a considerable amount of energy. Temperatures vary even more in temperate forests and are cooler on average than in tropical forests,

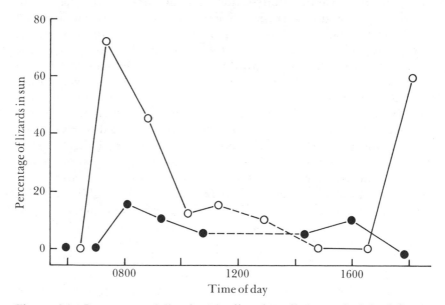

Figure 6.3 Percentage of lizards (*Anolis cristatellus*) perched in full sun during a day. Lizards in partial sun were counted as half in sun, half in shade. Open circles = lizards in open park; closed circles = lizards in forest. (Modified from Huey, 1974, copyright 1974 by the American Association for the Advancement of Science.)

which may help to explain the relative rarity of lizards there (Huey, 1974).

The tendency of snakes to lie on tarred roads at night (Porter, 1972) reflects their habit of taking heat from substrates of locally elevated temperature. Even some toads (*Bufo boreas*, for example) are notably heliothermic and show a tendency to climb into exposed positions (Lillywhite, Licht, and Chelgren, 1973). Other amphibians have also been reported to sunbathe (Brattstrom, 1963; Lillywhite, 1970).

It has been suggested that sunbathing is used in particular to increase the rate of digestion (Moll and Legler, 1971; Hamilton, 1973). Because digestive rates vary directly with temperature, raising the body temperature will increase the energy available to an animal if food is abundant. If the costs of breaking down food are high, as they are for herbivores, the gain can be substantial.

AQUATIC AMPHIBIANS AND FISH

Some aquatic amphibians and fish exploit temperature gradients. Fish (Fry, 1964) and newts (Licht and Brown, 1967) have been shown to seek particular thermal conditions within a body of water. Brown (1971a) and Brown and Feldmeth (1971) described the tendency of pupfish (*Cyprinodon* spp.) in areas around hot springs to concentrate their activities in water of a rather precise temperature range (42°C for one population), which appears to promote peak metabolic efficiency and to give them access to thermophilic blue-green algae. This temperature is almost lethal. When chased, the pupfish exhibit extreme reluctance to enter the higher thermal regime (Brown, 1971a; Brown and Feldmeth, 1971). Licht and Brown (1967) found that both adult and larval red-bellied newts (*Taricha rivularis*) show strong thermal preferences in breeding streams that contain hot springs, and that temperature choices were of considerable importance; newts that accidentally entered the spring effluents often died.

ARTHROPODS

Insects in several orders thermoregulate (Heinrich, 1974; Bartholomew and Casey, 1977; May, 1977). The behavioral thermoregulation of butterflies parallels that of reptiles (Clench, 1966). Both groups bask, regulate their contact with the ground, avoid wind, seek shade, and modify their orientation in the sun. However, butterflies do not change body color, and reptiles are not known to "shiver" (a method confined to the skippers among the butterflies). Their reliance on these microclimatic conditions almost certainly limits their ranges

(Clench, 1966). Nocturnal moths generate heat primarily by shivering and by the process of normal flight. Large, strong-flying moths generate such great heat loads that they may experience difficulties cooling during the day. The shade-seeking tendencies of many insects are probably an adaptation to this problem (Heinrich, 1974). Overheating is evidently not a problem for large, strong-flying diurnal moths such as clear-winged hawk moths (Sphingidae). Differences in heat generation may be related to the small degree of temporal overlap between the two groups. Some sphingid larvae also show a marked convergence with the lizards in their patterns of orientation to sunlight (Casey, 1976).

Sawfly larvae (*Perga dorsalis*) exhibit close parallels to the thermoregulatory patterns of birds and mammals. At low temperatures they aggregate, reducing individual heat loss; when temperatures exceed 30°C they expose themselves, increasing heat loss; and at temperatures over 37°C they spread liquid over their bodies, promoting heat loss through evaporative cooling (Seymour, 1974).

Many social insects regulate the temperatures of their living quarters (Wilson, 1971a). This is accomplished in three principal ways: by placing nests in carefully selected locations, by constructing them in particular ways, and in a few cases by behaving in ways that change temperatures within the nest. The shapes and exposures of nests may differ with the environment and geographic area. Ranges of many ants are severely limited by temperature. Their nests are considerably taller near the northern limit of distribution or in partially shaded situations, thus forming effective solaria (Brian, 1955; Pontin, 1969). This exposure apparently provides the extra warmth that makes survival possible in areas of low radiant energy. Some species rest only under rocks (which absorb heat) in thermally marginal areas (Brian and Brian, 1951; Brian, 1956).

The nests of some tropical termites are efficient solar heaters and coolers. Warm air produced by colonies of *Macrotermes natalensis*, an African species, is conveyed upward through their tall nests; reaching the top, the air spills into the outer layers, cools as it sinks to the bottom of the nest, and is again picked up and circulated through the colony. This system provides effective air conditioning with minimum exposure to the elements, at low energetic cost (Lüscher, 1961).

Short-term active thermoregulation takes place in some social species, the best known example being the regulation of hive temperature by the honeybee (*Apis mellifera*) (Lindauer, 1961). Honeybees keep the central hive temperature between 17°C and 36°C throughout the year by a complicated system of behaviors, including fanning and bringing in water when it is hot; when it is cold the bees group to form insulating layers around active individuals that produce most of the heat. A large proportion of the energy gathered by bees goes into regulating the temperature of the hive. Although some other bees and social insects perform certain of these acts, none are anywhere near as efficient as honeybees (Michener, 1974). Some ants have pursued a very different but equally effective strategy. Being wingless, ant workers cannot fan or bring water to their nests rapidly in the manner of winged insects. Instead, they move the eggs, larvae, and pupae up or down in the nest, taking advantage of conditions existing in different parts of the nest (Wilson, 1971a). This technique gives them a flexibility that even honeybees cannot match, since the bees place their young in immovable brood cells.

Humphreys (1974) reported that a large wolf spider (*Geolycosa godeffroy*) both sunbathes and moves its egg sacs about, incubating them in this way. This behavior is quite similar to that of lizards, although the spiders do not maintain their preferred temperature with the accuracy attained by lizards; this may be a limitation of their small size. The desert funnel-web spider (*Agelenopsis aperta*) is under severe constraints to build its web out of direct exposure to the midday sun. In direct sunlight it can often only tolerate thermal conditions existing on its web long enough to capture small prey (Riechert and Tracy, 1975). The spider realizes a greater energetic gain at shaded sites than at sunny ones, although contact of prey with the web may be lower at shaded sites (fig. 6.4).

Overnight reserves of endotherms

Small endotherms have to feed regularly and in the most extreme cases nearly constantly because of their high metabolic rates and small sizes. Obtaining adequate reserves to survive an inactive period (night or day) may be critical. These problems are most severe for small animals, but larger animals may also face them, particularly at high latitudes or during extended periods of inactivity. Behavioral techniques used to minimize energy loss during these periods include roosting communally using sheltered nesting areas, burrowing under snow, moving into caves at night, storing food, and extending activity into periods of low light.

COMMUNAL GROUPS

Communal groups may elevate their local temperatures, and this could be their principal advantage over individuals that roost only under severe conditions (Zahavi, 1971). Many small birds in high latitudes roost communally at such times, including treecreepers (*Certhia* spp.) (Löhrl, 1955), wrens (*Troglodytes troglodytes*) (Armstrong,

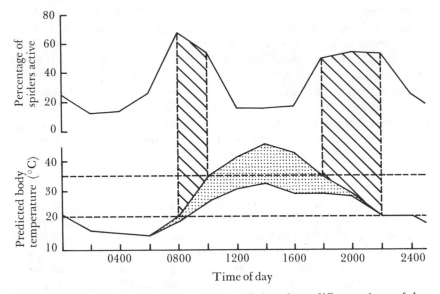

Figure 6.4 Percentage of spiders active on their webs at different times of day in July and August (*top*), and predicted spider temperature under existing conditions assuming a web-over-litter substrate (*bottom*). Hatched areas represent time periods during which over 50 percent of the individuals were active. Stippled area represents range of spider temperatures, exact temperature being dependent on the amount of exposure to solar radiation. Upper boundary of predicted temperature curve signifies spider temperature if in full sunlight, and lower boundary signifies spider temperature if in full shade. Area enclosed by dashed lines represents body temperature range within which over 50 percent of the spiders are active. (Modified from Riechert and Tracy, 1975, copyright 1975 by the Ecological Society of America.)

1955), and bushtits (*Psaltriparus minimus*) (Smith, 1972). Armstrong (1955) noted that the three smallest English winter birds—wren, long-tailed tit (*Aegithalos caudatus*), and goldcrest (*Regulus regulus*) —roost communally during that season.

Yellow-necked mice (*Apodemus flavicollis*) frequently shelter in groups outside the breeding season, especially when it is cold, resulting in a substantial reduction of energy demands (Fedyk, 1971). Bats conserve energy by roosting in aggregations, although the extent of their clumping may ultimately be limited by the distance that individuals must fly in order to obtain food (Herreid, 1967; Kunz, 1974). If satisfactory roosting places are scarce, one would expect a trade-off between these two factors.

Although a marked thermoregulatory effect may occur in forms that huddle together, thermal gradients within open roosts may be modest (fig. 6.5). Francis (1976) found that in a large roost (two and a half to three million birds) of New World blackbirds and starlings the greatest thermal gradient (maximum of 2.0°C) resulted entirely from the use of pine trees as cover; this benefit could have been obtained by birds that did not roost communally. High winds considerably reduced even this small thermal advantage, resulting in an overall gain of about 0.5°C. Kelty and Lustick (1977) and Yom-Tov, Imber, and Otterman (1977) obtained generally similar results at large starling roosts in Ohio and Israel and concluded that any energetic advantages resulted from the use of protected sites and not from the combined heat generation of the birds. However, they emphasized that the energy saved from choosing these sites may be substantial, resulting in estimated midwinter expenditures 12–38 percent less than those which would be experienced in adjacent exposed areas.

Although clumping behavior associated with thermoregulatory problems appears to be widely practiced by small endotherms, it is by no means confined to them. During severe weather such large animals as elk (*Cervus elaphus*) (Altman, 1956) and musk oxen (*Ovibos moschatus*) (Tener, 1965; Wilkinson, 1974) may aggregate. Huddling is commonly practiced by incubating emperor penguins (*Aptenodytes forsteri*) and, later in the breeding season, by their young during severe storms (Prevost, 1961).

USE OF SHELTERED AREAS

Animals may obtain a substantial thermoregulatory advantage by avoiding exposure during cold nights. Kendeigh (1961) calculated that at extreme temperatures (-30°C), house sparrows (*Passer domesticus*) in a nest box realize an energy saving of over 13 percent (fig. 6.6). Some birds, such as the wren (Armstrong, 1955) and eastern bluebird (*Sialia sialis*) (Frazier and Nolan, 1959), will even roost communally in cavities, which also lowers individual heat loss.

The thermoregulatory advantage must be weighed against the danger of predation to individuals who roost in a vulnerable place. In spite of the thermal advantages, a number of hole-nesting birds apparently do not use these cavities in the winter—for example, English great tits (*Parus major*) and blue tits (*P. caeruleus*) (Armstrong, 1955). During the breeding season, weasels prey on nest inhabitants at a high rate (Perrins 1965; Dunn, 1976), and it is likely that this would occur during the winter as well. In Germany Winkel and Winkel (1973) found that great tits use cavities at just two times

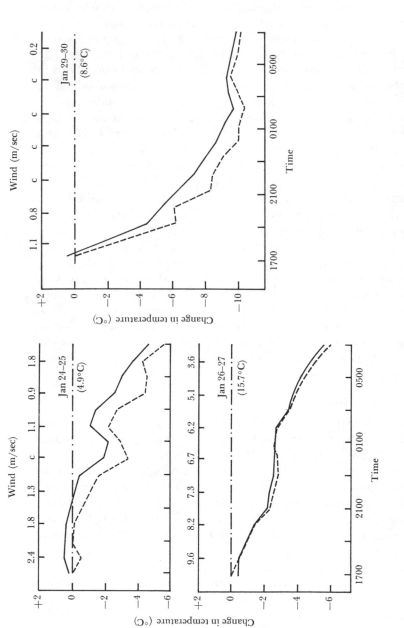

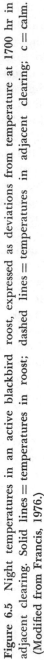

Figure 6.5 Night temperatures in an active blackbird roost, expressed as deviations from temperature at 1700 hr in adjacent clearing. Solid lines = temperatures in roost; dashed lines = temperatures in adjacent clearing; c = calm. (Modified from Francis, 1976.)

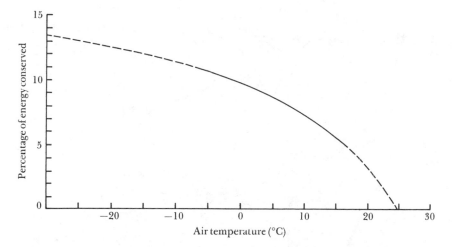

Figure 6.6 Relationship between percentage of energy conserved by house sparrows roosting in cavities and air temperature. (Modified from Kendeigh, 1961.)

other than the breeding season: in winter and during the molting period. These are both times at which energy losses can be large. Presumably only during these periods does the added energetic saving outweigh the vulnerability to predators. Perhaps the added vulnerability to predation explains why a number of other species roost communally only under severe conditions (Armstrong, 1955).

Pitts (1976) hypothesized that the odor of feces may attract predators to the nesting holes. In studies on the winter roosting habits of Carolina chickadees (*P. carolinensis*), he found that these birds frequently changed their roosting holes, were extremely secretive prior to entering them, and did not remove feces from these nesting sites. In contrast, downy woodpeckers (*Dendrocopos pubescens*) and white-breasted nuthatches (*Sitta carolinensis*), which often use the same roost cavity night after night, remove feces from the holes (Kilham, 1971).

Construction and positioning of nests and burrows may considerably reduce the energy requirements of small homeotherms. Several species of hummingbirds usually place their nests under overhanging limbs and in other sheltered locations, which significantly decreases heat loss by radiation, convection, and conduction (Calder, 1973, 1974) (fig. 6.7). Considerable heat escapes through the nest itself, but the hummingbirds seem to minimize this loss by building substantial nests. White, Bartholomew, and Howell (1975) and Bartholomew,

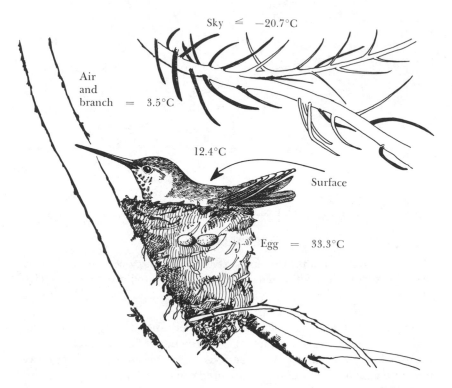

Figure 6.7 Mean predawn temperatures at a broad-tailed hummingbird (*Selasphorus platycercus*) nest in a spruce tree. The (synthetic) egg temperatures and air temperatures were recorded from thermocouples. The branch undersurface temperature (which did not differ significantly from the air temperature), the open-sky temperature, and the dorsal surface temperature of the hen were obtained with an infrared radiation thermometer. Original illustration by Lorene L. Calder. (Redrawn by Jaquin Schulz from Calder, 1973, copyright 1973 by the Ecological Society of America.)

White, and Howell (1976) found that the huge colonial nests of the sociable weaver (*Philetairus socius*) may ameliorate temperatures strikingly, as much as 18–23°C above ambient temperature. They also show periodic bursts of activity, which may raise the temperature, and function in a way analogous to nest thermoregulation by honeybees.

Lynch (1974) demonstrated experimentally that white-footed mice (*Peromyscus leucopus*) can be induced to build larger nests in the laboratory in response either to increased cold or decreased photoperiod. Presumably the decrease in photoperiod is a harbinger of decreasing temperatures under natural conditions. Several heteromyid rodents of western North America choose locations within their burrows that keep them at the lower end of their range of thermal neutrality (Kenagy, 1973).

The microclimate selected by procellariiform seabirds (albatross, shearwaters, petrels) has been investigated systematically by Mougin (1975). He found that in the Antarctic all species seek some form of shelter for their nest sites, but the requirements become progressively more exacting as the size of the species decreases. Wilson's storm petrel (*Oceanites oceanicus*), slightly larger than a swallow, requires nest holes that reduce exposure strikingly. The giant petrel (*Macronectes giganteus*), about the size of a typical albatross, nests in the open but only occupies nest sites that are conspicuously in the lee of prevailing heavy winds.

Grouse and their relatives often bury themselves under the snow during the winter. Examples include the ruffed grouse (*Bonasa umbellus*) (Darrow, 1947) and black grouse (*Tetrao tetrix*) (Pauli, 1974). Several other birds, including small passerines, may bury themselves in a similar way (reviewed by Sulkava, 1969; Haftorn, 1972; Novikov, 1972). Welty (1962) noted that snow buntings (*Plectrophenax nivalis*) may even spend part of the day under snow during extreme cold. Sulkava suggested that this behavior is correlated with strong cold, and Novikov noted that the more severe the conditions, the longer these birds remain in their snow burrows. Both of these observations are consistent with the idea that individuals run increased risks of predation as a result of this habit, and foxes are known to hunt for buried grouse (Bent, 1932).

Pearson (1953), French and Hodges (1959), Langner (1973), and Carpenter (1974) have documented one of the most improbable forms of behavioral thermoregulation, the roosting of hummingbirds and other species in caves in the high Andes at night, where the open air temperatures may drop below freezing. The temperatures of the caves may remain as much as 20°C above that of the surrounding environment (Langner, 1973). The hummingbirds travel substantial distances to these caves in some instances. Langner (1973) indicated that during the winter they may make daily round trips of up to 240 km to reach the nearest flowers.

Grubb's studies (1975, 1977) on the foraging locations of small, forest-dwelling birds during the winter indicate that less conspicuous behavioral changes than these may have significant thermal effects. As wind velocity increases, birds tend to forage at lower heights and in the lee of the wind. Chickadees and titmice, which typically forage in the exposed small branches of leafless deciduous trees when wind velocity is low, reduce their average foraging height considerably when winds are strong. Nuthatches, which normally forage on trunks, show smaller changes at such times.

OTHER METHODS

Three other behavioral methods for coping with thermoregulatory problems are available to diurnal species. First, such species may store snacks to be eaten at night. Certain small herbivorous birds store large numbers of seeds in their crops prior to roosting (Brooks, 1968; Evans, 1969). Redpolls (*Acanthis flammea* and *A. hornemanni*), which winter farther north than almost any other small birds in North America, have an esophageal diverticulum (Fisher and Dater, 1961) that presumably functions in this way. Second, diurnal species may extend their activity into periods of extremely low illumination, even feeding under laboratory conditions in virtual darkness (Brooks, 1968). Although it is doubtful that they could feed on a more dispersed natural food source at such a low light intensity, they may be able to move to and from food sources at this time, thus maximizing the time available for feeding (Brooks, 1968). Snow buntings, another arctic species, sometimes remain active at night during the winter (Morse, 1956). However, most diurnal birds have not evolved in this direction, probably because of visual difficulties. Light seems to be less critical for small mammals, which therefore have longer potential activity periods than birds. Third, diurnal species may simply reduce their levels of activity under unusually severe conditions. For example, Kessel (1976) found that black-capped chickadees (*Parus atricapillus*) visit bird feeders in central Alaska later on unusually cold ($-46°C$ to $-51°C$) than on normal winter days, and that their rate of visitation is also considerably lower. This suggests that under these conditions the birds suffer less energetic loss if they curtail activity than if they increase it by searching for food. It is also possible that they are resorting more often to hypothermia (Chaplin, 1974; Grossman and West, 1977). Kessel's temporarily missing birds (many of them banded) survived this severe period and resumed normal activity when conditions returned to normal. Grubb (1977) found that under the coldest winter conditions in Ohio, Carolina chickadees (*P. carolinensis*) vacated their usual feeding areas in a forest and retreated into dense brush. Evans (1976) reported that shorebirds cease foraging in very cold and windy winter weather and related this to conditions in which the energetic costs of foraging exceed the energy gained; similar behavior has been reported in geese under inclement conditions (Markgren, 1963; Raveling, Crews, and Klimstra, 1972). These species definitely do not use previously stored food.

DORMANCY AND MOVEMENT

Dormancy is a very common response to harsh conditions, and may yield substantial energetic savings (Tucker, 1965). The many adaptations involved in dormancy are primarily physiological, rather than behavioral-ecological (see Bartholomew, 1972). But their purpose is fundamentally ecological: to reduce the metabolic rate (and usually the body temperature as well), thus reducing the animal's energetic needs. As a strategy, dormancy depends on the predictable recurrence of favorable conditions, first to provide the extra resource needed for the period of dormancy, and then to provide those needed for a complete recovery. If dormancy is not feasible, the only alternative is to migrate, a life-style adopted by many birds and some other species that have the requisite mobility. For the most part, species adopt one option or the other, although a few make use of both (for example, bats that migrate considerable distances to winter roosting areas, and hummingbirds that enter nocturnal hypothermia on their summer breeding grounds).

DAILY TORPOR

Many small endotherms with extremely high energetic demands enter torpor daily; this is well documented in bats (Kunz, 1974), several groups of rodents (chapter 2), and a few birds (hummingbirds, goatsuckers, colies) (Haftorn, 1972). Daily torpor is commonest in small aerial feeders (birds and bats), owing to the heavy demands of both flight and endothermy. Some larger aerial feeders, such as swifts and goatsuckers, may also enter daily torpor. Shrews, the smallest terrestrial endotherms, are not known to do so. In many cases torpor only occurs under extremely unfavorable conditions (Hainsworth, Collins, and Wolf, 1977), as in incubating hummingbirds faced with unseasonably cold temperatures (Calder and Booser, 1973). This ability allows these birds to nest early in the season, which allows their young to become independent during a period when abundant nectar resources are available.

Bartholomew (1972) suggested that daily torpor has evolved in one other ecological context, that of small rodents in arid regions. These animals live mainly on seeds, which may be scarce and unpredictable. Bartholomew reasoned that this condition was an adaptation to unpredictable food resources. But it seems equally likely that these small rodents are sometimes unable to gather seeds rapidly enough to maintain a normal metabolic rate (N. E. Stamp, personal communication). During many years they take only a small fraction of the available seeds (Tevis, 1958; Pulliam and Brand, 1975; Nelson and Chew,

1977), and some species are known to enter daily torpor only when food supplies are restricted (Tucker, 1966). In some cases these daily cycles are supplemented by longer periods of dormancy (see Hudson, 1967).

Some animals can reduce body temperature without entering torpor. At cold winter temperatures black-capped chickadees can depress their body temperatures as much as 7°C during the night, reducing their energy demands by as much as 60 percent and thus maintaining an energy reserve overnight (Chaplin, 1974). Inca doves (*Scardafella inca*) (MacMillen and Trost, 1967) and vultures (Heath, 1962; Hatch, 1970) undergo modest nocturnal hypothermia as well. Kushlan (1973) hypothesized that the spread-wing posture often assumed by vultures in the morning helps them use radiant energy to regain normal daytime temperatures. This strategy seems particularly efficient in that vultures use thermals in hunting food, and thermals do not form until later in the morning when the land has warmed somewhat.

LONG-TERM TORPOR

As conditions worsen, long-term torpor (dormancy) replaces daily torpor in animals capable of both alternatives. Long-term torpor is often part of a circannual rhythm (Bartholomew, 1972), which implies that conditions are so predictable they can be tied to a clock. Some patterns of daily torpor show a circadian rhythm (Tucker, 1966). Bartholomew (1972) believed that dormancy evolved in response to such a wide spectrum of ecological circumstances that no simple generalizations could account for all cases.

MOVEMENTS

When unable to maintain themselves on existing resources or to enter torpor, animals must move. Although birds, bats, and some insects are capable of moving long distances to favorable areas, they may not be able to do so quickly enough to escape short periods of deprivation (chapter 2). In fact, small movments by less mobile animals may improve their circumstances more than equivalent movements improve the circumstances of most birds or bats. Confronted with shortages of moisture, salamanders simply move down in the soil column, probably through cracks and crevices (Jaeger, 1974); the resulting environmental change is as great as might be encountered on the surface over distances of many kilometers. Many small animals of the soil litter probably make similar movements (see Cole, 1946).

OTHER METHODS

Social animals often save energy by moving as a group through resistant media. The saving is greatest for animals that move rapidly or through relatively viscous media such as water. Schooling fish may obtain a considerable energetic saving in this way, but Radakov (1972) has argued that the energetic saving is less important than a number of biotic factors. Schools do not usually have leaders (see Breder, 1959), so the forward individuals are probably continually changing. Certain species of freshwater fish may only form schools when the water becomes extremely cold and thus somewhat more viscous (Hergenrader and Hasler, 1968).

Birds flying in groups obtain a similar energetic advantage. Lissaman and Shollenberger (1970) calculated the vees of twenty-five birds can fly as much as 71 percent farther in formation than singly, by exploiting the upward rising components of the wingtip vortex currents generated by individuals to the front and side. In order to maximize this gain, they must maintain a certain interindividual distance and angle of the vee, which will vary with size and velocity. Gould and Heppner (1974) measured these factors in Canada geese (*Branta canadensis*) and found them to be quite variable. Nevertheless, a substantial advantage is gained, which may allow long-distance fliers to make longer flights or to get to foraging areas more economically, reducing the necessary daily intake of food.

Mammals in deep snow also move through a resistant medium; deer, caribou, and moose form trails that may be used by many individuals (Peterson, 1955; Bergerund, 1974). Wolves travel single file through deep snow when not hunting (Mech, 1966; Kolenosky, 1972), taking turns breaking the trail.

In such cases the leader obtains little if any benefit. One might expect leaders to change, which they often do in formations of flying birds. Groups of closely related individuals (such as groups of wolves) ought to manage the rotation more cooperatively than do groups of unrelated individuals, especially members of different species (see Trivers, 1971). The members of unrelated groups would more often be expected to cheat, enjoying the benefits but not taking turns as leader.

Miscellaneous maintenance activities

There are many activities related to maintenance that make minor demands on an animal's time and energy. A few of these are briefly considered here.

CARE OF THE INTEGUMENT

Feathers and fur have great metabolic significance to endothermic animals (see Prosser, 1973) because they are major organs of insulation. They trap the most dead air, and thus insulate best, when lustrous and unmatted. Thus, it is highly advantageous for individuals to groom themselves regularly (Hutchinson, 1954). Bumblebees, which are largely endothermic (Heinrich, 1974), groom their furlike setae fastidiously (Free and Butler, 1959). Although it seldom demands very much time or energy, maintenance of insulation will be of most importance when energetic demands are great and time is therefore limited.

Birds must also care for their flight feathers properly. Inadequate feather care and stresses occurring during feather growth may damage the flight feathers, leading to losses of efficiency and maneuverability. The major energetic demands of molting may explain why most birds molt during periods (late summer and fall in temperate latitudes) when they are typically under no stress from reproduction or inclement winter weather.

In some birds, preening may be associated with the synthesis of vitamin D. Oil from the preen gland that has been smeared on feathers exposed to sunlight may produce the vitamin, which is gathered during later preening sessions (see Kennedy, 1968, 1969). This is analogous to the production of vitamin D in human skin exposed to sunlight.

Insects often spend considerable time grooming their tarsi and antennae. Efficient foraging may be difficult if they do not groom, because chemoreceptors critical for finding food are located there.

RESPONSES TO ECTOPARASITES

Grooming may also serve to remove ectoparasites, which plague animals of all kinds (including ectotherms). Some are not known to affect their hosts seriously, but many others do. For example, ticks may transmit debilitating diseases, remove substantial amounts of blood, and create sites for ordinary infection. Such effects surely justify the considerable amounts of time devoted to grooming by many animals, especially mammals. The bathing behavior of many animals, such as bathing in mud by ungulates and dusting by birds, may also aid in the control of ectoparasites.

Blood-feeding dipterans sometimes take significant amounts of blood from their victims, in addition to spreading disease. The exact extent of their physical damage is seldom apparent, but they annoy their hosts intensely, decrease host foraging time, and promote infection. They may occasionally even kill their hosts (Schaller, 1972). Behavior that mitigates the effect of insect attack is well developed in some animals, especially certain ungulates. The integumentary muscles of these animals are capable of an effective twitch response. Head-to-tail positioning has evolved in some species, allowing individuals to swat biting insects from their otherwise vulnerable anterior parts (see Geist, 1971).

The vulnerability of birds to mosquitoes differs strikingly in a species-specific way (Edman and Kale, 1971; Maxwell and Kale, 1977). Some are seldom bitten, whereas others have few if any defenses. Avoidance behavior may involve continually wiping the otherwise vulnerable legs and burying the head inside body feathers; such techniques all cost a certain amount of time and energy, but failure to engage in them can cost more.

STRUCTURAL MAINTENANCE

Animals build a wide variety of structures associated with food gathering, predator avoidance, and reproductive activities. The construction and maintenance of burrow or runway systems by moles, shrews, and mice require substantial expenditures of time and energy. The burrows of moles (*Talpa europaea*) are kept in repair for long periods of time, and the animals move regularly through them picking up food items such as earthworms and insect larvae that fall through the walls of the burrow (Godfrey and Crowcroft, 1960). Similarly, the runways of voles, such as the meadow vole (*Microtus pennsylvanicus*), facilitate movement through the thick grasses in which they usually live, at the same time giving them ready access to food sources and some measure of protection from predators (Hamilton, 1939). Some mammals and birds make special nests not associated with reproductive activity, and these may require sizable investments of time and energy to build and maintain. The summer tree nests of squirrels (Hamilton, 1939) and nightly tree nests of a number of primates such as the gorilla (*Gorilla gorilla*) (Schaller, 1963) are good examples.

Synthesis

The importance of behavioral thermoregulation relative to feeding and defense from predators differs markedly from species to species, and across environments within a species. Endotherms and ectotherms differ fundamentally as a result of their vastly different heat budgets, but it is of interest that they nevertheless show many similarities (for example, sunbathing). In general, responses depend on the microclimatic conditions in which individuals find themselves, and these may differ markedly within a small area.

I have concentrated primarily on the behavioral aspects of thermo-regulation, but it is clear that these form only part of a physiological-behavioral adaptive complex, which presumably evolved as a whole. This is nowhere more strikingly apparent than in the adaptations of some lizards to desert conditions. Color shifts, combined with postur-ing and the selection of appropriate sunning sites, permit such lizards to raise their body temperatures quickly, in this way extending their periods of activity during the cool early morning. Foraging time is thus increased, and rapid warming minimizes the period during which individuals are highly vulnerable to predators. When conditions be-come hot later in the morning, changes of color and of orientation reduce the uptake of radiant energy. By moving between direct sun-light and shade, individuals can maintain relatively constant tempera-tures. Finally, individuals can to some extent store heat during the day, and by dissipating it passively as ambient temperatures decline, extend their periods of activity. Clearly these behavior patterns and their corresponding physiological modifications are mutually de-pendent.

7 Reproduction

BEFORE AN ANIMAL can reproduce, it must accumulate nutritional reserves beyond those required for its own maintenance. Females must be able to create large amounts of new protoplasm, and males must typically be able to court, fend off other males, or assist in caring for the young.

Finding a mate may require substantial time and energy, particularly in solitary species with low population densities, such as mammals with large nonoverlapping home ranges (for example, large cats, prosimians, certain large rodents). Considerable effort may go into courtship, aggressive display, or both. Finding and preparing a place in which to raise the young (a burrow for many mammals, a nest for most birds) may be particularly costly (White and Kinney, 1974); it can take weeks to construct burrows or nests in some instances. The nest site, and perhaps also a large surrounding territory, may require defense against conspecifics and even the members of other species. Care of the young may create extraordinary demands (West, 1960; Kendeigh, 1969) and at the same time restrict the foraging behavior of the parent. If individuals reproduce and winter in different areas, failure to accumulate the large additional energy stores required for migration in either direction can prevent an individual from reproducing.

Accumulation of energy stores

Animals that put a large investment into gametes must accumulate substantial energy stores before even beginning to reproduce. They may surmount this obstacle in a variety of ways. Many species lay down extra protein and fat prior to breeding (Ward, 1965b; Jones and Ward, 1976); this reserve may be used directly in development of the gonads, and also in the incubation of eggs or embryos (see King, 1973; Ricklefs, 1974). If parental care occurs, it may allow the parent to channel directly to the young a high proportion of food gathered during their dependency. Some of these adaptations are spectacular. The emperor penguin (*Aptenodytes forsteri*), which undertakes extended periods of incubation, may lose over 40 percent of its preincubation weight during the breeding season (Prevost, 1961). Protein and fat reserves fall quickly in queleas (*Quelea quelea*) during the first few days of the reproductive effort (fig. 7.1), sometimes to a dangerously low level (Jones and Ward, 1976). Mammals, pigeons, and a few fish (see Noakes and Barlow, 1973) have minimized the short-term strain on resource acquisition by producing nutritious secretions that permit the young to be fed directly from resources gathered by the adults at an earlier time. However, subsequent to the production of eggs or young most animals can exploit their bodily resources only indirectly, by using them for their own maintenance while channeling newly gathered food directly to the young.

Where there is significant seasonality, young are usually born during the warm months (Schoener, 1971); food is usually more abundant then, and endotherms can minimize their heat losses. Endotherms are severely restricted by low temperatures if only one parent raises the young, since it must leave them to forage. As temperatures decrease, the rate of heat loss by unattended eggs or young increases, as do the energy demands of the incubator itself (White and Kinney, 1974). Some species store enough food as fat to incubate without feeding—for example, some female eiders (*Somateria mollissima*) (Palmer, 1976). But this option is not available to most species, especially small ones. Eiders line their nests with their own feathers (famous for their insulating qualities), and if a female must leave her nest, she covers the eggs with these feathers.

Caching food in advance might seem to be a good strategy. Surprisingly few birds and mammals cache food for use during the breeding season, despite the fact that many do so for use during the winter (chapter 2). A few squirrels (*Tamiasciurus* spp.) (C. C. Smith, 1968) and other rodents appear to use stored seeds during their reproductive

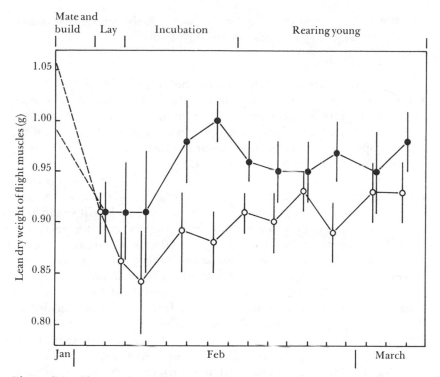

Figure 7.1 Changes in the mean weight (±2 SE) of flight muscle protein of breeding adult queleas. Black circles = males; white circles = females. The dashed lines indicate the probable decline in protein during the first few days before sampling. (Modified from Jones and Ward, 1976, with permission of the *Ibis,* journal of the British Ornithologists' Union.)

periods, and four species of birds—two nutcrackers, a jay (Swanberg, 1951; Balda and Bateman, 1971; Bock, Balda, and Vander Wall, 1973), and a tit (Haftorn, 1973)—are known to use stored food in this way. Seeds stored by corvids permit them to nest early in the season, and their young consequently have a long time to develop before the onset of severe conditions (Balda and Bateman, 1971; Vander Wall and Balda in Bock et al., 1973). Stored food forms the major resource of both adult and young nutcrackers (*Nucifraga* spp.). However, in the piñon jay (*Gymnorhinus cyanocephala*) it makes up only a minor part of the food of the young, although it forms the major food of adults during the reproductive period. Perhaps the difficulty of protecting and maintaining a store of food while raising young eliminates this strategy in most cases. Many types of food, particularly the rich sources of protein sought by breeders, do not store well

(Haftorn, 1956) , and this may be why some of the young tits studied by Haftorn (1973) sometimes refused such items.

Care of young

Parental care is widespread in the animal kingdom and universal among birds and mammals. Varying amounts of parental care occur among reptiles (Bellairs, 1969) and amphibians (Noble, 1931; Wells, 1977) , and it is highly developed in some fish (Marshall, 1965) . It also appears sporadically throughout the invertebrates, for example, in social insects and spiders (Wilson, 1971a) .

Parents can expend their effort all at one time or in a sequence of reproductive acts spread over most of a lifetime. They may either produce large numbers of gametes simultaneously, which will suffer high rates of mortality, or produce and subsequently care for relatively small numbers of young. Although parental care will restrict the activities of the parent (s) for a relatively long period, it may allow successful reproduction where it would otherwise be impossible.

Parental care has other behavioral and ecological implications. It provides an excellent opportunity for the transmission of specified or difficult foraging techniques. This may permit the exploitation of resources that would otherwise be unavailable, such as sparse, unpredictable, and hard-to-manipulate resources (chapter 2) . Young may thus learn complicated behavioral patterns that are not readily incorporated into an innate repertoire.

AMOUNT OF PARENTAL CARE

A heavy initial energetic investment in the young will shift the major foraging effort of parents toward the beginning of the breeding season. This should be advantageous where particularly large but brief pulses of available food occur, pulses so sudden or unpredictable that the parents cannot synchronize the needs of their offspring to them. Many animals feed their young over substantial periods, but a brief period of superabundance should facilitate the production of large eggs or offspring with good opportunities for rapid subsequent growth. Also, a large initial investment could reduce the impact of predation on the young since the time of high vulnerability would be shortened.

High initial investment ought to be most common where predator pressure is much higher on young than on adults because danger to the gravid, relatively immobile females will be minimal. This condition is most frequently met in moderate- to low-productivity ecosystems where, however, accumulating a large resource surplus may be difficult.

The extreme precocity of many plains-dwelling ungulates appears to be a direct consequence of the high risk of predation experienced by the young (Estes, 1966; Schaller, 1972). Most ungulates are precocious, but the condition is developed most strongly in species that live where there is little or no concealing cover such as the open savanna, or in species that cannot securely defend their young from certain predators, for example, small and medium-sized antelopes.

Many placental mammals that give birth to altricial young either hide them effectively or have few natural enemies. Primates and the related tree shrews provide a good example. Certain tree shrews hide their young and spend little time with them, feeding them as seldom as once every other day and spending the rest of their time apart from them. The young fend for themselves at an early age (Martin, 1968). On the other hand, parental care is prolonged in most primates, reaching its extreme in the pongid apes, where the young depend on their mothers for years (Schultz, 1969). The social systems and large body size of pongids make them relatively invulnerable to predators.

Predator pressure also modifies the reproductive patterns of birds in a variety of ways. Small numbers of young are produced by many tropical birds (Skutch, 1949, 1976); among other effects, this reduces the frequency with which a brood must be visited. Careful nest placement and the adoption of cryptic behavioral patterns may further reduce danger to the offspring. Foster (1976) noted that long-tailed manakins (*Chiroxiphia linearis*) produce clutches of one or two and that they nest in small forks far out on the branches. This position probably reduces loss to a regular predator, the coati (*Nasua nasua*). However, it also exposes these birds to increased rates of nest loss from wind.

Korschgen (1977) believed that the twenty-six-day incubation period of female eiders, who eat little or nothing during this time, is an adaptation to minimize nest predation. Korschgen found that these birds occasionally starve while incubating, which indicates the apparently great intensity of nest predation on this species. He also observed that outbreaks of avian cholera in eider populations were concentrated during the latter part of the incubation period, when the birds were in the worst physical condition.

Some birds build their nests preferentially in thorn bushes; it takes longer to negotiate this cover, which degrades feeding efficiency, but it also deters predators. Brosset (1974) noted that tropical forest birds avoid building their nests near pathways normally used by predators, concentrating them between tree buttresses, at tips of twigs overhanging open spaces or water, and in small isolated trees. Hoopoes

(*Upupa epops*) eject feces into the face of a would-be predator attempting to enter a nesting hole (Löhrl, 1977). These birds also produce foul, musky-smelling secretions from their oil glands, which Löhrl believes may be defensive in nature. All these methods of concealment and defense cost something in terms of time and energy.

The movement patterns of ungulates with young may be highly modified as a consequence of predator pressure. If the parents or the group to which they belong are unable to defend the young against common predators, the young may adopt a cryptic existence. This presumably accounts for the coloration of young deer, who may be left while the parents are feeding. If surprised, they generally do not have any effective defense, at least not against their largest predators. The foraging patterns of some predators appear to be designed to find these young ungulates (Schaller, 1972). For example, African wild dogs (*Lycaon pictus*) will fan out while hunting in a way that increases their chance of finding a fawn. Wildebeest (*Connochaetes taurinus*) give birth to young that are able to move with the herd within a few minutes, thus enjoying some protection by being in the group. Large species such as buffalo (*Syncerus caffer*) and eland (*Taurotragus* spp.) are able to protect their young actively, owing to the parents' large size (see Jarman, 1974).

Unusual reproductive behavior may also occur where predator pressure is relaxed. Brush turkeys (Megapodiidae), Australasian galliform birds, incubate their eggs in warm soil or in mounds of fermenting vegetation. The eggs are extremely large, numerous, and slow to develop; the young hatch and usually leave the incubators without ever seeing their parents (Frith, 1959). Other galliform birds have fewer, smaller eggs with shorter incubation times, and most give the newborn at least a minimal amount of subsequent parental care. The megapodes appear to be channeling an unusually large effort spread over an unusually large amount of time into the production of highly precocial young. Males are responsible for the care of the mounds and the eggs within, and this consumes all of their attention over a period of several months. The mound must be constantly tended and added to in order to keep the eggs at a relatively constant temperature. A single egg may represent 18 percent of a female's body weight, and the entire clutch laid by a female may exceed three times her weight (Lack, 1968). During years of poor resource availability, the interval between successive eggs increases, suggesting that the number of eggs is a function of the ability of the female to collect enough food to form an egg (see Frith, 1959).

Many of these birds live in tropical forests where there are few periods of temporary superabundance (see Morton, 1973). Because of its size and conspicuousness, a mound would seem to be very attractive to egg predators, and indeed there are reports of predators digging out eggs (Lincoln, 1974). But predation within this group's isolated range (Australia, New Guinea, and surrounding islands) is probably nowhere near the intensity of that experienced in other zoogeographic regions, making this a feasible method of tending the eggs.

Bob-white quail (*Colinus virginianus*) and some medium-sized terrestrial passerine birds such as meadowlarks (*Sturnella* sp.) have rather broadly overlapping feeding patterns. Quail hatch as active young from large eggs and can almost immediately perform most of their own feeding with only some initial assistance from the adults. However, the passerine birds hatch blind and helpless from much smaller eggs and are heavily dependent on the adults for at least a few weeks. Clutch size of the quail may be twice that of meadowlarks, yet meadowlarks often raise two broods and quail often only one (Bent, 1932, 1958). Incubation of the quail eggs takes somewhat longer than incubation of the meadowlark eggs. Uncovered young are especially vulnerable because of their own activity and the activity of parents feeding them. Thus the rate of predation on quail nests at this point should be lower than that on meadowlark nests, since the former are at this especially vulnerable stage for only a short period. The actual time of incubation and care of young in the nest is nevertheless similar in the two species. Although the two strategies might appear to be adapted to rather different temporal patterns of food abundance (for a high pulse and a relatively constant resource level, respectively) or to different predator pressures, both may exist within a single grassland.

When compared with tree-nesting passerine birds, ground-nesting passerines appear to have converged somewhat with quail; their eggs are larger than those of related tree-nesting passerines, and their period within the nest is abbreviated (Nice, 1962). Meadowlarks belong to the family Icteridae (New World blackbirds), which are typically tree, bush, or marsh nesters. The meadowlarks have larger clutches than other icterids, except for bobolinks (*Dolichonyx oryzivorous*), which are also ground nesters (Bent, 1958).

ONE PARENT CARES FOR THE YOUNG

Parental care in birds and mammals is often performed solely by the female or, more rarely, by the male. The young may be either precocial or altricial. Parental care in other taxa (arthropods, fish, amphibians, reptiles) is usually carried out by only one parent (see

Brown, 1975a), but more often than in birds and mammals this is the male—for example, leptodactylid frogs (Martin, 1970). Here I concentrate on birds and mammals.

Uniparental care of young in birds is most common where resources are abundant because incubation and provisioning of young are apt to take up a high proportion of the care giver's time. This problem can be alleviated by nesting close to a predictable food supply, by having the male provision the female, or by better insulating the nest (White and Kinney, 1974).

In altricial birds, uniparental care is fairly common in extremely productive habitats where there is little development of the vertical dimension (marshes, grasslands) (Orians and Willson, 1964). Uniparental care is absolutely more common in precocial species, again most frequently in open-country habitats (Orians, 1969c). Precocial animals are to some degree preadapted for uniparental care, because the initial input of energy into a large infant or egg minimizes the subsequent input required, at least for early care (Orians and Willson, 1964).

Uniparental care is often associated with polygynous or promiscuous mating systems but is by no means confined to them (Orians, 1969c). For a polygamous system to be advantageous for the female, it is necessary that she successfully rear more young than she would by being mated to a monogamous mate (fig. 7.2). This requirement could be met if certain males were able to occupy territories in the richest habitat large enough so that two or more females could find adequate food for themselves, or if the females could establish similarly favorable conditions themselves. This seems most likely to be the case in open-country areas (Verner and Willson, 1966), which often show great variation in productivity and a tendency to change rapidly over time (Orians, 1969c).

Mammals are more frequently polygynous or promiscuous than are birds (Eisenberg, 1966; Orians, 1969c), a direct consequence of the minimal parental role played by most male mammals. This follows logically from mammalian physiology, which gives females the functions of both producing the young and initially feeding them. Few mammals are monogamous (Kleiman, 1977); a high proportion of these are carnivores, who often have great difficulty obtaining sufficient meat to nourish their young. Monogamous males may provision the females or young and may assist in training the young. But male lions (*Panthera leo*) (Schaller, 1972), tigers (*Panthera tigris*) (Schaller, 1967), and mountain lions (*Felis concolor*) (Hornocker, 1969) do not contribute to the maintenance of their young. Male care should

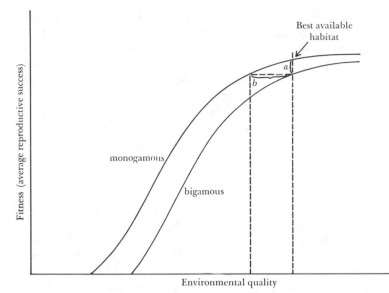

Figure 7.2 Conditions necessary for the evolution of polygynous mating patterns. Average reproductive success is assumed to be correlated with environmental differences, and females are assumed to choose their mates from available males. The distance *a* is the difference in fitness between females mated monogamously and females mated bigamously in the same environment; the distance *b* is the polygny threshold, the minimum difference in quality of habitat held by males in the same region sufficient to make bigamous matings by females favored by natural selection. (Modified from Orians, 1969c, copyright 1969 by The University of Chicago.)

permit occupation of marginal areas where resources are available but hard to secure.

BOTH PARENTS CARE FOR THE YOUNG

Monogamy, with some care by both parents, is the rule in birds but the exception in mammals (Orians, 1969c). Even in birds, the reproductive morphology and physiology of the male and female dictate qualitative differences in their contributions to rearing offspring, although these differences are usually less striking than those of mammals. In either case, the production of eggs or fetuses and the nourishment of young make substantial demands on females. Males can mediate this drain by bringing food to their mates, and this may be the functional basis of courtship feeding in many birds (Royama, 1966b).

Defense is often the specialty of male birds (see Hinde, 1956) and may demand considerable time and energy (Black, 1975). Even where males regularly participate in direct parental care, wide variation exists in the extent of the participation. In the common grackle (*Quisculus quiscula*) some monogamous males feed young more than females do, some less, and some not at all. Even polygynous males may feed young at one or more nests (Howe, 1976). The contribution of males to feeding the brood may differ with the age of the offspring. Male song sparrows (*Melospiza melodia*) feed older offspring at higher frequencies than they do younger offspring (J. N. M. Smith, 1978); this species has several broods per year, and this mate behavior may help pairs to start the next brood as soon as possible.

Some male wood warblers, such as the yellow warbler (*Dendroica petechia*), regularly feed their females on the nest (Morse, 1966). This species tends to live at relatively low densities where studied. Males of the closely related black-throated green warbler (*D. virens*) seldom if ever assist their incubating females directly, and instead spend much of their time displaying on the territory (Morse, 1968b). This species lives at high density, and constant display may be required to maintain a territory under such conditions. Male black-throated green warblers do assist in feeding their young after the young have fledged and territories begin to break down. The effects of density could be studied by looking at high- and low-density segments of a single population. Isolated male black-throated green warblers sing in a stationary position much less than individuals in dense populations (Morse, 1970a) (fig. 7.3). Since this singing is a major component of territorial display, they should be able to make a more direct contribution to the rearing of their young than can those in dense populations, but the time budgets of these birds are not known well enough to test this prediction.

Female warblers forced to obtain all their own food during incubation remain off the nest only about 15 percent of the time but feed at extremely high rates while off the nest (Morse, 1968b) (fig. 3.5). They probably lose weight during incubation, judging from the amount of time they incubate and the voracious way in which they forage when off the nest, a pattern also seen in black-throated blue warblers (*D. caerulescens*) (Black, 1975). Parental weight loss occurs in blue tits (*Parus caeruleus*) while the young are being fed (C. M. Perrins, personal communication), but females apparently maintain their weight during incubation. However, the male brings food to her on the nest. During incubation queleas actually reverse some of the weight loss they suffer while laying (Jones and Ward, 1976).

Paying inadequate attention to one's offspring, even if they are

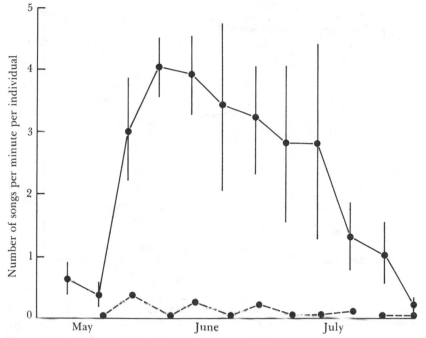

Figure 7.3 Frequency of stationary singing (±1 SD) by male black-throated green warblers in dense populations (solid line) and by isolated individuals (dashed line). (Redrawn from Morse, 1970a.)

adequately fed, may lead to increased mortality of the young. Murton and Isaacson (1962) reported cases in which woodpigeons (*Columba palumbus*) encountered feeding conditions that were adequate for egg laying but which did not leave sufficient time for proper incubation. Unguarded eggs suffered high rates of predation, and there were few surviving progeny. Hunt (1972) and Yom-Tov (1974) noted that as adult gulls (*Larus* spp.) (fig. 7.4) and carrion crows (*Corvus corone*) spent more of their time away from the nest foraging, their young suffered increasingly high rates of predation, largely from conspecifics.

RELATIVE CONTRIBUTION OF THE TWO SEXES

The relationship between male and female parents should be completely cooperative if a single strategy permits each member of the pair to maximize its reproductive success. But as Trivers (1972) has pointed out, the interests of males and females may be quite different with respect to reproductive patterns and care of the young. Nowhere is this fact clearer than in polygamous relationships where one sex, fre-

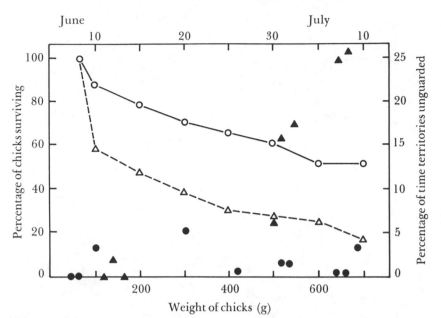

Figure 7.4 Survivorship of herring gull chicks to a given weight on Goose Rock (solid line) and Little Green Island (dashed line); percentage of time territories left unguarded by parents on Goose Rock (black circles) and Little Green Island (black triangles). Goose Rock is located near a garbage dump, providing an easy source of food for the adults; Little Green Island is not close enough to a garbage dump to provide an easy food source. (Modified from Hunt, 1972, copyright 1972 by the Ecological Society of America.)

quently the female, provides most or all of the care of the young. Under some conditions the best option for the male may be to have several mates and for the female to have a single mate. The result may be a compromise. For example, Downhower and Armitage (1971) have shown that monogamy is the optimal condition for female yellow-bellied marmots (*Marmota flaviventris*), and a harem of two to three females is optimal for the male. The mean observed harem size is intermediate. Males provide territorial space within which the young may be reared, and to some extent reduce aggressive interactions among females. The fewer females there are, the less aggression. Females appear to be competing for food.

Under conditions where one partner can rear the offspring, it may behoove the other to desert and to mate with as many other individuals as possible. Conflict arises if this polygamous pattern lowers the fitness of the deserted individual as a result of the added work and danger that may befall it as the sole provider; any behavior on the part of the potentially deserted sex that acts to increase the other sex's investment in the young should be favored. For example, the vulnerable sex (most frequently the female) may require its prospective mate to perform elaborate courtship rituals, to feed it, to build a nest, or to perform a variety of other activities. Any one of these may require so much effort that to desert simply becomes unprofitable, particularly if the next prospective mate is certain to demand the same.

Subtle forms of cheating may occur, as when one parent ceases caring for the young sooner than the other or makes a smaller investment than the other. Parental investment is any investment in an offspring that decreases the parent's ability to invest in other offspring (Trivers, 1972). Trivers has pointed out that at various times within a reproductive cycle, commencing with fertilization, one sex or the other may be particularly vulnerable to desertion as a result of having made a greater investment in the young than its partner, up to that particular point. If the deserted partner is able to rear the offspring by itself from that point, its best option may be to continue rearing them without the assistance of its partner. However, the investment already made by the deserted partner should not completely dictate its own subsequent behavior; the deserted partner should "ask" only whether it can rear more offspring by continuing to care for these young by itself or by abandoning the offspring and making a second reproductive effort with a new mate (Dawkins and Carlisle, 1976; Maynard Smith, 1977).

Most of the evidence on parental care in monogamous birds seems interpretable without recourse to Trivers' argument. He cited evidence suggesting that male monogamous birds invest somewhat less than females, but it is difficult to evaluate these studies because many possible variables are involved in the overall parental investment made by the two sexes. In several instances these male birds showed strong shifts in the type of contribution made, and these were associated with environmental differences. Although they paid less direct attention to the young when in dense populations, the other activities they performed (such as increased display) were probably essential to territory maintenance in this setting. However, these observations were not made with Trivers' proposition in mind; they should be repeated with an eye toward quantification of the relative contributions of the two sexes. Emlen and Oring (1977) suggested that birds are unusual, in that males and females often share rather equally in parental care.

COMMUNAL BREEDING

Sometimes parental birds or mammals are assisted by other individuals, usually of the same species (Skutch, 1961, 1976). It is widely believed that these associates (often called helpers) may increase the number of offspring reared, even permitting reproduction where it would otherwise be impossible. But the relative reproductive success of pairs with and without associates is known in only a few cases (Brown and Balda, 1977). Pairs of superb blue wrens (*Malurus cyaneus*) and Florida scrub jays (*Aphelocoma coerulescens*) with associates have success rates two or even three times that of pairs without associates (Rowley, 1965; Woolfenden, 1975).

This difference could result from several causes. Most frequently mentioned are improved nutrition of the young, reduced predation, and improved territory defense. Seldom can these alternatives be unequivocally distinguished in field studies. Brown and Balda (1977) suggested that increased success may result simply from the fact that large groups (containing associates) tend to occupy larger or higher-quality territories than groups without associates.

The feeding contributions of associates differ markedly. Some broods of Mexican jays (*Aphelocoma ultramarina*) were fed more by associates than by their parents (Brown, 1970, 1972). But Zahavi (1974) believed that associates at Arabian babbler (*Turdoides squamiceps*) nests often make only a minimal contribution to the food requirements of the young and that overall they may have a negative influence on nesting success by competing for food and attracting predators. However, the data presented by Zahavi do not permit firm conclusions (Brown, 1975b).

It is not possible to measure the effect of associates simply by recording the amount of food that they bring to the young. For instance, if protection from predators is best accomplished by groups of birds, the extra individuals may provide a major benefit, especially if the predators do not seriously threaten adults. The mere presence of extra individuals about the nest could serve this function, since absence from the young may lead to high nestling mortality (Murton and Isaacson, 1962; Hunt, 1972; Yom-Tov, 1974). Woolfenden (1975) believed this to be the major benefit of associate scrub jays. Relatively large, aggressive species such as jays and babblers often attack or mob predators, so associates in these species may substantially lower predation on nests. Unfortunately, this factor is difficult to quantify (Emlen, 1978) and difficult to distinguish from territorial defense (Brown, 1978).

If resources are rather constant and hard to procure, associates may permit breeders to accumulate the additional energy required for rearing young where two individuals could not do so by themselves. Associates seem to be relatively common in areas of this kind (Skutch, 1961, 1976). Gaston's data (in Emlen, 1978) on the success of jungle babblers (*Turdoides striatus*) in less favorable habitats is consistent with this interpretation. Groups with associates raised an average of 1.0 young; those without associates raised an average of only 0.3 young. But the frequency of groups with associates did not differ from that in favorable habitats, where success of pairs both with and without associates was higher than in the less favorable habitats. Helping should be advantageous where resources are present in substantial amounts but require a considerable expenditure of time or energy to exploit. Moehlman (1978) suggested this explanation for black-backed jackals (*Canis mesomelas*), which have access to large numbers of prey (rats) that require considerable time to hunt. Associates may enhance survival in areas where an unassisted pair could breed by reducing the strain on parents or predation on the young. Stallcup and Woolfenden (1978) found that parent jays with associates had considerably higher survival rates than those without helpers.

These observations raise the question why associates should spend their time and energy raising someone else's young when they might be raising their own. However, they may not actually have the option of breeding. Nesting associations usually occur where nesting territories are at a great premium and the breeders are dominant over the associates (Stallcup and Woolfenden, 1978). Wynne-Edwards (1962) explained the participation of associates by a group-selection argument (that is, the group is benefited at the expense of the individual). However, the difficulties of group selection have been pointed out repeatedly (Lack, 1966; Wiens, 1966; G. C. Williams, 1966). There seem to be no data on this phenomenon that specifically support group selection, and the presence of associates can be explained without it.

Hamilton (1964) pointed out that close relatives may help to perpetuate a large proportion of their own genes by assisting in the raising of such young. Hence, if existing conditions prevent them from breeding, nonbreeders may improve their inclusive fitness (Hamilton, 1964) by helping their relatives to raise offspring. Maynard Smith (1964) has called this process "kin selection." Unlike group selection, kin selection is not a qualitative departure from natural selection, because seemingly altruistic behavior may in fact benefit the actor as well as the recipient. However, the benefits obtained from helping others with their young should decrease rapidly as relationships become more remote (Hamilton, 1964), if this is a primary basis for these associations. Close relatives, such as older siblings, should thus be the major contributors to such a system.

Are associates in fact closely related to their beneficiaries? The high frequency of helper caste systems found in social Hymenoptera may be the result of their unusual haplodiploid system, which results in sisters being more closely related than parents and other sibs (Hamilton, 1964; Trivers and Hare, 1976) and more closely related than in conventional breeding systems. However, exact degrees of relatedness are seldom known for vertebrates, although it appears that most relationships in these social groups are close. For example, helping offspring have been reported in several vertebrates, such as superb blue wrens (Rowley, 1965), Mexican jays (Brown, 1970, 1972), scrub jays (Woolfenden, 1975), Arabian babblers (Zahavi, 1974), common babblers (*Turdoides caudatus*) (Gaston, 1978a), lions (Schaller, 1972), African wild dogs (van Lawick and van Lawick-Goodall, 1971), and black-backed jackals (Moehlman, 1978). In the scrub jays (Stallcup and Woolfenden, 1978) and common babblers (Gaston, 1978a), male associates make significantly more trips with food to the nest than female associates. This difference is consistent with the fact that, in common with most communally nesting birds (Gaston, 1978b), both species are female outbreeding; that is, female offspring tend to disperse from their natal groups, whereas males remain. Thus, male associates are usually more closely related to the young than females. In many of these groups only one pair reproduces directly, and the others contribute work only.

Groups with associates often have skewed sex ratios (Brown, 1978). An unbalanced sex ratio accompanied by low resource availability should favor an associate system if members of the more abundant sex are unable to raise young on their own or with only a small effort from their mates, thus making polygamy impossible (see Orians, 1969c). In most cases there appears to be little sign of promiscuity within these groups, although this is hard to monitor. Polyandry does occur in the Tasmanian native hen (*Tribonyx mortierii*), a large terrestrial rail, where two male sibs often form a group with a single female (Ridpath, 1972). Promiscuity occurs in the unique breeding system of the noisy miner, an Australian honeyeater (*Manorina melanocephala*) (Dow, 1977).

The basis for differential survival of the sexes may be related to the associate system. Sex ratios subsequent to birth often diverge markedly from unity, sometimes favoring the male, sometimes the female. If females are smaller, they are often socially subordinate to males. Under these circumstances, they have little chance of dominating a group.

Even if a breeding female disappears, a resident female's chance of becoming the breeder is slim—father-daughter and sib matings are virtually unknown in scrub jays (Woolfenden in Emlen, 1978). Therefore, her chances of breeding at the natal site are very low. Thus, she must disperse, and dispersal is believed to result in a relatively high mortality rate (Woolfenden in Emlen, 1978).

By helping to raise genetically related individuals, usually younger sibs, associates might enhance their inclusive fitness, especially if they are unable to reproduce themselves. If helping improves their chances of becoming successful parents later, the advantage of being an associate should be even greater. The first reproductive efforts of birds are often significantly less successful than subsequent ones (Lack, 1954, 1968), and in the marginal conditions frequently experienced by these animals, the problem may be particularly severe. By allowing their relatives to help and thus gain experience, parents in turn may increase their own inclusive fitness.

It might even be advantageous under certain conditions to assist unrelated breeding individuals (Trivers, 1971). However, if associates are unrelated to the breeders, there is no reason why the breeders should tolerate them unless they are making a contribution to the breeders' success. Failure to discriminate against the occasional outsider that has broken into a family group should not, however, be considered strong evidence against the existence of this advantage for the parents, particularly in light of the apparently high frequency with which their own offspring actually turn out to be the associates.

Ligon and Ligon (1978) have studied some associates unrelated to breeders. They suggested that unrelated green woodhoopoe (*Phoeniculus purpureus*) associates may contribute as much care to a brood as closely related individuals. In one instance unrelated individuals presented considerably more food to an offspring than Ligon and Ligon felt was required for its successful fledging. Associates regularly displayed to the offspring. They interpreted this and certain other behavioral patterns as indicating that associates are competing for the attention of the offspring. The latter might subsequently help the helpers in their own future breeding activities. Unfortunately, Ligon and Ligon did not indicate whether the relationships described are typical of the population as a whole. However, mortality in these woodhoopoes is considerably higher than in some other populations of communally nesting species studied to date (40 percent yearly versus around 15 percent for scrub jays) (Stallcup and Woolfenden, 1978). On the basis of this relatively high mortality rate alone, one might predict a more frequent mixing of unrelated individuals than that characterizing populations with lower rates. Although Ligon and Ligon provided no information on the frequency and importance of helping by unrelated individuals, they emphasized that these systems may operate outside the confines of a tightly knit family group. Cases such as this could not be accounted for through kin selection. The loose colonial system of noisy miners (Dow, 1977) may bear some resemblance to the woodhoopoes, although Dow found it impossible to ascertain the relationships of most group members. Caution should be used in interpreting these studies on communal nesting; without exception they demonstrate only a correlation between the presence of associates and enhanced reproductive success of the breeders in these groups. The standard interpretation is that the associates make real contributions, but it is not possible to eliminate logically consistent alternatives. Zahavi (1974) was the first to suggest that if large groups (which normally have the most associates) can survive only on good territories, the high success rate of these groups may be a function of habitat quality rather than of the activities of nonbreeding individuals. His study provided no information on critical habitat variables, but Gaston (1976) and Brown and Balda (1977) subsequently reported a correlation between independently derived criteria of habitat quality and group size in jungle babblers and Hall's babblers (*Pomatostomus halli*), respectively (fig. 7.5). Brown and Balda pointed out that virtually none of the studies on associates measured habitat quality, so that the standard interpretation of enhanced reproductive success cannot be considered proved. Brown and D. D. Dow (unpublished work cited in Brown, 1978) apparently were the first to test experimentally whether groups really do have an advantage. Using another Australian species, the grey-crowned babbler (*Pomatostomus temporalis*), they found that groups of six to eight that were reduced to three (the breeding pair plus one associate) had a nesting success similar to that of natural groups of three, whereas undisturbed groups of six or eight produced nearly twice as many fledglings at the same time. Other studies are now needed to test the generality of this finding.

Despite Brown and Balda's warning (1977), it still seems parsimonious to attribute enhanced reproductive success to these groupings, given the close genetic relationships that characterize most of them. If the associates are not contributing, it is difficult to see why they are tolerated. Of course, the arguments invoking habitat quality and those invoking help by associates are not mutually exclusive.

Ricklefs (1975) and Brown (1978) suggested that cooperative breeding has evolved in a particular demographic context, where annual recruitment of yearlings into a population is much greater than the

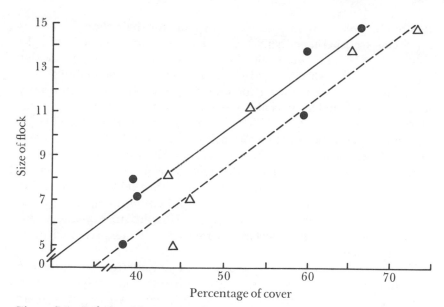

Figure 7.5 Relationship between flock size and percentage of cover in Hall's babblers. Circles and solid regression line = percentage of flock range covered by trees; triangles and dashed regression line = percentage of flock range covered by grass and forbs. (Redrawn from Brown and Balda, 1977.)

death rate of older individuals. The consequence is a saturated habitat. Associate systems have been reported most frequently from low-latitude areas, and many of these occur in relatively unchanging environments. However, it is difficult to draw up a simple set of ecological conditions associated with these groups. Some occupy extremely unpredictable areas (for example, Australian babblers), and some occur at higher latitudes or altitudes (for example, piñon jays). It is likely that resi-dent species form such alliances more frequently than migratory spe-cies, given the much greater likelihood of separation suffered by migrants and the long life span of many residents. Only a small minor-ity of communally breeding birds is distinctly migratory (Fry, 1977).

Argument rages as to whether helping associations can be explained through individual selection without resort even to kin selection (see Brown, 1978). It is clear that there are some direct individual ad-vantages. For example, if space or mates are limiting, as often appears to be the case in these groups, remaining on a parental territory might be the best way to survive until one is eventually able to obtain a mate of one's own. Breeding is delayed until a point considerably after morphological maturity in many species of birds that do not have helping associations. Lack (1968) and others have attributed this to the inability of young individuals to provide for their young as well as for themselves, perhaps in part from lack of experience. Given tight conditions, the same could hold for helpers. Rowley (1978) made this argument for the white-winged chough (*Corcorax melanorhamphus*), an Australian crowlike bird (Corcoracidae) that lives in a notoriously unpredictable climate. It is also possible that there are overall advantages to group living that could account for the associations, but this does not explain the tendency of young birds to remain in their natal groups. None of these explanations adequately accounts for the substantial amounts of help frequently given by associates. And, associates are not chased away in several species where they seldom if ever bring food to the nest (Brown, 1978). Although helping might be favored solely for the experience that a helper gains, in superb blue wrens success does not differ between experienced and inexperienced birds (Rowley, 1965).

Trivers' arguments (1974) about the manipulation of mates can be extended to parent-associate relationships. In some ways the interests of breeders and associates will be different. In particular, there should be conflict over who the breeders in a group will be. The presence of well-defined hierarchies in all such groups implies the existence of competition. The basis for competition to be a breeder should exist in any situation where nonbreeders associate with offspring that are less than full sibs.

Manipulation is also important in the communal breeding system of groove-billed anis (*Crotophaga sulcirostris*), a tropical American cuckoo. As many as four monogamous pairs may share the same nest, and all of the females lay eggs in it. Vehrencamp (1977) demonstrated a dominance hierarchy among the females and found that the lowest-ranking individual was the first to start laying eggs. Females often rolled eggs out of the nest before they commenced to lay. By being late in the nesting sequence a female's chance of having her eggs removed was lower than that of individuals commencing earlier. The early-laying individuals laid larger clutches than the late layers but still reared significantly fewer young than the late ones (fig. 7.6). Since the early layers initiated incubation, this limited the egg-laying opportunities of late layers. However, the late layers reduced this possible disadvantage by producing eggs in a more rapid sequence than the early ones. Vehrencamp also found that early-laying females contributed more parental care than late-laying females. However, mates of the late layers incubated and fed young more than any other group members, which is consistent with their having the greatest reproductive investment.

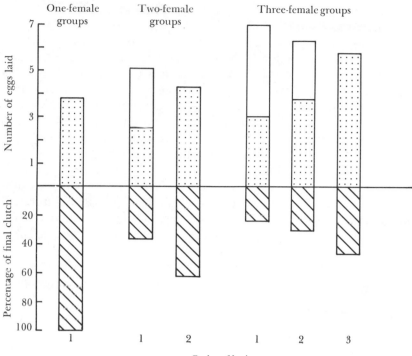

Figure 7.6 Order of laying and ownership of eggs in communal nests of groove-billed anis. White areas = eggs removed; stippled areas = eggs retained; hatched areas = percentage of final clutch owned by different individuals. (Modified from Vehrencamp, 1977; copyright 1977 by the Ecological Society of America.)

LENGTH OF CARE BY PARENTS

Long periods of parental care represent one possible response to poor resources. Examples include diets based on low-protein fruit (Morton, 1973), foods that are difficult to manipulate—for example, oystercatchers (*Haematopus ostralegus*) feeding on bivalves (Norton-Griffiths, 1968)—or foods that are hard to subdue—for example, tigers and lions that feed on large prey (Schaller, 1967, 1972). The ability to perform certain feeding tasks with adequate precision may take an extremely long time to develop, making the young dependent on their parents for a large part of this time (chapter 2).

Trivers (1974) has pointed out an important consideration regarding the length and intensity of parental care. It is to an offspring's benefit to obtain as much investment as possible from its parents, up to a point that is typically beyond the point at which the parents are selected to cease investing. If the offspring succeeds, it does so at the expense of its parents and its present and future siblings, if any. The offspring's demands will be of particular importance if the parents might raise an extra brood (as in birds occupying changeable habitats), or if the parents might breed again as soon as they regain reproductive condition (as in primates weaning their young). At a certain point, when the probability of the offspring's survival is sufficiently high, parents should be selected to commence another brood or litter, if time and resources permit. But the offspring might enhance its survival (and its inclusive fitness) by receiving additional care from its parents, because it is twice as related to itself (1.0) and its own future offspring (0.5) as it is to its siblings (0.5) and their future offspring (0.25) (fig. 7.7). The parents, being equally related to each of their offspring (0.5), are selected to invest equally in them, other things being equal.

Conflict of the sort predicted by Trivers' model is easily observed in young birds and mammals. In birds it often takes the form of incessant, even aggressive begging, even after the young are able to forage

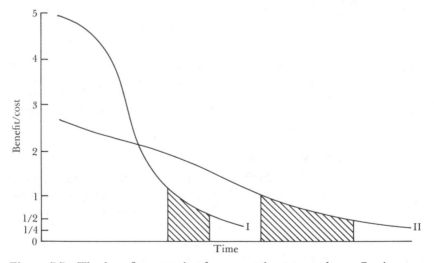

Figure 7.7 The benefit-cost ratio of a parental act toward an offspring as a function of time. Benefit is measured in units of reproductive success of the offspring and cost in comparable units of reproductive success of the mother's future offspring. Two species are plotted. In species I the benefit-cost ratio decays quickly; in species II, slowly. Hatched areas indicate times during which parent and offspring are in conflict over whether parental care should continue. (Modified from Trivers, 1974.)

for themselves. Sometimes the young follow their parents about during the parents' foraging—for example, juvenile herring gulls (*Larus argentatus*) (Drury and Smith, 1968). In response, parents in many species eventually drive the young away.

Spatial and temporal factors

Resource availability may actually define the limits of the breeding range (see El-Wailly, 1966; White and Kinney, 1974), and other correlated factors may do so as well. However, West (1960) found that tree sparrows (*Spizella arborea*) did not improve their overall energy balance by migrating from winter to summer areas (Illinois to northern Manitoba), because the energy required for existence and incubation would be lower in Illinois than in Manitoba during the summer, and resident birds would not have the added expenses of migration. Although the longer days in Manitoba would permit longer foraging periods, they would not make up for opposing factors such as added costs of living in a cooler climate and the costs of migration. West concluded that energetic factors were not responsible for migration in these tree sparrows. However, many of the critical data for his conclusion were obtained from confined birds and hence did not allow several important environmental variables to be assessed. For instance, finding food may be a formidable problem in the field. West's data do not deny the possibility that energetic considerations set ultimate limits to breeding ranges but suggest that other factors intervene before this point. He suggested the possible importance of competitors, who may be a primary reason for migration in the first place (Lincoln, 1950; Cox, 1961). In a similar study Cox (1961) concluded that sedentary tropical finches would obtain little direct energetic advantage by migrating to higher latitudes to breed.

Factors such as competition and predation probably do modify energy budgets under natural situations. However, the importance of an energy budget in defining a breeding range cannot be assessed until the budget is broken down in a way that permits assessment of the most critical period, such as the incubation or fledging period. The factor being selected should be the ability of an individual to accumulate temporary surpluses at these critical times, not simply a minimization of energetic expense; it is these surpluses that permit reproduction or improve the probability of survival through critical periods.

The reproductive season, a time of extremely high energy demand, should typically coincide with the point at which the largest energy surplus can be accumulated (see Kendeigh, 1969; Schoener, 1971). However, adequate surpluses may be so unpredictable or of such short duration that their accumulation represents a major challenge. Consider a desert habitat where food is a direct function of rainfall, which is itself very infrequent and unpredictable. Desert animals employ remarkable combinations of physiological and behavioral adaptations in order to shorten the time required for breeding. In Australia, desert birds respond quickly to the presence of rain, and the breeding season may commence almost immediately (Serventy, 1971). Certain species with major adaptations to desert conditions also have populations in more humid areas, which allows us to compare their reproductive behavior under the two kinds of conditions. In the desert both male and female zebra finches (*Taeniopygia castanotis*) construct the nest, but only the female does so in more humid regions (Immelman, 1963). Desert populations of this species may also start breeding at a very young age (as early as two months) when environmental conditions are favorable, and they continue breeding as long as favorable conditions exist within a local area (Serventy, 1971). Budgerigars (*Melopsittacus undulatus*) can start to breed at nearly as early an age as the zebra finches (Vaugien, 1952, 1953). The finches are extremely nomadic, as are many other desert dwellers (Serventy, 1971), and this improves their chances of finding satisfactory companions.

Some migratory species bring part of the energy reserves needed for reproduction with them from other areas, in spite of the costs of transport incurred during migration. Female Ross' geese (*Chen rossii*) arrive on their breeding ground in northern Canada after a trip of over 3,000 km, with the reserves required to lay a complete clutch of eggs (Ryder, 1967). Goslings are hatched early enough to take advantage of the major pulse of secondary productivity later in the summer. If the females were to gather on the breeding grounds the energy required to produce their clutch, the young would hatch too late; even if a second pulse occurred, the end of the brief arctic summer would arrive before the young could be fledged.

Synchronization of the breeding season to periods of high energy demand may lead to patterns that are at first rather puzzling. The emperor penguin broods its eggs through the middle of the Antarctic winter, and as a result food is most plentiful when the chicks' demands are greatest, the following spring (Prevost, 1961).

Most species that have several periods of high energy demand within the year (breeding, molt, migration) clearly separate these activities in time. West (1968) found that willow ptarmigans (*Lagopus lagopus*) living under simulated wild conditions maintained a remarkably constant energy demand throughout the year. The outstanding exceptions to this pattern occur in species dependent on highly unpredictable resources. Several birds of the Australian desert will interrupt a molt cycle to breed when favorable conditions suddenly appear (Serventy, 1971), as will piñon jays when an unpredictable seed crop suddenly becomes available (Ligon and Martin, 1974).

Foster (1974a, 1975) described another situation in which extensive overlap occurs between breeding and molting of feathers, this being the case in many species of tropical birds that characteristically suffer high rates of nest predation (cf. Willis and Oniki, 1978). Foster proposed that in response to this heavy predation rate, individuals may attempt to reproduce many times but with reduced effort at each attempt. Foster (1974b) argued that some tropical species do not raise clutches as large as they are capable of raising.

Even in relatively predictable situations, individuals may not breed every season. The long periods of gestation and care seen in many large mammals such as elephants and apes preclude yearly breeding (see Eisenberg, 1966). Even some birds such as large albatrosses (*Diomedea* spp.) will skip the breeding season following a successful one, since the fledging of chicks will take over a year (see Lack, 1966). However, species with shorter breeding periods may miss years as well, probably as a result of being in poor condition for reproduction, or poor environmental conditions, or not securing a mate (see Lack, 1966, 1968).

Other limiting factors

Breeding populations often appear to be restricted by food limitations, predators, or environmental factors. Some animals may have additional requirements, however. Here I briefly discuss one special requirement of several species: nest holes. The importance of nest holes in some species has been clearly demonstrated by several European investigators. By putting up large numbers of nest boxes in forests, severalfold increases in the numbers of hole-nesting tits and Old World flycatchers have frequently been obtained (Enemar and Sjostrand, 1972; Hogstad, 1975). These are species that appropriate holes rather than make them. Since these species have close relatives not dependent on these special requirements, one might ask why such traits are retained. Lack (1966) argued that old-growth forests, to which these species are presumably adapted, would have adequate numbers of old trees so that nest-holes would not be limiting. He assumed that evolutionary adjustments have not yet been made to the immature forests characteristic of modern Europe under the influence of human activity (see Lack, 1965).

Synthesis

The relationships among parental care, foraging behavior, and endothermy have many interesting implications. I have stressed the point that extended parental care has some important implications for the young, making possible the learning of complex behavioral foraging patterns, which provide access to resources that would otherwise be difficult to master, if available at all. It is of interest that parental care is ubiquitous among truly endothermic animals, namely birds and mammals. This condition could be influenced strongly by the large energetic demands placed on parents by endothermic young. These demands may be so great that it is impractical or impossible to provide at one time all the resources necessary to permit the young to reach independence. The one notable exception is the case of the brush turkeys of Australasia. Typically, the young are never cared for directly by the parents, although the great amount of effort and the time that the male puts into tending the nest of fermenting vegetation is certainly analogous to true incubation. This is still quite different from the situation in taxa that deposit eggs and leave them, and the demands on the female, who must produce the huge eggs required to form offspring capable of fending immediately for themselves at hatching, result in one of the most prodigious reproductive outputs in birds. It is probably not accidental that this adaptation is found only in a zoogeographically isolated area, where neither predation pressure on the eggs nor competition with other birds is as intense as it would be in other parts of the world. This "exception" tends to prove the general rule.

It should be noted, however, that parental care is by no means confined to birds and mammals. In most other taxa parental care is usually associated with protection from predators or from unfavorable environmental conditions. Direct interaction between parents and young is rare, with the great exception of social insects. In animal groups other than birds and mammals, parental care usually represents a collection of novelties, not a fundamental characteristic of a large taxonomic group.

Conflicts of interest that arise in reproductive relationships are now being studied by many workers, following the initial contributions of R. L. Trivers based on the inclusive fitness theory of W. D. Hamilton. This development is placing social relationships clearly into the framework of natural selection and seems to be likely to account for many complex relationships among individuals that had previously been difficult to explain. It is also creating an urgent need for data adequate to test the critical central issues, such as whether many seemingly cooperative systems can be explained solely by individual selection without the introduction of kin selection. The problem is complicated greatly by the fact that groups chosen for study tend to be composed of long-lived individuals with low reproductive rates, which makes it necessary for investigations to be conducted over periods of many years. There seems to be no easy way around this problem, since the development of complex social systems is functionally related to the long life spans and low reproductive rates. It may be that such studies should be planned as cooperative enterprises, permitting the accumulation of large data based over long periods of time.

8 Competition, Especially for Mates

ANY RESOURCE in limited supply, such as food, space, or mates, is likely to become a focus for competitive interactions. Competition for mates may have secondary ecological effects remarkably similar to those produced by competition for food or space. In this chapter I concentrate on mechanisms associated with competition for mates.

Interference and exploitation

Two basic types of competitive interactions may be recognized, interference and exploitation. These terms were introduced in the context of interspecific competition by Elton and Miller (1954) and Park (1954). *Interference* is defined as "any activity which either directly or indirectly limits a competitor's access to a necessary resource or requirement," and *exploitation* as the "utilization of a resource once access to it has been achieved" (Miller, 1967). In interference the individuals denying others access to a potentially limiting resource will be the competitively successful ones; in exploitation the first individuals to use a limiting resource or those that use it most efficiently will be the competitively successful ones. Being primarily concerned with the role of behavioral interactions in determining patterns of resource use, I shall deal mainly with interference mechanisms. However, since interference and exploitation provide two alternative mechanisms of competitive interaction, it is appropriate to assess their relative importance in communities and to determine the circumstances under which selection will favor one or the other.

190

Miller (1967) suggested that interference is the dominant mode of competition in higher vertebrates, with exploitation prevailing among simple metazoa. But this conclusion might simply reflect a bias inherent in the literature, since interference mechanisms have been studied mainly in animals whose behavior, usually based on vision or hearing, is particularly conspicuous to human observers (mammals, birds, fish, certain crustaceans, and insects) (Miller, 1967; Morse, 1974). The argument that exploitation predominates in other metazoa is based largely on the absence of relevant information and should therefore be considered tentative, particularly in light of growing information about the importance of pheromones in the communication of these groups (Wilson, 1971a). A fact implicit in Miller's argument (1967) is that interference is more efficient than exploitation; if so, it should result in lower overall mortality. In a time of diminishing resources, only a few individuals may succumb if an interference mechanism such as a hierarchy exists. Although dominants profit most, if low-ranking individuals are induced to disperse their chances of survival may also be better than if they remain and fight. Unfortunately, there are few data relevant to this hypothesis, but Murton, Isaacson, and Westwood's studies (1971) on woodpigeons (*Columba palumbus*) suggest that this pattern occurs (fig. 8.1), and Woolfenden et al. (1976) reported that survival of dominant cattle egrets (*Bubulcus ibis*) is higher than that of other individuals when these birds are faced with an inadequate food supply.

Interference should be most prominent (relative to exploitation) in relatively predictable conditions, for here the assurance of future access to resources is highest. The cost of maintaining such behavior under these circumstances may be low. Once two individuals contest a given resource, they are unlikely to do so again in the future (Morse, 1971c, 1976b; Diamond, 1978) (fig. 8.2); additional contests are likely to have the same outcome as the first, and the striking decline in aggression often seen after contests (Morse, 1971c, 1976b) is presumably a consequence of this. Interference may benefit both the high and low ranking, although benefits to the high ranking should be greater than those to the low ranking, and advantages to the latter will only be a consequence of the interaction. To the extent that resources are apportioned in an orderly way, such that little time and energy are wasted in aggressive behavior, they can be used mainly for the essential function of maintenance. It should also be noted that the potential for interference may be greatest under constant conditions where population densities are likely to be high, leading to levels of resource exploitation high enough to necessitate direct competition.

If selection for interference occurs in both fluctuating and constant

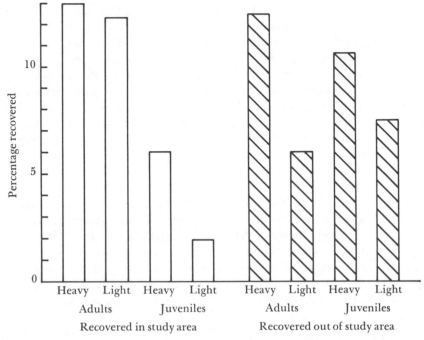

Figure 8.1 Probable survival of woodpigeons depending on initial age, weight (heavy = >450 g, light = <450 g), and distance of recovery. The figure includes records for birds known to be alive one month or more after the date of initial marking. (Modified from Murton, Isaacson, and Westwood, 1971, with permission of the Zoological Society of London.)

environments, where does selection for exploitation occur? In highly unpredictable situations (as opposed to regularly fluctuating ones) interference mechanisms should confer relatively little advantage, for they will not provide a dependable reward. Whether the resources of animals not reported to practice interference are relatively unpredictable is not clear. Forms not known to practice interference are often smaller than animals that do so, and they are generally more vulnerable to environmental fluctuation than the latter because they have a smaller margin of safety on which to operate (Slobodkin, 1961). But this impression may result from a lack of appropriate observations on animals not known to exhibit interference mechanisms.

Populations specializing on temporarily superabundant resources should not experience selection for high levels of interference. Fruit subject to rotting or seeds that will fall from trees into the litter are cases in point for birds. The frequency of hostile interactions in tit flocks decreased markedly during seasons when large amounts of beech

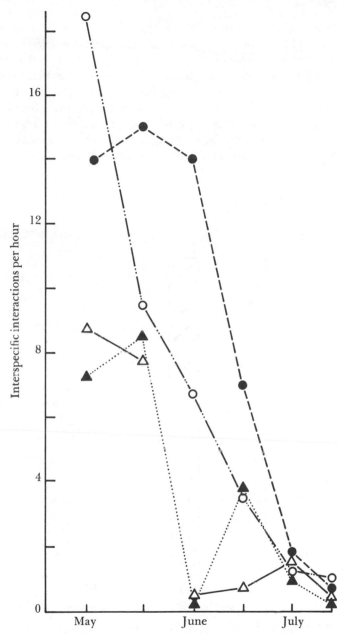

Figure 8.2 Frequency of intraspecific interactions between warblers during the breeding season. White circles = black-throated green warbler; black circles = yellow-rumped warbler; white triangles = blackburnian warbler; black triangles = magnolia warbler. (Modified from Morse, 1976b.)

mast were available (Gibb, 1954). However, during a bumper crop of longleaf pine (*Pinus australis*) seeds, aggressive interactions between pine warblers (*Dendroica pinus*) and brown-headed nuthatches (*Sitta pusilla*) increased (fig. 8.3), probably because exploitation of the food brought the species together (Morse, 1967b). Carpenter (1978) found that hummingbirds that were almost always nectar-limited did not curtail their normal high frequency of aggression in response to a superabundant resource, but that Hawaiian honeycreepers (Drepanididae) and Australian honeyeaters (Meliphagidae),

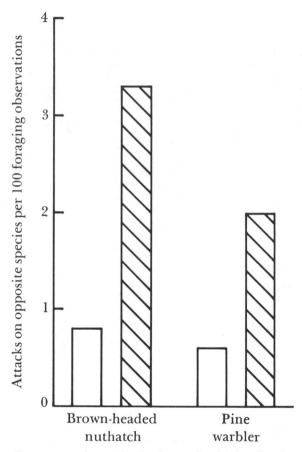

Figure 8.3 Frequency of attacks by brown-headed nuthatches and pine warblers on the opposite species when no longleaf pine seeds were available (white bars) and when a heavy seed crop was available (hatched bars). (Modified from Morse, 1967, copyright 1967 by the Ecological Society of America.)

which have superabundant nectar sources much of the time, did show levels of aggression that fluctuated in response to resource availability.

In some species with cycling population densities, aggressive behavior seems to have a direct genetic basis. Tamarin and Krebs (1969) and Gaines and Krebs (1971) showed correlations among population densities, proportions of aggressive individuals, and certain genetic markers such as transferrins (also see Krebs and Myers, 1974).

Types of interactions

Although behaviors associated with interference competition may be categorized as hostile, they take a wide variety of forms, from direct contact to highly ritualized encounters. In this section I survey the nature and distribution of hostile interactions and also consider their secondary effects, which may have considerable ecological significance.

DIRECT CONTACT

Direct contact is the simplest form of interaction, and one might therefore expect it to represent the primitive condition.

Frequency of direct contact. In the popular literature direct contact often appears in the form of a life-or-death struggle between two desperate combatants. In fact, no such unrestricted encounters occur in most groups of animals, even in highly competitive situations (Wilson, 1971b), but they do occur regularly in a few groups.

Dominant individuals in some species may physically evict their young or those of closely related individuals from social groups, and the outcasts are frequently injured. Subadult male toque monkeys (*Macaca sinica*) are often severely mauled during eviction, and the wounds contribute to their high mortality at this time (Dittus, 1977) (fig. 8.4). Levels of aggression toward offspring may differ greatly in closely related species, for example, species of ground squirrels (*Spermophilus*). Michener (1973) believed that these differences reflect the social systems of the squirrels. Where males are dominant, members of social groups are highly aggressive toward juveniles; where males are subordinate to females, aggression toward juveniles is mild or nonexistent.

Mortality frequently accompanies interactions among ants, bees, and other social hymenopterans, especially in intercolony affairs (Brian, 1955; Free and Butler, 1959; Pontin, 1961). The unique haplodiploid system (males from haploid eggs, females from diploid eggs) of the aculeate Hymenoptera, resulting in sisters that are more closely related

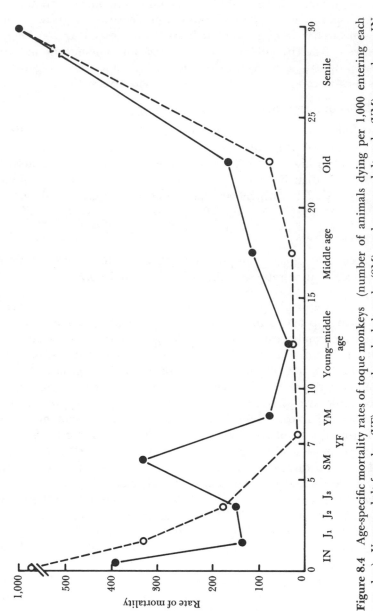
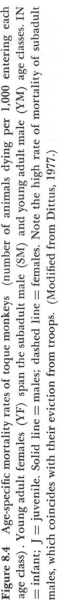

Figure 8.4 Age-specific mortality rates of toque monkeys (number of animals dying per 1,000 entering each age class). Young adult females (YF) span the subadult male (SM) and young adult male (YM) age classes. IN = infant; J = juvenile. Solid line = males; dashed line = females. Note the high rate of mortality of subadult males, which coincides with their eviction from troops. (Modified from Dittus, 1977.)

to each other than to mothers and daughters (Hamilton, 1964, 1972), may account for the high frequency of mortal combat among these insects. The outcome of successful group combat may favor genes very similar to those of the wounded or killed individuals on the successful side. The extent of "commitment" to combat is perhaps best seen in certain species with grotesquely developed warrior castes that are capable of doing little else but fighting. Wilson (1971b, p. 183) provided a graphic description of the soldier caste in the myrmicine genus *Pheidole:* "These individuals have mandibles shaped approximately like the blades of wire clippers, and their heads are largely filled by massive adductor muscles. When clashes occur between colonies the soldiers rush in, attack blindly, and leave the field littered with severed antennae, legs, and abdomens of their defeated enemies."

Potentially lethal altruism is relatively easy to explain in haplodiploid social insects, where workers are highly related to their reproductive sisters and are usually sterile anyway. Potentially mortal combat in diploid species typically involves the purely selfish interests of two individuals disputing access to a resource. Geist (1971) contended that mortality from fighting may be substantial in several species of ruminants, although the use of antlers and horns makes the outcome of a contest primarily a matter of strength (Geist, 1966a). Male deer not infrequently gore each other severely in competition for females during the rutting season. In some species over 10 percent of the adult males appear to die from male intraspecific combat—for example, European moose (*Alces alces*) and red deer (*Cervus elaphus*). Between 5 percent and 10 percent of the bull musk oxen (*Ovibos moschatus*) on Banks Island in Canada were killed during one yearly rut (Wilkinson and Shank, 1976). Mortality would be much higher if these animals did not have such effective defenses. There seems to be little support for the notion that animals rarely kill or maim one another in intraspecific fighting (Lorenz, 1966). Studies on lions (Schaller, 1972) record overtly aggressive behavior leading to severe injury or death, most often when members of different prides interact. Even superficial wounds may become infected and subsequently fatal, as Stamps (1977) found in lizards (*Anolis aeneus*).

Substantial physical combat occurs in several other groups of animals but usually with less extreme consequences, owing in many cases to the evolution of key fighting structures. For example, the horns of bighorn sheep (*Ovis canadensis*) spiral in a way that makes them useless as daggers. Instead, they have become enlarged into battering rams, and most contact between males consists of butting and pushing (fig. 8.5). Although this behavior might have serious consequences, the skulls of these sheep are greatly thickened in areas of impact, such

Figure 8.5 Hornlike structures. *A,* mountain sheep *(Ovis canadensis)*; *B,* mountain goat *(Oreamnos americanus)*; *C,* titanothere *(Brontotherium platyceras)*; *D,* hadrosaurian dinosaur *(Parasaurolophus tubicen)*. (Drawings by Jaquin Schulz; titanothere redrawn from illustration in Osborn, 1929, *U.S. Geological Survey Monograph* 55; hadrosaur based on outline in Lull and Wright, 1942 [Geological Society of America] and personal communication from Gregory Paul.)

that a match is usually terminated when the weaker individual becomes so exhausted that it is unable to continue the contest (Geist, 1966a).

Fighting over food can also lead to death. Rose and Gaines (1976) found an increase in the frequency of wounding in prairie voles *(Microtus ochrogaster)* during the winter, measured by the frequency of perforations of the skin; large individuals have significantly fewer wounds than smaller individuals. Voles may also increase their rates of aggression during the breeding season (Lidicker, 1973), much as in ungulates. Most work on small mammals has been done on rodents, but this pattern may be general, as suggested by changes in male aggressiveness with age noted in a small Australian marsupial, Macleay's marsupial mouse *(Antechinus stuartii)* (Braithwaite, 1974) .

Advantages of avoiding direct contact. Intraspecific combat involving direct contact appears to be the exception rather than the rule. There are several good reasons for this. First, where individuals may become severely injured, even the winner may be a loser (Geist, 1971) . This may account for the relatively high ratio of displays to attacks in species with dangerous weapons but ineffective defenses— for example, mountain goats *(Oreamnos americanus)* (Geist, 1965) (fig. 8.5) . Even in rigidly stratified systems where one male does most of the mating—for example, turkey *(Meleagris gallopavo)* (Watts and Stokes, 1971) —there are clear advantages of fighting in such a way that contestants are not likely to be severely injured. Second, even if one individual kills another, this might help a third individual more than the victor (Dawkins, 1976) . Third, overt aggression diverts time and energy away from other activities such as avoiding predators or raising young. Fourth, exhaustion itself may increase vulnerability to predators (Estes, 1966) . Fifth, the conspicuous activity of combat may attract predators, making the individuals immediately vulnerable (Wiley, 1973; Hartzler, 1974) . Clearly, to be retained such behavior must lead to benefits sufficient to overcome strong counter-selection.

The number of cases in which overt aggressive behavior is associated with sexual selection suggests that reproductive considerations are often overriding. Collias (1944) and others noted that aggressive behavior may be commoner and more intense in polygamous than in monogamous forms. This is not surprising when one considers the size of the stakes in polygamous systems, particularly those in which harems are very large (see Trivers, 1972; LeBoeuf, 1974) .

Where hierarchical relationships occur, dominant individuals must be able to assert themselves in direct combat, for if their dominance were based only on sham, this would soon be revealed through the constant testing performed by less dominant individuals. This testing should be especially frequent if the stakes involved are large. One or a few bull elephant seals *(Mirounga angustirostris)* may succeed in fertilizing most of the females in a colony (LeBoeuf, 1974) , and dominant stags of several deer species that undergo a communal rutting period may enjoy extremely high reproductive success as well (Geist, 1971) .

Instances of unrestrained aggression between species have also been described (Ripley, 1959, 1961) and have been termed "aggressive neglect" (Hutchinson and MacArthur, 1959) . In one example a species of honeyeater (*Myzomela obscura*) on the Moluccan Islands so persistently attacked a species of sunbird (*Nectarinia sericea*) that it (the honeyeater) tended to neglect its own young. The scarcity of such examples in the literature suggests that they may be maladaptive results of historically recent interactions.

DISPLAYS

Hostile interactions are often ritualized (see Blest, 1961). Little if any bodily contact occurs between participants in many encounters, and long-distance communication may even be the principal means of interaction. Information of several kinds may be communicated: interspecific and intraspecific, intersexual and intrasexual, and individual. Although several possible modes of communication exist, most attention has been devoted to visual, vocal, and olfactory modes.

Visual displays. The very presence of an individual is a display to other individuals capable of visual perception. The presence of another individual would be particularly conspicuous to a normally solitary animal; in the case of typically gregarious species, further information might be required before significant display value was attained.

An animal's size may indicate much about its potential status as a competitor where large animals are able to dominate smaller animals (see Guhl, 1962). For example, size is an important factor mediating the interactions of male pinnipeds on their breeding grounds. Small males do not often challenge larger ones (Bartholomew, 1970). Geist (1966b, 1971) noted that the horns of bighorn sheep have considerable signal value, with large size apparently acting as a symbol of dominance, thus minimizing frequency of attack. Although this signal is slightly more complex than body size, it is one that can readily be perceived by the animals, perhaps more readily than body size. Furthermore, the main fighting structures are the focus of the display. Topinski (1974) demonstrated the importance of antlers in displays of red deer, which shed these structures at different times. When individuals lose their antlers they also lose dominance. This relationship probably holds in other antlered forms as well (Beninde, 1937).

Geist's hypothesis may also apply to the hornlike structures of extinct mammals and reptiles (fig. 8.5). Stanley (1974) suggested that the blunt horns of titanotheres (giant extinct mammals) served as butting organs, and Barghusen (1975) suggested a similar function for some of the bony elaborations of the skulls of some therapsid reptiles. Hopson (1975) surveyed cranial modifications of hadrosaurian dinosaurs and concluded that although some probably served as butting organs, others were also important as visual signals of the sort described for mountain sheep. The displays presumably communicate an individual's ability to prevail in a physical encounter. Physical contact should be adequate to prevent the success of an imposter for long, although selection for deception may nonetheless occur (Dawkins, 1976).

A frequent consequence of sexual selection is that one sex (usually the male) is larger than the other. But in many cases large size differences do not occur, and in some the sex apparently experiencing the more intense sexual selection is actually smaller than its mate. For example, in duikers (paleotropical forest-dwelling antelopes) and achouchis (neotropical forest-dwelling caviomorph rodents) the males are smaller than the females (Ralls, 1976). More than one selective pressure may be responsible for this difference. Ralls suggested that the large size of newborns, intense resource competition by females, and only mild competition among males for mates might account for the size differences noted in duikers and achouchis. Large newborns may enjoy higher survival than small ones, and this may strongly select for larger mothers. Male guppies (*Poecilia reticulata*) are under intense sexual selection for their bright coloration (Endler, 1978), yet they are much smaller than the females, who produce large broods of live young.

Secondary sexual characteristics other than size usually do not enhance the physical competence of their owners and may even degrade it. Presumably such traits exist because females prefer males who exhibit them, employing the traits and indices of some aspect of the quality of prospective mates. Extreme examples of such traits include the "tails" of the paradise whydah (*Steganura paradisaea*), a weaver finch (fig. 8.6), and the quetzal (*Pharomachrus mocinno*), a trogon. In both cases the male's tail (actually upper tail coverts growing over and past the true tail feathers) may reach a length several times that of the body. The feathery ornamentation of some birds-of-paradise (Paradisaedidae) is equally extreme, especially compared to that of the closely related but drab bower-birds (Ptilonorhynchidae), which display in ways strikingly different from the birds-of-paradise.

Possession of these large and often ungainly structures virtually dictates that the males will differ in other ways from their females. Selander (1966) demonstrated that female boat-tailed grackles (*Cassidix major*) can feed in ways unavailable to males, who are larger and have much larger tails than the females. Males are unable to pick food off the surface of the water while in flight, a strategy frequently used by females. The large tail is used in the sexual display of the males. This is a case in which adaptations associated with display and directly independent of resource competition nevertheless have

Figure 8.6 Paradise whydah. *Top,* male during breeding season; *middle,* male outside of the breeding season; *bottom,* female. (Drawing by Jaquin Schulz.)

a secondary effect on patterns of resource use. Male whydahs and quetzals are probably even more severely restricted in their patterns of resource use than the grackles, since their plumes are larger than the grackles' tails.

Up to a certain point it may be advantageous for males and females to feed on different resources, in that way reducing competition, but the resources present may limit the extent to which the two sexes can diverge. Increased prowess in display may be directly correlated with mortality—for example, grackles (Selander, 1965). If the most extreme males, assumed here to be the most attractive to females, suffer much higher rates of mortality than less extreme males, the latter may on balance be favored by selection. Such compromises are probably reached in many sexually dimorphic species, as evidenced by the high mortality rates of males (Selander, 1965; Geist, 1971). Nevertheless, sexual selection has gone to extraordinary lengths. It may well be that severe environmental shifts, triggering rapid changes in productivity or in the types of resources available, have led directly to disaster for some species—for example, extinction of Irish elk (*Megaceros giganteus*) (Gould, 1974). One might predict that sexual selection will generally produce its most extreme results under relatively constant conditions. Whydahs molt their tail feathers after the breeding season, but the quetzal and birds-of-paradise do not, perhaps because of longer periods of sexual activity. Although predation rates on males and females of the latter two species have not been reported, it is almost inevitable that predation on adult males is higher than it would be if they did not possess these secondary sexual characters. In forms with strong sexual selection, overall male mortality is higher than in related sexually monomorphic species (Selander, 1966). Selander (1965) studied overall mortality rates of male and female grackles over a winter and found the rate of the males to be approximately twice that of the females. The sex ratio of nestlings is nearly even (50.8 percent males), but males make up only 42.8 percent of the population in October and 29.2 percent in March, just prior to the breeding season. Presumably this difference results both from a lower energy-gathering efficiency and a greater risk of predation for males than for females.

A species living in a community that contains interspecific competitors may be restrained in the extent or type of morphological change that it can undergo, for incipient changes in size or other structural changes may increase similarities to a prospective competitor, thus halting such a trend. Any display structures (plumes, long tail feathers) that impede an individual's ability to compete successfully will also be selected against strongly if similar species are present.

Colors and patterns may also serve as the basis for displays. These characteristics would not seem to have the strong ecological consequences associated with size or other morphological differences, but they do allow for individual or group recognition, which is a basis for differential response to individuals resulting from past experience.

Rhijn (1973) has reported differences in behavior of male ruffs (*Philomachus pugnax*), a large sandpiper of the Old World, that depend on coloration. During the breeding season male ruffs develop elaborate collars of erectile feathers about their necks and similar tufts of feathers on their heads. The ornamental feathers are generally either black or white, although considerable individual variation occurs (fig. 8.7). Males with black collars are aggressive and territorial on the communal display grounds (leks), while white-collared birds are not aggressive and generally move about the display areas, tolerated by other males attending territories.

A similar pattern holds in white-throated sparrows (*Zonotrichia*

Figure 8.7 Variation in head plumage of male ruffs during the breeding season. Birds with black head plumage are dominant. (Drawing by Jaquin Schulz.)

albicollis) , in which both the males and females have bright and dull morphs (differences in head striping) (Lowther, 1961), the bright morph being dominant (Lowther and Falls, 1968; Ficken, Ficken, and Hailman, 1978) . These differences have a direct cytogenetic basis (Thorneycroft, 1975). Continuous plumage variation occurs in the closely related Harris' sparrows (*Z. querula*) (fig. 8.8). The amount of black about their faces determines their position in the hierarchy. These differences are to some degree related to age, but this accounts for only part of the overall variability (Rohwer, 1975; Rohwer and Rohwer, 1978). More commonly, simple age-related color differences serve as indicators of dominance.

Cases involving sexual dichromatism may also be attributable to sexual selection, which could operate even where conditions do not favor structural dimorphism. Visually conspicuous individuals are preyed on more often than inconspicuous ones, even in the absence of sexually selected characters (see Kaufman, 1974a, b) . Males possessing

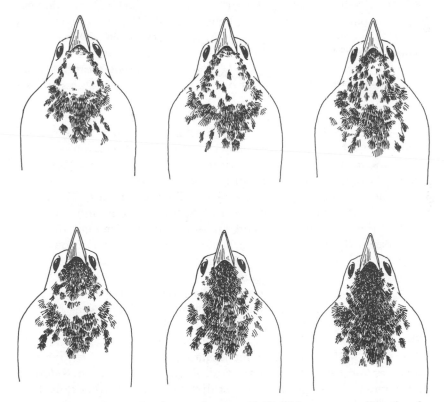

Figure 8.8 Variation in throat plumage of Harris' sparrows. (Drawing by Jaquin Schulz.)

bright colors may have to display them for effective results, and if so this behavior should make them conspicuous to predators as well as to prospective mates or competitors, resulting in increased predation. Perhaps for this reason monomorphic, dichromatic birds tend to contribute less to nesting activities than do closely related monomorphic, monochromatic species, even if monogamous (see Lack, 1968). In these cases the brightly colored males would presumably lure predators to the nest.

One ecological consequence of visual displays is that they may influence the locations in which individuals spend much of their time. Displaying individuals typically frequent conspicuous locations (Morse, 1966, 1968b). If they spend considerable time in such areas, it may often be efficient for them to forage there. The existence of suitable display locations is also important in habitat selection. Kendeigh (1941) and Persson (1971) have both suggested that some species of small songbirds, such as the yellow warbler (*Dendroica petechia*), will not set up territories in areas lacking an adequate display perch, even if they spend most or all of their foraging time below this elevated perch. Resource exploitation in several other *Dendroica* warblers is probably influenced by the tendency of the males to spend most of their time displaying in prominent locations when high densities of individuals (of one or more species) are present (see chapter 2).

Auditory displays. Vocalizations may function in a manner analogous to that of visual displays and may also be acted on by sexual selection. Morton (1977) has pointed out that on the whole large animals have the ability to produce lower-frequency vocalizations than small ones, a simple consequence of the difference in size. Thus, low-frequency vocalizations might reinforce displays communicating the sender's (large) size. Davies and Halliday (1978) demonstrated that male toads (*Bufo bufo*) are much more likely to try to dislodge amplexed males that give relatively high-frequency vocalizations than those giving low-frequency vocalization.

Song is used in the courtship and aggressive activities of many birds, frogs, and toads. Often females do not sing, although song is quite common in female tropical passerine birds with permanent pair bonds (Skutch, 1976). In the latter case, other functions of song may be involved as well, such as pair maintenance (see Thorpe, 1972). Song and other vocalizations may serve additional functions, such as defense of a territory. A single type of vocalization can hypothetically both attract females and repel males (Armstrong, 1963); in other situations separate songs may perform these distinct functions (Morse, 1970a). Regardless of their function, however, if different vocalizations are used in different contexts, this gives rise to the possibility that they will be given from different sites. This separation may affect the foraging location of the vocalizer and perhaps facilitate ecological segregation between the sexes.

From considerations of sound attenuation and degradation alone, vocalizations and other auditory displays should be tailored to the habitat (Ficken and Ficken, 1962; Chappuis, 1971; Morton, 1975; Wiley and Richards, 1978). The ideal vocalization in open habitats will differ from that in forested habitats, and vocalizations in the tree canopy will differ from those in the understory. This has important implications because certain parts of a habitat will be more favorable for broadcasting displays than others. This factor should localize the display sites used and favor foraging near them. Of course, selection on the acoustical properties of vocalizations has probably often been in part a consequence of the resource exploitation patterns of the singers, but once perfected, these vocalizations should act as a conservative force tending to slow changes in niche dimensions.

Displays may attract predators as well as mates and competitors. If predation is heavy, it should exert pressure on auditory signals. However, given the acoustical sensitivity of some predators (Shalter, 1978a), it is doubtful that displays can be made completely cryptic to them. Increased predation probably occurs in sage grouse because of both auditory and visual displays (see Wiley, 1973; Hartzler, 1974).

Howard (1974) suggested that the size of mockingbird (*Mimus polyglottus*) repertoires is a product of sexual selection. Individuals with the most songs attain the best territories (those with the most adequate food supplies), which should permit them to produce more young than others, or at least to raise them more quickly (fig. 8.9). The size of the repertoire is expanded by taking in elements of other birds' songs as well as elements contained within the typical repertoire. The mockingbird is monogamous, monomorphic, and monochromatic and defends a territory permanently, making it one of the last species for which strong sexual selection might be anticipated. Howard's study makes clear that we are only beginning to understand some of the variables responsible for distributing animals in space and time.

Although communication often operates in both visual and auditory channels, in some situations one mode predominates. This is seen clearly if one compares birds of the floor and of the canopy in tropical forests. Visibility on the floor can be extremely poor, with illumination only a tiny fraction of that penetrating the canopy (see Richards,

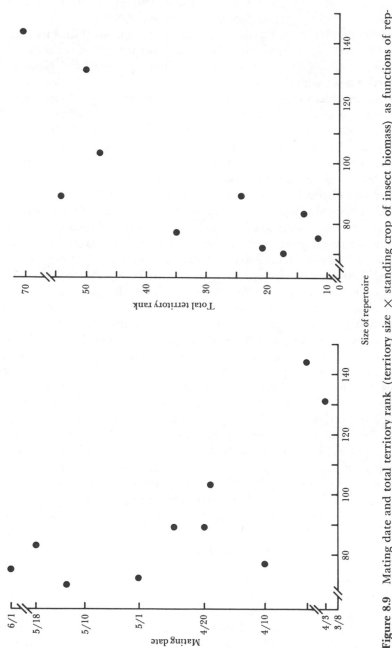

Figure 8.9 Mating date and total territory rank (territory size × standing crop of insect biomass) as functions of repertoire size in mockingbirds. (Modified from Howard, 1974.)

1952). Here bright plumage would probably be almost useless. Brown, gray, and black birds in fact predominate, and they typically give loud, distinct vocalizations. On the other hand, the bright-colored birds for which tropical forests are famed typically occur either on the forest edge or in the canopy, where visibility and illumination are much greater (see Meyer de Schaunsee, 1963). Appropriate detailed comparisons do not seem to have been made, but the vocalizations of edge and canopy species appear to be simpler and less specifically distinct (at least to the human ear) than are those of the floor species —for example, descriptions of the vocal repertoire of several antbirds (Formacariidae) of the forest floor (Willis, 1967, 1972).

Olfactory displays. Chemical releasers, or pheromones, are now known to provide an extremely important means of communication, probably serving as the major medium in many mammals and invertebrates. In fact, E. O. Wilson (1970) stated that chemical communication is the paramount mode in most groups of animals. As in visual and auditory communication, chemical communication has several functions beyond our immediate interest. However, it does clearly act in aggressive situations associated both with competition for mates and for resources.

Chemical displays differ fundamentally from auditory ones and most visual ones in that they have a longer life. Use of chemical cues, as in scent marking, should function effectively in identifying the territory of an individual that is unable to move rapidly about it at all times. Many rodents and canids mark their territories in this way (Ralls, 1971; Eisenberg and Kleiman, 1972), although some nonterritorial animals scent mark as well (Eisenberg and Kleiman, 1972; Johnson, 1973). Scent marking is frequently associated with aggressive situations, and when fights break out between an actively marking animal and an intruder, the individual marking typically wins (Ralls, 1971). Thus, this behavior may act in a manner analogous to some visual displays, serving as an indicator of dominance. But scent marking is not always associated with aggressive situations (Eisenberg and Kleiman, 1972; Johnson, 1973), and it is only one of several types of olfactory communication (Eisenberg and Kleiman, 1972).

Olfactory communication also differs from visual and auditory communication in that it is not possible to direct it actively to the extent that the other displays can be directed. This is desirable where blanket coverage is advantageous. Given their greater permanence, olfactory displays do not have to be given as often as other displays. As endotherms, with high energetic demands, mammals must use relatively large areas. But unlike birds, they have low mobility, which means they cannot continually monitor their essential local environments. Thus, a lasting display should have definite advantages, even if it cannot invariably be backed up with force.

Synthesis

The argument that it is advantageous to be different from one's mate is supported by the observation that islands with low species diversity often exhibit high frequencies of structural sexual dimorphism (Selander, 1966; chapter 2). This suggests that such divergence is hampered in saturated communities with high species diversity. It may well be that in ecologically crowded situations a high degree of specialization has occurred in auditory and chemical signaling. One can even imagine competition for communication channels (see Chappuis, 1971; Morton, 1975).

Competition for mates causes selection within a sex, driving the divergence between the sexes. Simple competition for resources such as food would typically pit all members of a population against one another; the consequence should be an extremely variable population, but one exhibiting continuous variation. Discontinuous differences may lead to foraging differences, but they probably continue to exist in their observed forms only because of ongoing competition for mates. Although sexual selection for body size may be of great significance in groups exhibiting indeterminate growth, its secondary ecological significance will be greatest in forms exhibiting determinate growth, because in the former a continuum of body sizes exists.

E. O. Wilson (1975) has asked whether sexual foraging differences are a consequence of sexually selected size dimorphism or whether, at least in some instances, these differences are a direct result of selection for diversification of foraging itself. If sexual selection is not a major factor, we should frequently find ecological morphs specialized for different patterns of resource exploitation, but with members of both sexes belonging to each morph. The fact that so few feeding morphs of this sort exist (at least in vertebrates; see chapter 2) is consistent with the argument that dimorphisms are typically a consequence of sexual selection, rather than the result of primary selection for diversification in foraging.

9 *Territoriality*

HOSTILE INTERACTIONS are an important proximate cause of spacing patterns. The ultimate purpose of such patterns is often to secure access to resources contested by other individuals. We may distinguish cases in which an individual or group (1) defends a large area containing all its required resources, (2) defends a point source containing a particular resource, or (3) establishes a priority for the use of a resource. Territories and social dominance hierarchies can profitably be thought of as the opposite ends of a spectrum along which are many situations combining aspects of both. For the sake of convenience I initially treat them as alternative responses to a number of key variables such as resource abundance, defendability, and patchiness.

One way to compete for resources is to defend a space against some or all other individuals. Such a space is referred to as a *territory* and is defined adequately for most purposes as "any defended area" (Noble, 1939). Several authors have added refinements to Noble's definition. For example, Emlen (1957) defines a territory as a space within which an animal is aggressive toward and usually dominant over certain categories of intruders. In different situations those intruders may be members of the same sex, the same species, or other species. This definition is rather similar to Willis and Oniki's (1978) concept of a territory as a dominance space. Pitelka (1959) defined a territory as any exclusive area, while Eibl-Eibesfeldt (1970) considered any space-associated intolerance as territoriality. As Brown and Orians (1970) pointed out, the flexibility of Noble's definition is one of its strengths.

211

Pitelka's definition is very similar and has the virtue of emphasizing the function of territoriality—use of an area without undue disturbance. A territory may be held by an individual, a pair, a family, or even a group of unrelated individuals.

Characteristics of territories and their holders

Overt defense (attacking, chasing, threatening intruders) or displays (overt or ritualized) identify the territory holder and make it conspicuous to rivals (Brown and Orians, 1970). Territory holders are usually adults if sites are scarce, but if enough spaces become available the most dominant of the young may also secure territories (Knapton and Krebs, 1976).

The essential characteristics of a territory are: (1) it is a fixed area (which may shift over time), (2) it is actively defended, and (3) the holder has exclusive use of it (with regard to a given set of individuals). The area is usually a site set in space, such as a plot of land, but it may also be an area set in relation to its mobile defender, such as a space of 10 m around an individual. That territories are actively defended implies that they do not overlap. This assumption is embodied in the many published diagrams showing contiguous but nonoverlapping territories (for example, Smith, 1968; Dhondt, 1971). In fact, wolves (*Canis lupus*) show little tendency to overlap and may even maintain substantial buffer zones (Mech, 1977). This separation is probably maintained by the high level of aggression among packs; resident wolves will often attempt to kill trespassers. Territorial overlap is often temporal; that is, boundary areas may be used by two adjacent individuals, but at different times (Brown and Orians, 1970).

ORIGIN OF TERRITORIES

Territoriality probably arises in evolution from the widespread habit of maintaining individual distance. It may develop in response to the distribution of a critical resource, the holder securing exclusive right to the resource by its immediate presence. Selection would occur for defense of the site immediately surrounding the resource, which becomes a territory. Progressively enlarged territories might arise if they give improved access to more dispersed or mobile resources.

Brown emphasized that the many diverse manifestations of territoriality are quite similar with respect to the economics of their defense (fig. 9.1). An animal can afford to defend a territory only when the gains obtained from this activity exceed the expenses incurred. That is, it would be profitable to hold a territory only when the calories gained more than offset those used defending the territory. Brown suggested that conditions favoring territoriality in birds usually occur

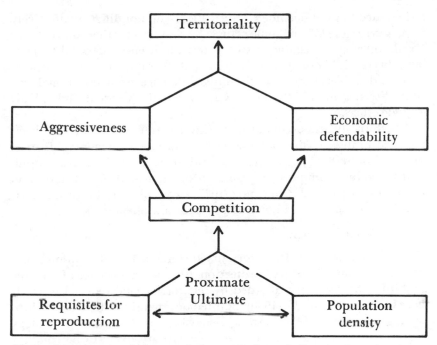

Figure 9.1 A general theory of the evolution of diversity in avian intraspecific territorial systems. (Redrawn from Brown, 1964.)

where food is predictable and relatively evenly dispersed. Although territoriality at highly concentrated food sources might be advantageous, energy spent defending such a source may well render it infeasible. Problems associated with predation and reproduction may also determine whether a territory will be established.

What animals defend territories?

Among the vertebrates territorial behavior has been reported in bony fish, frogs, salamanders, lizards, crocodilians (Cott, 1961; Greer, 1970), birds, and mammals. References to all these groups except crocodilians may be found in Klopfer (1969) and Brown and Orians (1970). A great proportion of the literature on territoriality is devoted to birds (see Nice, 1941, and Hinde, 1956, for examples), probably because of the frequency of the phenomenon in birds and their popularity as study groups. Territorial behavior is generally assumed to be much more widespread in birds than in mammals, yet territorial systems occur in small carnivores (Lockie, 1966). Group territories are frequently reported for rodents (C. C. Smith, 1968; Archer, 1970) and monkeys (Crook, 1970b). Many mammalian "home ranges"—sites regularly visited but not defended (Burt, 1943) —do not differ qualitatively from territories. When home range holders meet other individuals, hostile interactions characteristic of territorial encounters take place. This fits Eibl-Eibesfeldt's definition (1970) of territoriality (space-associated intolerance), and the only difference from traditional territoriality lies in the efficiency with which the resident individuals exclude others.

Territoriality also occurs in many invertebrate groups and has been reported in insects (dragonflies, crickets, various hymenopterans, true flies, and butterflies), crustacea (mantid shrimps, fiddler crabs, amphipods), mollusks—limpets (Stimson, 1970; Branch, 1975), chitons, and octopuses (Kyte and Courtney, 1977) —and phoronid worms. It is clearly a widespread and potentially important phenomenon.

Costs and benefits

Major benefits of territoriality include increased assurance of a good food supply, improved protection (through knowledge of the area and from the dispersion afforded by territories), and reduced interference in breeding. Hinde (1956) pointed out that since the relationships among selective forces governing behavior, structure, and physiology are extremely complex, the realized benefits are likely to be complex and to differ greatly from species to species. For example, territories are often but not always associated with reproductive activities. Winter territories of migratory birds are not directly associated with breeding (Lack, 1968), but they are in the case of permanent residents such as tits (Hinde, 1952; Krebs, 1971), where winter maintenance of a territory may be critical to reproductive success in the next breeding season.

The costs of territories are also large. Defending a territory always demands time and energy, often in substantial amounts. Territoriality may increase the risk of predation, although territory holders probably often reduce their vulnerability somewhat by selecting safe areas and getting to know them well. It is clear, however, that male African antelopes who set up breeding territories increase their vulnerability to predation (Estes, 1969).

How are territories defended?

Although the means of defense are highly varied, visual, auditory, or chemical display usually serves to maintain the territory; Physical contact is rarely used except possibly when the territory is first being established. For example, rates of overt fighting between male warblers (*Dendroica* spp.) are about an order of magnitude greater when territories are being established than afterward (Morse,

1976b) (fig. 8.2). Klomp (1972) and Barash (1973a) give similar examples for other birds and for pikas (*Ochotona princeps*). Peek (1972) proposed a three-level territorial defense system for species such as these. For example, in red-winged blackbirds (*Agelaius phoeniceus*) vocalizations serve as a first line of defense against potential intruders, visual displays appear when other individuals have already trespassed or are near the boundaries of a territory, and the territory holders resort to chases and attacks only when all else fails.

By removing individuals and playing recordings of their songs on their territories, Göransson et al. (1974) and Krebs (1977a) have demonstrated that song is actually a mechanism of territorial defense. Working with great tits (*Parus major*), Krebs found that territories were occupied within eight to ten hours after the original owners were removed, unless recorded vocalizations were played, in which case the territories were not completely occupied until twenty to thirty daylight hours had elapsed. Göransson and coworkers found that playbacks may prevent thrush nightingales (*Luscinia luscinia*) from establishing territories at a site for an entire season. Both studies are based on small samples of birds, but they clearly show a real repellent effect of song. It is not clear why the effect is shorter-lived in tits than in nightingales, but it could be a function of the pressure on territories or of the period within which a territory must be set up in order to rear young successfully. These times might differ in tits and nightingales, the former being a permanent resident, the latter a long-distance migrant.

Defending individuals may spend great amounts of time in areas where encounters with invading individuals are most likely. Lockie (1966) noted that weasels (*Mustela* spp.), which extensively patrol the boundaries of their territories, concentrate their activity where pressure from adjacent animals is greatest (fig. 9.2). They also supplement their actual presence with scent markings.

Owners usually prevail in territorial encounters. The asymmetry may occur because the owner has invested more time and energy in the territory than the intruder has. It knows the best foraging and hiding places and has established stable relationships with its neighbors. Thus it can count on its ability to exploit the site effectively in the future (Dawkins and Krebs, 1978). The territory is to some degree unknown to the intruder. Being unable to make an accurate assessment of its worth, the intruder may be less willing than the owner to escalate a conflict. The owner has already made a considerable investment in the territory, which it would have to duplicate were it to give up this territory and establish another one (assuming that territories still were available).

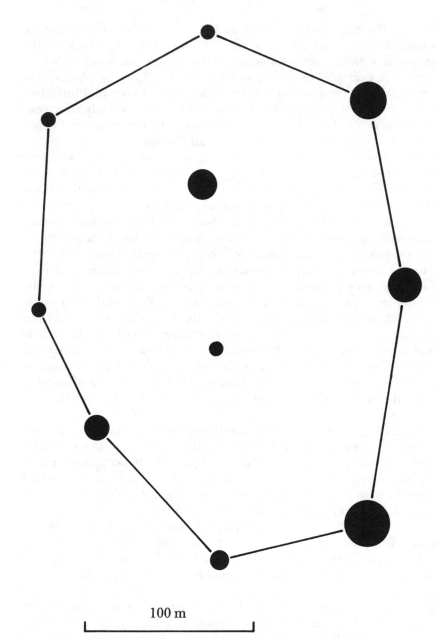

100 m

Figure 9.2 The use made of a territory by a male weasel as shown by trapping results. Dots represent traps, and the lines enclose one weasel territory. The sizes of the dots represent the frequency with which the weasel was caught in the different traps. (Modified from Lockie, 1966, with permission of the Zoological Society of London).

Krebs (1976, 1977b) has argued that singing a wide variety of songs may help territory holders to defend their space more effectively than singing only a few songs. He suggested that the varied repertoire creates the illusion that several birds are present, rather than just one. Intruders, being under time constraints, might not bother to test such an area further. However, if it were not backed by the ability to defend the site, this guise might be exposed by the intruders and fail. Howard's studies (1974) on mockingbird (*Mimus polyglottos*) vocalizations (chapter 8) suggest an alternative explanation for large repertoires: the production of many songs may simply identify an individual's quality as a mate or defender of a site. That characteristic in turn might be a consequence of its age or tenure. It is impossible to distinguish between these alternatives without experimentation.

Territories differ with respect to who is excluded from them. The owners may evict conspecifics only (except the mate and young or social group), or they may also evict potentially competing species. Territoriality has not been much studied from an interspecific point of view (but see Orians and Willson, 1964; Murray, 1971). It is clear that at least in some cases territories are defended against other species that are ecologically similar and thus potentially strong competitors.

Considerable variation exists with respect to who defends the territory. In breeding territories females perform much of the reproductive effort, and territorial defense is usually carried out by the male. But females in many monogamous species also participate in defense. Examples include the Carolina wren (*Thryothorus ludovicianus*) and several *Dendroica* warblers (Morton and Shalter, 1977; Morse, 1976b). Most of the females' attacks in these studies were directed at other females. In the warblers at least, this could have resulted from the ecological specialization of males and females. Where individual territories are set up, the females naturally defend these—for example, red squirrel (*Tamiasciurus hudsonicus*) (C. C. Smith, 1968). In the case of territorial polygynous species, females may do much of the defending against other females, especially where defense occurs against other females of the same harem—for example, red-winged blackbirds (Nero and Emlen, 1951; Holm 1973). In group territories defense may be carried out by all individuals, as is probably the case with some jays (Brown, 1963a). But large males perform most defensive activities in some baboons (Crook, 1970b).

Types of territories

Territories may differ greatly in size, depending on the uses to which they are put. Mayr (1935), Hinde (1956), and others have distinguished several types of territories: type A, a large breeding area within which nesting, courtship, mating, and most foraging take place; type B, a large breeding area that does not provide most of the food; type C, a colonial nesting territory; type D, a territory used only for pairing and mating (for example, a lek). Although overlap exists, these types constitute useful points of reference. Where defense may be limited to a nest site, as in nesting sea birds, individuals obtain their nourishment elsewhere. Sometimes they have separate feeding territories. Drury and Smith (1968) have published observations on colonially nesting herring gulls (*Larus argentatus*) indicating that they defend areas both during and after the breeding season.

The ultimate packing of territories occurs when individuals touch one another. This would not seem likely in species where individuals display aggression toward one another, but it may be approached in some colonial species. For example, royal terns (*Thalasseus maximus*) on their nests may nearly touch one another in some cases (Buckley and Buckley, 1977). The fact that these tightly packed nesting sites show almost perfect hexagonal packing is consistent with the idea that some degree of defense is involved (fig. 9.3). Hexagonal packing maximizes the average distance between neighbors (see Grant, 1968; Barlow, 1974a). Territories range from such minimal sizes up to many square kilometers. The territories of several raptors in western North America reach this latter extreme (Craighead and Craighead, 1956).

Clearly, a territory that provides all necessary resources for an animal will be considerably larger than a territory used only for reproductive purposes. Its actual physical size may also differ with the resources present, with territories in areas containing high food densities being considerably smaller than they are elsewhere (Orians, 1971; Slaney and Northcote, 1974). Studies on nectar-feeding sunbirds (Gill and Wolf, 1975b), Hawaiian honeycreepers (Carpenter and MacMillen, 1976), and hummingbirds (Gass, Angehr, and Centa, 1976; Kodric-Brown and Brown, 1978) all show close relationships between numbers of flowers present (energy available) and territory size. This has been seen in a variety of other animals as well. Territory size of red grouse (*Lagopus lagopus scoticus*) varies inversely with the density of new green shoots, a preferred food (Miller and Watson, 1978). Lance (1978) found an inverse relationship between territory size and the nutrient content of heather on one of Miller and Watson's study areas. In studies in *Sceloporus* lizards, Simon (1975) found a similar inverse relationship between the amount of food and territory size and was able to manipulate sizes of territories by provisioning them with food (fig. 9.4). She also found that territory sizes varied predictably as a function of sex, with males defending areas about twice as large as those of females of similar size. Large individuals, requiring more

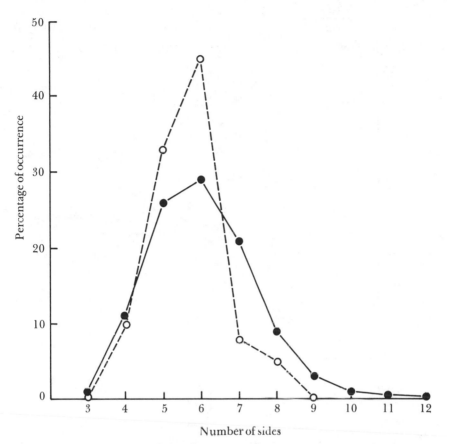

Figure 9.3 Frequency distribution for number of sides of randomly generated Voronoi polygons (solid line) and of royal tern nest sites (dashed line). The number of hexagonal nest sites is significantly greater than the number predicted by chance. (Redrawn from Buckley and Buckley, 1977.)

resources than smaller ones, tend to have larger territories, other things being equal (Schoener, 1968b). Similarly, carnivores require much larger territories than herbivores of a similar size (Schoener, 1968b) (fig. 9.5). This is explained by the manyfold-of-magnitude drop in available energy per trophic level (Slobodkin, 1961; Turner, 1970), assuming that the populations being compared are both food-limited, and by the difficulty of searching for mobile prey.

Territory size and structure may differ strikingly even within a population, depending on resource availability. Seaside sparrows (*Ammospiza maritima*) (Post, 1974) occupied territories in artificially ditched salt meadows that averaged seven times the size of those in

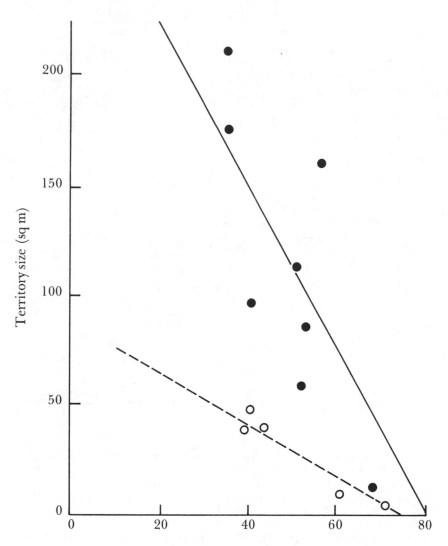

Figure 9.4 Relationship between territory size and food abundance in lizards (*Sceloporus jarrovi*). Food abundance = the number of prey items captured on sixteen "sticky" boards, each 10 × 15 cm in area and exposed for four hours. Black circles = lizards one year or older; white circles = lizards under one year. (Redrawn from Simon, 1975; copyright 1975 by the Ecological Society of America.)

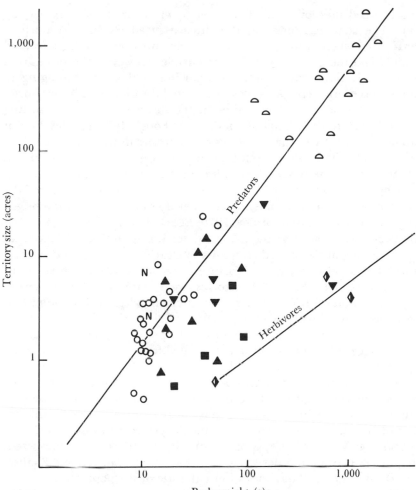

Figure 9.5 Relationship between territory size (acres; 2.47 acres = 1 hectare) and body weight (grams) for birds of varying diet. Omnivores (10–90 percent animal food) are black; herbivores, half-black; predators, white. N = nuthatch species. Circle = 90–100 percent animal food (mainly arthropod); half-circle = 90–100 percent animal food (mainly vertebrates; upright triangle = 70–90 percent animal food; square = 30–70 percent animal food; upside-down triangle = 10–30 percent animal food; diamond = 0–10 percent animal food. (Modified from Schoener, 1968b; copyright 1968 by the Ecological Society of America.)

undisturbed meadows (8,781 ± 2,435 sq m versus 1,203 ± 240 sq m, se). The large territories in the ditched meadows were all-purpose territories, whereas the small ones did not provide an adequate food supply and the birds had to forage outside. However, the birds from small territories spent a similar proportion of their time foraging to those in the large territories. Post attributed the difference in territories to a scarcity of suitable nest sites in the disturbed habitat, resulting from the greater likelihood of flooding. Although fledging rates did not differ significantly in the two types of territories during the three years of the study, the results suggest that in poor years the birds in disturbed areas would be more successful than those in the undisturbed areas.

Songbirds of several species defend considerably larger areas at the time of territory establishment than they do during most of the breeding period (Brown and Orians, 1970). Lack (1966, 1968) argued that in some species, such as the great tit, territories are not defended at all after this initial period. But this is clearly not the case in several species of wood warblers, and failure to record territorial interactions may reflect the fact that territories were stabilized during earlier fighting (Morse, 1976b). Patterson and Petrinovich (1978) found that territories of white-crowned sparrows (*Zonotrichia leucophrys*) remained extremely constant in size over the period of a breeding season. They suggested that this constancy resulted from a relatively low population density with few nonbreeders.

In the fall some permanent resident birds in the temperate zone actively defend territories that are larger than those they will use during the next breeding season. It may be that these fall territories are adjusted to satisfy the greatest demands expected within the period of a year (see Verbeek, 1973; Morton and Shalter, 1977). Carolina wrens require territories in order to survive the winter (Morton and Shalter, 1977), and their fall territories limit the size of the population.

The presence of both intraspecific (Klomp, 1972; Armitage, 1974) and interspecific (Yeaton and Cody, 1974; Morse, 1977a) competitors may strongly affect territory size. Yeaton and Cody found that territory sizes of song sparrows (*Melospiza melodia*) varied up to thirteenfold on small islands off the northwest coast of North America (fig. 9.6), depending on the number of other small nesting birds present. As the number of other species present increased, so did the sizes of individual song sparrow territories. Similar results were obtained on islands off the Maine coast by Morse (1977a) for some species, but not all. Song sparrows showed the same type of compaction reported by Yeaton and Cody, as did parula (*Parula americana*) and yellow-rumped (*Dendroica coronata*) warblers. However, black-throated green warblers, the socially dominant warbler species on these islands, did not

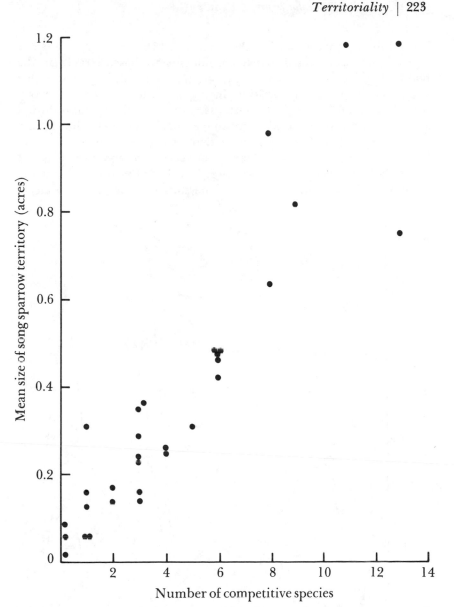

Figure 9.6 Relationship between territory size (acres) of song sparrows and number of potentially competing species. 2.47 acres = 1 hectare. (Modified from Yeaton and Cody, 1974.)

occupy islands smaller than the territory sizes occupied by them in large mainland areas with many other species of warblers present.

Although these correlations suggest that food is a limiting factor in the determination of territory size, it is very difficult to demonstrate this beyond doubt. Indeed, Lack (1966) suggested that territory size is not typically resource related, but his arguments are based largely on the population size of great tits during one unusual breeding season (fig. 9.7), and on a similar situation in a population of song sparrows studied by Tompa (1962). In both cases, breeding populations during a single season were considerably higher than they were during any other year, leading Lack to suggest that populations could not have been limited by resources in those other years. If a large percentage of a population is composed of new breeders that all settle at about the same time, the advantages of prior ownership are much less of a factor

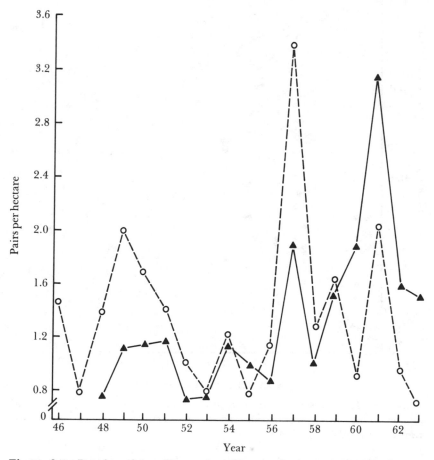

Figure 9.7 Density of breeding pairs of great tits in two English forests: Marley Wood (solid lines) and Forest of Dean (dashed lines). Not depicted is the density of pairs in Marley Wood in 1947 (0.3 pairs per hectare). (Modified from Lack, 1966.)

and the result may be a density of territory holders far greater than usual. Reproductive success per individual in the year of high density was in fact lower than in other years, in both species. Inexperienced breeders usually have low success rates, but Tompa's comments (1971) on the high frequency of fighting among neighboring birds suggest that the low success rate was more than just the result of their being first-time breeders.

Some workers believe that the primary purpose of breeding territories is to attract mates, and that territory size is therefore not expected to show obvious relationships to variables such as the food supply. This may be particularly likely where good sites are scarce. Studying savanna sparrows (*Passerculus sandwichensis*), McLaren (1972) and Welsh (1975) concluded that mating opportunities constituted the greatest territorial advantage for males. Some individuals were able to secure two mates, and this did not appear to decrease the success of their females. In fact, the birds performed much of their foraging in undefended areas off the territories. Polygynous species are particularly likely to be found in areas that contain plentiful food but a restricted number of nest sites (Orians, 1969c). Brown and Orians (1970) suggested that male nonsocial insects frequently employ territorial strategies that directly increase their access to females. Male grasshoppers (*Ligurotettix coquiletti*) advertise acoustically from desert bushes and also defend these bushes against other males (Otte and Joern, 1975). Similarly, sphecid wasps may vigorously defend favorable perches located near areas in which females feed and nest (Alcock, 1975).

Noting the frequent lack of close correspondence between resource supply and territory size, Verner (1977) proposed that territories are held in order to restrict the access of other individuals to resources, whether food or mates. Holding an oversize territory might indeed reduce the fitness of one's competitors, but it seems likely to reduce the fitness of the territory owner even more, since the costs of defense go up (probably often more than linearly) with territory size. One individual's territory may of course depress the fitness of a neighbor, but this is more likely to be an incidental consequence of the territory than one of its evolutionary causes. A plausible alternative explanation for Verner's observation is that an individual tries to defend a territory large enough to provide all of the resources necessary under most conditions. Although rare, extremely harsh conditions, such as an extended rainy period causing low insect productivity (Morse, 1967b, 1976a), can be devastating. Defending a territory large enough to exceed normal requirements will decrease the chance of adult or juvenile mortality and will have the consequent effect of restricting resources otherwise available to competitors, just as Verner suggested. In both instances nonterritorial individuals are unlikely to rear as many off-spring as the territory holders in the next generation.

Similarly, it is possible that territory holders are unable to measure the potential food resources with great precision, yet they must often attempt to do just this at the beginning of the breeding season in cases where the critical period for food comes much later (Brown and Orians, 1970). In this case an individual might play it safe, maintaining a large enough territory to raise young successfully through the worst parts of most years. The exact level of resources will depend on several factors, including their degree of predictability and both interspecific and intraspecific pressure.

Spacing of territories

The spacing of territories depends very much on the particular purposes they serve. Where all resources are contained within the territory, relatively even spatial distributions are found, except where satisfactory habitat exists only in patches. Where only certain required resources are associated with the territory (for example, nesting and display sites), considerably more clumped distributions may occur. Colonially nesting seabirds typically do not feed on their territories, and it has been suggested that they clump in part to facilitate synchronization of the breeding cycle (Darling, 1938), which presumably minimizes the vulnerability of newly hatched young. However, many studies have failed to demonstrate this effect (Klopfer and Hailman, 1965). It may be that only large colonies can successfully withstand predation on eggs and young by gulls and other predators. On the other hand, if prey, even in large numbers, are ineffectual against predators, thorough dispersion may be the best strategy, as in Brewer's blackbirds (*Euphagus cyanocephalus*) and great tits, which are vulnerable to snakes and mammals (Horn, 1968; Krebs, 1971). Post (1974) noted that seaside sparrows tended to have spaced territories where pressure from mammalian predators appeared to be high and clumped territories where predation appeared to be low.

In species forming leks (communal display grounds), clumping of territorial display areas should cause more females to be attracted to the males than would be attracted if they acted as individuals (see Crook, 1965; Lack, 1968). This effect appears to account for the existence of leks, despite the fact that some individuals mate much more frequently than others. Predation may strongly affect the form that leks assume (Lack, 1968). Lekking males may be subject to severe predation, as Estes (1969) found in wildebeest (*Connochaetes taurinus*) and Hartzler (1974) in sage grouse (*Centrocercus urophasianus*). This implies that predation strongly affects the spatial and temporal distribution of leks. Different species of grouse show widely varying spatial distribution, ranging from sets of small, immediately adjacent territories (Wiley, 1973) to individual display areas located considerable distances apart, but often within visual or auditory range of one another (Wiley, 1974). The females attracted to these display areas are dispersed during the nesting season and may even defend territories against one another. Snow (1962a, b) noted that lek-forming birds of tropical forests are primarily frugivorous (manakins, cotingas, birds-of-paradise), and that insectivorous birds rarely form leks. The relative ease with which males can obtain fruit may be an important factor permitting the large expenditure of time on the leks. The absence of males from the female's foraging ground may conserve her supplies as well (Crook, 1965). In addition to fruit the females typically take highly proteinaceous foods (primarily insects), which are required to produce and to feed the young. Males may initially have been able to escape parental duties because of the tendency of these birds to raise small broods, which apparently lowers nestling mortality (see Morton, 1973; Foster, 1974b).

Like many other lek-forming species, grouse are precocial; after hatching the female leads the young to food, rather than bringing it to them. Male lek-forming mammals could not contribute directly to the nourishment of their young; thus, there are fewer constraints on the development of such systems in mammals than there are in birds. But it is probably more than an interesting coincidence that, in mammals, leks are also confined to herbivorous species.

Feasibility of holding a territory

Territoriality may be an excellent way for certain favored individuals to gain access to resources, but it has its limits. Whether a territory can be defended depends on the mobility of the animal, the degree of clumping, and the predictability and quality of the resource. For example, when food becomes scarce it may become impossible to obtain adequate resources within an area of defensible size. As the shortage develops, competition for resources will intensify as intrusions by other individuals expanding their own foraging in response to the scarcity increase. At a certain point it will become necessary to cease defense, or even to emigrate, as expenditures in defense exceed the benefits gained (see Brown, 1964).

Hinde (1952) studied the time course of territoriality in sedentary great tits in England and found that when winters are unusually mild, established individuals tend to remain on their breeding territories. If conditions become more severe, they cease territorial defense but remain in the same area in social groups with tits and other insectivorous species, maintaining a dominance hierarchy. When conditions improve they resettle their territories. On warm days in late winter they will at least temporarily settle within their territories; this may result from the lowered energy demands of mild days. This strategy allows birds to remain near their territories without defending them when it would be energetically infeasible to do so. If conditions become very severe, the birds may quit the area entirely. Apparently English great tits seldom, if ever, make long-distance movements, but some of the Dutch great tits studied by Kluyver (1972) do so regularly.

Carpenter and MacMillen (1976) developed a model permitting them to predict when nectar-feeding Hawaiian honeycreepers (*Vestaria coccinea*) would shift from territorial to nonterritorial states (fig. 9.8). This shift occurred in response to the richness of the flower sources and competitive pressures. The birds became nonterritorial when resource availability was either low or high. When resources are superabundant there is no advantage to defending them. Gill and Wolf (1975b) demonstrated that nectarivorous golden-winged sunbirds (*Nectarinia reichenowi*) defend territories containing flowers with high nectar rewards but not those containing flowers with low rewards.

Highly mobile animals should be able to defend territories much more effectively than less mobile ones. This may be one reason why territoriality appears to be more widespread among birds than mammals (see Brown and Orians 1970); the difference may also be due in part to the high proportion of insectivorous birds (Morse, 1975a). Hence comparisons should be made between similar feeding types (for example, between insectivorous birds and insectivorous mammals). But even when this correction is made, territoriality is much more frequent in birds (see Fisler, 1969). The comparison between birds and reptiles is of interest because the problems of territoriality defense in reptiles may be relatively similar to those in birds. Reptiles are not very mobile relative to birds, but their nutritional demands are much lower because of their ectothermic condition. Many reptilian territories are simply display or sun-basking areas and are thus not typical of bird territories. But some species of reptiles, including several iguanids, defend all-purpose territories. The fact that mammals are especially unlikely to form territories (at least rigorously defended ones) may result from their need for large areas in relation to their mobility. Hamilton, Buskirk, and Buskirk (1976) noted that defense of space and resources by Chacma baboon (*Papio ursinus*) troops was far more effective in small swamp territories than in large territories in the desert. They attribute this difference to the probability of encountering other troops. However, problems remain in linking mobility and demand. For instance, why is territoriality more frequent in carnivorous

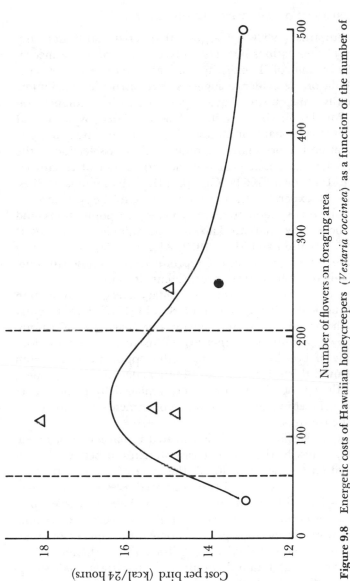

Number of flowers on foraging area

Figure 9.8 Energetic costs of Hawaiian honeycreepers (*Vestaria coccinea*) as a function of the number of flowers in their foraging area. The model predicts that the honeycreepers will be territorial at the intermediate flower (and food) densities enclosed by the dashed lines. Triangles = territorial birds; white circles = nonterritorial birds; black circles = aggressive but nonterritorial birds. Curve fitted to eye. (Modified from Carpenter and MacMillen, 1976, copyright 1976 by the American Association for the Advancement of Science.)

mammals than it is in others, since carnivorous forms typically use much larger areas than do herbivorous forms of similar size?

CAN TERRITORIALITY LIMIT POPULATION SIZE?

Wynne-Edwards (1962) has argued that territoriality not only affects the sizes of populations but has evolved specifically as a mechanism of population control. The distinction between cause and effect is critical. The bulk of the evidence suggests that territorial encounters are fundamentally antagonistic, that is, they reflect direct competition for resources. Population size may indeed be regulated by territorial interactions, but it is probably unnecessary and unwarranted to argue that such regulation is more than a secondary effect of selection at the individual level for traits that promote the acquisition of certain resources (see J. L. Brown, 1969; Klomp, 1972). Numerous studies, including removal experiments, have documented the presence of nonbreeding individuals (most frequently males) in populations, and in some cases these individuals are known to be capable of breeding if given the chance (Watson and Moss, 1970; Klomp, 1972). J. L. Brown (1969) has pointed out that unless reproductively competent non-breeding individuals of both sexes are present, it cannot be assumed that territoriality has any effect, even a secondary effect, on population size. Knapton and Krebs (1974) and S. M. Smith (1978) have recently demonstrated that surplus male and female song sparrows and rufous-crowned sparrows (*Zonotrichia capensis*) occur in some populations and are capable of breeding if they have the opportunity. Knapton and Krebs removed thirteen males and nine females in an experiment carried out in the spring. These birds were replaced by fifteen males and ten females. Breeding success of the replacements did not differ from that of the controls.

Some removal studies have not demonstrated replacement of breeding individuals; it may be that in these cases territorial behavior is not limiting population size, at least not at the time of the experiments. When Cederholm and Ekman (1976) removed crested tits (*Parus cristatus*) and willow tits (*P. montanus*) from their territories, new individuals did not occupy the spaces until after the breeding season. It is possible that regulation took place at some other time of the year, as seems to occur in the Carolina wren (Morton and Shalter, 1977). Rippin and Boag (1974a) found that only when territorial male sharp-tailed grouse (*Pediocetes phasianellus*) were removed early in the breeding season, and in substantial numbers (at least half of the territorial males), would other males colonize the communal display areas. Presumably the established territorial males were so aggressive that unless many of them were removed the nonbreeders could not establish themselves.

Fretwell and Lucas (1969) discussed the effects of territoriality on population size where habitats differ in suitability. They hypothesized that a characteristic "free" distribution will occur where each individual is free to choose its territory site without regard to other individuals, and that this differs from the dominance distribution that occurs where settlement is restricted by other individuals. Under the dominance distribution at equilibrium an individual should be indifferent as to whether it moves to the next less suitable habitat. In companion papers, Fretwell (1969c) and Fretwell and Calver (1969) argued that if birds in high-density habitats have consistently higher success rates than those in low-density habitats, one could conclude that territoriality was acting to limit density. Klomp (1972) pointed out possible difficulties in this approach. For example, if individuals in high-density habitats exhibit consistently higher success rates than other individuals, it is only possible to state that territorial behavior limits density after other density-limiting mechanisms can be excluded. Also, it is necessary to treat experienced and first-year breeders separately because they usually have different success rates. Klomp argued that an enormous amount of field work would be required to establish population regulation in this way and that the technique is not a substitute for removal experiments.

The remarkable fluctuations reported by Lack (1966) (fig. 9.7) and by Tompa (1962) for a few populations of songbirds are consistent with the idea that territoriality limits the densities of breeding populations (Krebs, 1971). Fluctuations in density might reflect environmental variations important to the breeding biology of the population. Great tits settle onto breeding territories from winter flocks in a sequential manner, with prior territory holders establishing territories before first breeders. But if extremely warm weather occurs in late winter, it may lead to a more rapid (and more nearly simultaneous) settling, causing a high breeding density. If a large proportion of individuals are first breeders, the effect would be greatly enhanced. This would also favor synchrony, a factor accounting for the high breeding density of song sparrows reported by Tompa (1962) following a period of heavy mortality among territorial adults. Based on his studies of sticklebacks (*Gasterosteus aculeatus*), van den Assem (1967) suggested that settling patterns may be of major importance in determining breeding densities, and Armitage (1974) reported a similar effect in yellow-bellied marmots (*Marmota flaviventris*). These examples suggest that such variation is widespread among territorial animals.

Knapton and Krebs (1974) have actually tested the settling effect in song sparrows. They used three areas, one which remained a control, a second from which they removed song sparrows sequentially, and a third from which they removed a considerable number of individuals simultaneously. They found that only in the area of simultaneous removal did the territory size and population density change significantly (fig. 9.9). This is consistent with the settling effect hypothesis. They also noted that of four species of birds, for which there are long runs of data on territory size, the two that have relatively low adult mortality have relatively constant territory sizes from year to year—oystercatcher *Haematopus ostralegus* (Harris, 1970) and tawny owl *Strix aluco* (Southern, 1970)—whereas the two with relatively high adult mortality have considerable year-to-year variation in territory size—red grouse (Watson and Moss, 1970) and great tit (Lack, 1966).

Settling rate may remain as a variable in these populations because it is a consequence of variable environmental and demographic patterns. Further, although a low nesting success accompanied simultaneous settling in the great tits and song sparrows, individuals that reproduced may have enhanced their fitness (relative to individuals not

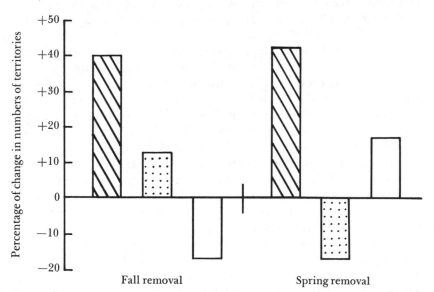

Figure 9.9 Changes in numbers of song sparrow territories after simultaneous (hatched areas) and successive (stippled areas) removal of sparrows. Unfilled areas = controls. (Modified from Knapton and Krebs, 1974; reproduced by permission of the National Research Council of Canada from the *Canadian Journal of Zoology*, 52, pp. 1413–1420, 1974.)

reproducing). Large numbers of offspring were produced, although individual output was low.

Mobile territories

There may be little functional difference between fixed territories and mobile defended spaces. Samson (1976) found that male Cassin's finches (*Carpodacus cassinii*) defended their females and the areas around them, rather than any fixed space. Since the population that Samson studied was composed of many more males than females, females were very likely to be the most important limiting factor. Mobile defense of females has also been reported in a number of other cardueline finches and is usually accompanied by male-biased sex ratios (Newton, 1972). Male hamadryas baboons (*Papio hamadryas*), usually polygynous, guard their females in a similar way (Kummer, 1971).

Analogous mobile territories may be associated with feeding. Ingolfsson (1969) and Prŷs-Jones (1973) reported that glaucous gulls (*Larus hyperboreus*) defended feeding eiders (*Somateria mollissima*) against encroachment by other gulls. The gulls pirated food items from the diving eiders when they brought food to the surface. Jutro (1975) found that laughing gulls (*Larus atricilla*) defended sites around picnickers and bathers on the beach. If the birds were not fed within five or ten minutes, the territory broke down.

Individual distance

Some animals, the "contact" species of Hediger (1955), tolerate extreme crowding. But most animals allow other individuals to approach only up to a certain critical distance without responding agonistically. The minimal distance (individual distance) tolerated without attack or retreat may be considered a smaller version of the mobile territories just discussed. The actual size of this space may depend on what the animal is doing. When pine seeds are superabundant, brown-headed nuthatches (*Sitta pusilla*) forage so close together that they occasionally touch each other. At other times they do not tolerate such close approach (Morse, 1967b, 1970b). Species in which individuals huddle for warmth include some that maintain considerable spacing in their daily activities—for example, the wren (*Troglodytes troglodytes*) (Armstrong, 1955). Some swallows (Hirundinidae) only permit contact during unusually severe weather (Grubb, 1973; Mescrvey and Kraus, 1976).

Role-dependent differences in individual distance may be striking. Lockley (1961) found that dominant and subdominant male rabbits (*Oryctolagus cuniculus*) maintained a distance of nearly 1 m, whereas males and females stayed only about 30 cm (a body's length) apart. Sizable differences also exist between closely related species. For example, blue tits (*Parus caeruleus*) permit much closer individual distances than do several congeners (Morse, 1970b, 1978) (fig. 9.10). As with brown-headed nuthatches, these differences are related to the patterning of resources (clumped or spaced) used by each species. In blue tits there is considerable social facilitation of feeding (Morse, 1978). Maintaining minimal distances while searching for food may increase the efficiency of resource exploitation and reduce the frequency of hostile interactions. Condor (1949) and Hediger (1955) have discussed individual distance in greater detail.

Home range

A home range is an area regularly traversed by an individual in its activities (see Burt, 1943; Jewell, 1966). This includes territories

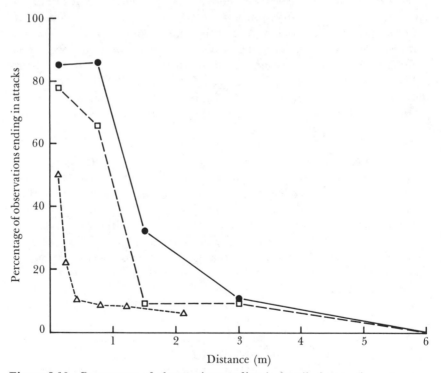

Figure 9.10 Percentage of observations ending in hostile interactions at varying intraspecific distances. Circles = Carolina chickadee; squares = tufted titmouse; triangles = blue tit. (Modified from Morse, 1970b, copyright 1970 by the Ecological Society of America; Morse, 1978c, with permission of the *Ibis,* journal of the British Ornithologists' Union.)

as well as undefended areas, but the term usually refers implicitly to the latter. Relationships between the sizes of home ranges, their characteristics, and the characteristics of the animals themselves are quite similar to those for territories. The size of a home range is a function of the animal's body weight, and the area occupied by a carnivore is much larger than that occupied by a herbivore of similar size (McNab, 1963). Areas formerly considered undefended are now known to be defended to some degree, even if in a ritualized manner or only occasionally. The primary difference between territories and home ranges may lie in the effectiveness with which defense is accomplished. Although it may be advantageous to keep all intruders out all the time, occasional defense may also provide significant benefits. Such benefits may mean the difference between being able or unable to occupy an area. Eisenberg (1966), Fisler (1969), and Brown and Orians (1970) tended to treat territories and home ranges as the ends of a range of related phenomena, and this seems to make much better sense than considering them to be qualitatively distinct.

It has traditionally been believed that many species do not defend territories or that they defend very small ones (for example, the immediate vicinity of a burrow), while roaming over much larger areas (Brown and Orians, 1970). But it is becoming clear that although individuals may not defend an area regularly or actively, they may nonetheless exert dominance when they meet other individuals. For example, flocks of small birds occupying winter ranges traverse the boundaries of these areas quite carefully, and if a group straying outside its usual area meets the resident group, it is usually repulsed (Morse, 1970b). Fossey (1974) reported protracted agonistic encounters (largely vocal) between groups of gorillas (*Gorilla gorilla*) when approaching each other on home ranges. Both cases strongly suggest territoriality, if of a relatively loose kind. Even where mutual avoidance occurs, it may reflect past interactions of the sort not likely to be seen in short-term studies (Morse, 1976b). There is probably no simple way to distinguish operationally between home ranges and territories (see Burt, 1943, Brown, 1966, and Jewell, 1966, for comparisons of territory and home range).

Brown's (1964) hypothesis of defendability is relevant to the question of how frequently an area will be defended. Mammals, because of their low mobility relative to resource demand, are unlikely to defend territories rigorously. C. C. Smith (1968) argued that rigorously defended territories have developed in red squirrels but not in other tree squirrels because red squirrels have a localized, easily defendable resource (pine cones), which is also abundant, dependable, and readily stored. Other tree nest sites are rigorously defended

(C. C. Smith, 1968). Significantly, Layne (1954) did not find territoriality in red squirrels living in deciduous forests, where resources are not as readily defended. Primates show an analogous relationship, with frugivores being nonterritorial and those feeding on more abundant and predictable resources (leaves and buds) being territorial (Carpenter, 1958; Marler, 1969). These studies show considerable variation in the degree of space-related defense among closely related species, or even within a single species.

10 *Spacing Patterns: Dominance Hierarchies*

SOCIAL DOMINANCE usually establishes priority of access to resources and is usually attained by those individuals who most successfully fight, chase, or otherwise supplant the other members of their group (Morse, 1974). Dominance is distinguished from territorial behavior (which may also be considered a form of dominance) by the absence of a clear reference point in space. The concept was developed by Schjelderup-Ebbe (1922) from his studies on flocks of domestic fowl (*Gallus domesticus*), but hierarchical relationships were recognized long before this time by a variety of workers (see E. O. Wilson, 1975). However, it remained for Schjelderup-Ebbe to establish it as a subject for intensive investigation in vertebrates.

The terms *dominance hierarchy* and *peck order* are often used synonymously. But as pointed out by Etkin (1964), dominance hierarchy is more general and usually more appropriate; peck order implies that the interactions between individuals are overtly hostile, but in fact, dominance relationships often lower the frequency and intensity of overt hostility.

A distinction is often made between peck-dominance systems, in which an individual is dominant over another one in a statistical sense (wins exceed losses), and peck-right systems, in which dominance is absolute (unilateral). Some of this variation may be more apparent than real. Individuals that have previously used an area intensively may exert dominance there over others that have used it less in-

237

238 | *Behavioral Mechanisms in Ecology*

tensively. Reversals may also occur while hierarchies are being established, and animals in reproductive condition are often dominant over nonreproductives—for example, female mangabey monkeys (*Cercocebus* spp.) (Chalmers and Rowell, 1971). Thus if care is not taken to record the temporal sequence of interactions, dominance may appear to be less regular than it really is.

Hierarchies may be either linear or nonlinear. In linear hierarchies a high-ranking individual dominates all lower-ranking individuals; these are characteristic of small flocks of fowl and of several other species as well. But in some species such as pigeons (*Columba livia*) (Masure and Allee, 1934a) and black-capped chickadees (*Parus atricapillus*) (Hamerstrom, 1942; Glase, 1973), hierarchical relationships are often not perfectly linear. "Triangles" and even "squares" may occur; for example, A dominates B, B dominates C, and C recognition.

Like territoriality, hierarchies may be viewed as the outcome of dominates A. As the sizes of groups increase the opportunities for nonlinearity also increase, owing in part to difficulties of individual aggressive competition for resources. A major difference is that territories are readily defendable, while the areas or resources allocated by means of hierarchies are not. Dominance hierarchies have been reported in mammals, birds, reptiles, amphibians, fish, arthropods, and mollusks (Gauthreaux, 1978). The phylogenetic limits of hierarchical behavior remain uncertain largely because it is difficult to determine the importance of such factors as olfactory communication (Eisenberg and Kleiman, 1972). Nonetheless, it is clear that hierarchies are widespread, although Itô (1969) and others have questioned whether all the hierarchical relationships observed in the laboratory, particularly among fish and invertebrates, also occur in the field.

Measures of dominance

Different workers use different criteria to rank animals, and as a result confusion exists in the literature, particularly in the primate literature (Richards, 1974). It is essential to state the criteria being used to rank the members of a hierarchy. For example, Rowell (1974) has noted that dominance relationships in primates may differ markedly from one context to another; dominance recognized among individuals in their regular daily interactions may bear little relation to that seen in food competition tests or in other measures of dominance. Eaton (1976) and others (see Kolata, 1976) have noted that aggressive dominance ranking of both male and female macaques (*Macaca fuscata* and *M. mulatta*) in large enclosures is not correlated with their frequency of mating, although it is uncertain how strongly reproductive success correlates with overall mating frequency. It is known that low-ranking male anubis baboons (*Papio anubis*) mate frequently with females coming into reproductive condition, but that they are not allowed to mate with females at the time of ovulation (DeVore, 1971). Hausfater (1975) found that a statistically significant correlation existed between dominance ranking and number of matings in a troop of yellow baboons (*P. cynocephalus*) in Kenya, although the alpha male in the troop did not mate as often as predicted. However, Hausfater believed that the dominance ranking changed so rapidly that no reproductive advantage accrued to dominants. Bernstein (1976) noted that alpha males in macaque groups may be chosen by females as much for their social characteristics as for their fighting abilities. It is becoming increasingly apparent that female dominance, female choice, and individual relationship between males and females complicate the formation of hierarchies and also their subsequent effects (see Cheney, 1978). Much of the work on primates consists of laboratory or corral studies, in which the animals are always at artificially high densities. Deag (1977) emphasized that studies of caged animals must be interpreted with extreme caution, but found evidence that hierarchies do exist in wild, unprovisioned Barbary macaques (*M. sylvanus*). Cheney and Seyfarth (1977) found that an alpha male baboon (*P. cyanocephalus*) in a free-ranging troop fertilized six of the eight females that became pregnant during the course of a fifteen-month study in South Africa. Cheney (1978) cited numerous other reports of dominant male monkeys preventing subordinate males from copulating with estrous females.

There is no evidence that the equivocal relationship between aggressive behavior and reproduction described in many primate studies holds for most other animals. Dominant male galliform birds on leks almost always perform the greatest amount of mating; examples include sage grouse (*Centrocercus urophasianus*) (Wiley, 1973), sharp-tailed grouse (*Pediocetes phasianellus*) (Rippen and Boag, 1974b), and greater prairie chicken (*Tympanuchus cupido*) (Ballard and Robel, 1974; Robel and Ballard, 1974). Male elephant seals (*Mirounga angustirostris*) that control a mating area perform most of the matings (fig. 10.1), although they may be unable to retain high rank for the entire breeding season (LeBoeuf, 1974). Dominant male dragonflies (*Plathemis lydia*) in communal areas enjoy preferential access to females (Campanella and Wolf, 1974). Dow and Schilcher (1975) also found a close correlation between success in fights and in mating in fruit flies (*Drosophila melanogaster*); many encounters were settled by display (wing threats).

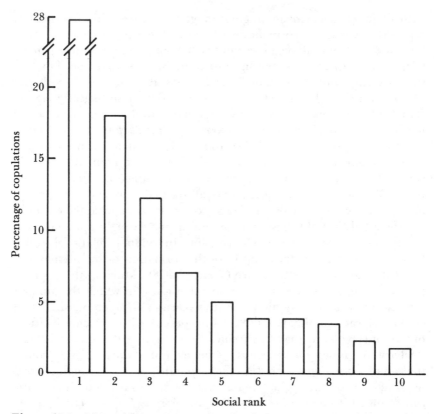

Figure 10.1 Mean yearly percentage of copulations by ten top-ranked male elephant seals in harem (from six-year totals). Individuals were assigned the social rank they held for the longest time during the six-week period that females were in estrus. (Modified from LeBoeuf, 1974.)

The contrast between some of the primate studies and those on other species warrants further attention. Direct comparisons are difficult, because many of the primate studies have been concerned with individuals in cages or corrals. It is tempting to dismiss the results of these studies as artifacts of crowding or food superabundance. Under such circumstances access to food may not be a relevant measure of dominance, although it might be in the field. On the other hand, the idea that populations are usually food-limited may have created a misleading set of expectations in the minds of ecologically oriented behaviorists. It would be profitable to initiate studies of primate dominance patterns in natural systems where food is demonstrably not a limiting factor, as well as in systems where severe resource shortage occurs periodically. Comparisons of hierarchical behavior during times of shortage and of plenty would be illuminating.

It remains possible that primates have significant attributes not shared by other groups of animals whose dominance relationships have thus far been studied. Certainly kinship and other alliances could complicate simple analyses, and these factors should be accorded extra attention.

Advantages of hierarchies

Hierarchies are not merely an alternative to territories; they have distinct costs and benefits which are closely associated with group membership. Social groups provide advantages in defense against predators, in defense of feeding areas, and in exploitation of resources (see Brown and Orians, 1970). Hierarchical relationship within a group may facilitate efficient feeding by reducing aggression and may also reduce vulnerability to predators. By giving certain individuals priority of access to resources or mates, hierarchies distribute these advantages unequally. Even when resources and mates are in adequate supply, dominant individuals may obtain the most satisfactory ones; if they are in limiting supply, rank may mean the difference between obtaining them or not. Even so, the indirect advantages to low-ranking members of groups may be sufficient to prevent the system from breaking down.

How are hierarchies established and maintained?

Physical interactions are extremely rare in some groups. Casual observers have frequently reported hierarchical relationships unaccompanied by overt hostility, as have the authors of intensive (but brief) studies. For example, Austin and Smith (1972) studied winter flocks of small insectivorous birds for an entire season without noting hostile interactions, although hostility is known to occur in these species and in closely related species when hierarchies are being established. In jungle babbler (*Turdoides striatus*) groups, aggression is rare and is confined almost entirely to first-year birds (Gaston, 1977).

Most long-term studies show that relatively high levels of hostile interactions occur at the time rankings are being established, after which such interactions are rarely seen (Morse, 1970b). Dominance systems that developed without hostile interactions would be open to interpretation in terms of group selection (Wynne-Edwards, 1962), but there is no convincing evidence that such systems exist (see Lack, 1966; Wiens, 1966). Systems in which encounters are rare can easily be explained in terms of individual selection, on the assumption that most individuals obtain advantages from sociality at one time or another. Once established, hierarchies are often extremely stable. Winter bird flocks typically remain stable for the season (Morse, 1970b), and primate troops or babbler flocks may remain stable for several years (Sade, 1967; Gaston, 1977).

CUES USED

If dominance systems are to operate efficiently, individuals must either respond to cues highly correlated with dominance, or they must use individual recognition. In studies on cockroaches (*Nauphoeta cinerea*), Ewing and Ewing (1973) demonstrated that these animals do not use individual recognition. The cue is thought to be related to physiological changes resulting from some unknown stress factor. Ewing and Ewing first ran pairs of individuals, then paired these individuals with others of different ranks, and finally rematched the original pairs. In this way reversals could be obtained predictably.

Predictable dominance orders found in some hymenopterans are directly dependent on their degree of ovarian development (Pardi, 1948) alone, and individual recognition plays no known role. Ungulate horns and antlers (Geist, 1971) probably serve as cues in a comparable way, as may the amount of black coloration in the facial regions of Harris' sparrows (*Zonotrichia querula*) (Rohwer, 1975).

Individual recognition might allow for particularly efficient relationships. It is known to occur in fowl and operates effectively up to a critical group size of about ten (Schjelderup-Ebbe, 1922). Apparently this is as far as the animals can count! More complicated relationships such as triangles and squares become common at larger group sizes, and the hierarchy becomes relatively unstable. Many mammalian groups, particularly ones that are stable over considerable periods of time, are probably based on accurate individual recognition. Fisler (1976) summarized the widespread existence of individual recognition in small rodents and showed that it depends on both olfaction and vision. In the antelope squirrel (*Ammospermophilus leucurus*) both of these sensory modalities clearly play a role in structuring the hierarchy.

The system with the greatest potential for accurate recognition would be one in which there are several different castes or generic types. These could be age groups, sexes, or other morphs. If an individual could distinguish its own caste from others, as well as the other individuals within its own caste, it would be able to increase its efficiency of recognition considerably at a modest cost. Wood-Gush (1971) has demonstrated such relationships among domestic fowl.

CHARACTERISTICS OF DOMINANT ANIMALS

Large individuals are usually dominant over small ones, and old individuals are usually dominant over young ones, even where determinate growth occurs. These relationships occur in both wild and domesticated animals (Fretwell, 1969b; Guhl, 1962; Gottier, 1968). Other characteristics may be important as well, such as condition, sex, time with the group, experience, and reproductive state (Gottier, 1968; Ewing and Ewing, 1973).

Size. In studies on juncos (*Junco hyemalis*) Fretwell (1969b) found a good relationship between size and dominance, tendency to participate in groups, and habitats frequented. Large dominant birds were often found in habitats different from those used by smaller birds and birds not in flocks. The individuals not in flocks were known in some cases to be low in the hierarchy. Survival rates of these solitary birds appeared to be much lower than those of flock members.

Why does a wide range of size persist in a junco population? Presumably the smaller birds are more efficient exploiters than the large ones under some conditions, or their lower energy demand allows them to survive where the larger individuals cannot. This relationship has interesting implications for Bergmann's rule that races of homeotherms from cool climates are larger than those from warm climates (Mayr, 1963; also Rosenzweig, 1968; James, 1970; McNab, 1971).

In studies on black-capped chickadee flocks, Glase (1973) also found that large individuals are usually dominant over small ones. Separate male and female hierarchies exist, but males as a group generally dominate females.

Age and sex. Within the hierarchies described by Glase (1973) adult chickadees usually dominated juveniles (birds under a year of age). However, this factor was not as important as that of sex; even juvenile males were typically dominant over adult females. In some species, such as juncos, it is possible that differences in dominance rankings are partly responsible for the tendency of males to winter farther north than females (Ketterson and Nolan, 1976). Males in most passerine species are generally dominant over females in encounters and typically are as large or larger than the females. But exceptions occur, as in purple (*Carpodacus purpureus*), house (*C. mexicanus*), and Cassin's (*C. cassinii*) finches (Samson, 1977), where the females often dominate the males, although they do not exceed them in size. The reason for this reversal is unknown. When females are the larger sex, as in hawks and owls, they are typically dominant over the males (Brown and Amadon, 1969). Differences in habitat use between certain immature and adult birds (both age classes being of similar size) are sometimes related to the higher dominance ranking of adults; examples include red grouse (*Lagopus lagopus*) (Jenkins, Watson, and Miller, 1967), great tit (*Parus major*) (Krebs, 1971), and American redstart (*Setophaga ruticilla*) (Ficken and Ficken, 1967; Morse, 1973a). Even this age-related dominance does not inevitably hold, however. Harrington and Groves (1977) report that migrating immature semipalmated sandpipers (*Calidris pusilla*) dominated adults more frequently than the adults dominated them. Where age classes are of different sizes, as in many mammals and fish, the smaller juveniles are usually subordinate to the adults.

Length of tenure within a group. Individuals typically enter a new group at the bottom of a hierarchy, even if they have come from a situation in which they held a high rank. Usually they work their way to the top of such a group slowly. For the most part, reports on this progression have been based on laboratory studies, but a similar relationship may exist under natural circumstances. Most new members of natural groups are young individuals, who are typically not as experienced as older ones, and they typically have no alternative to accepting a low rank.

Relationship to other individuals. In primates, rank is often inherited matrilineally, with daughters assuming ranks similar to those of their mothers. The ranks of males, who typically are outbreeding, are more variable than those of females, depending more on factors such as age, length of tenure within a troop, and the formation of coalitions with other males (see Cheney, 1978).

RELATIONSHIPS WITHIN AND AMONG GROUPS

The relationship of an individual to other members of a group may be of considerable importance in determining its position in that group. In the Carolina chickadee (*Parus carolinensis*) (Dixon, 1963), black-capped chickadee (Glase, 1973), and junco (Sabine, 1955, 1959), the position of a male in the hierarchy determines the position of his mate. On the wintering ground, the rank of a male Canada goose (*Branta canadensis*) determines the rank of his entire family (Raveling, 1970). In this case, an individual's dominance affects the fitness of its offspring well into the future. Rank is sometimes space dependent, with individuals having a high rank in some parts of the group's area and a low ranking in other places. This was first noted in the laboratory in flocks of parakeets (*Melopsittacus undulatus*) and pigeons (*Columba livia*) (Masure and Allee, 1934b; Ritchey, 1951) and was later seen in the field in Steller's jays (*Cyanocitta stelleri*) (Brown, 1963b), chickadees (*Parus* spp.) (Minock, 1972), ant-following birds (Willis, 1967, 1968), house sparrows (*Passer domesticus*) (Watson, 1970), and tassel-eared squirrels (*Sciurus aberti*) (Farentinos, 1972). Thus, the effect appears to be widespread. Under natural circumstances a shifting pattern of dominance may be related to the earlier activities of individuals on their group sites. Members of some chickadee flocks have a high dominance rank in the parts of their winter range that contain their breeding territories and a low rank elsewhere (Minock, 1972). Relationships such as these actually combine certain aspects of stationary territoriality and spatially mobile hierarchies.

HIERARCHIES WITHIN FAMILY GROUPS

Members of broods fed by parents may show hierarchical relationships that are associated with competition for food. For example, young magpie geese (*Anseranas semipalmata*) are aggressive toward one another, especially when begging for food from their parents (Davies, 1963). In domestic chicks, which feed themselves, dominance among family members may be established at an early age (Guhl, 1958). Herons and raptors commence incubating when the first egg is laid, which gives the first young hatched a considerable lead in development over its siblings (Lack, 1954; Kear, 1970). Unless food is presented in adequate amounts for all, the larger young get most of it, with the others eventually starving. The aggression of some nestling raptors is allegedly so strong that older individuals regularly kill the younger ones (see Lack, 1954). Dominance relationships occur routinely among littermates in mammals, and in some cases the rewards of this dominance are substantial. Newborn piglets (*Sus scrofa*) quickly establish a teat order, which is typically retained through weaning. The competition is for anterior positions, and it has been demonstrated that these sites provide more milk than posterior positions and also carry a lower risk of trampling by the mother (Gill and Thompson, 1956). Considerable aggressive interaction occurs, including the use of teeth (McBride, 1963). If the mother were in marginal condition the results of this scramble would probably mean the difference between life and death for the young. Even if all young survive, those weaning in best condition should have the highest probability of subsequent survival (see chapter 7). Although the tendency of mammals to give birth to littermates simultaneously reduces the scope for early dominance relationships, such relationships almost always develop anyway.

Relationships of hierarchies and territories

Members of hierarchies experience difficulties similar to those confronting individuals who compete for territories. This is especially true if favored habitat is patchy, in which case territorial groupings will resemble hierarchical groups. In threespot damselfish (*Eupomacentrus planifrons*) it can be shown that territorial individuals prefer to live in such groups, that some occupy more favorable sites than others, and that an individual's ability to improve its location is related to dominance (Itzkowitz, 1978). The typical result of territorial encounters is exclusion of all but the successful contestant or contestants. If territories overlap, however, hierarchical relationships may exist within common boundaries, as in areas occupied by bird flocks (Morse, 1970b). Here it is difficult to distinguish between territoriality and hierarchical relationships.

Changing resource patterns may cause animals to switch from a territorial to a hierarchical system, or vice-versa. Such shifts have been reported in great tits (Hinde, 1952) and house mice (*Mus musculus*) (Crowcroft, 1955). The mice set up hierarchical systems in corn storage bins, which provide abundant and concentrated resources, but are territorial where resources are poorer. American elk (*Cervus elephas*) also hold territories in some habitats but not in others (Altmann, 1952). Jaegers (*Stercorarius* spp.) are solitary nesters when they are associated with a defendable food source (a concentration of lemmings), but are loosely colonial where they feed by harrying other seabirds to disgorge their food (Lack, 1968). Verbeek (1973) has established strong correlations between spatial patterning of the food supply and nesting dispersion (individual territories or group nesting) in several species of corvids. Colonial nesters appear to occupy areas in which food sources are rather evenly dispersed (fig. 10.2). Crook (1965), Lack (1968), and Brown and Orians (1970) provided additional examples in which differences in sociality depend on resource distributions.

CHANGES BETWEEN TERRITORIALITY AND HIERARCHY

Inability to emigrate, absence of escape cover, and high population density can all induce a change from territoriality to hierarchical relationships in rodents (Archer, 1970). These factors are associated directly or indirectly with the degree of crowding and the ability of individuals to defend an area, or alternatively, with their ability to defend themselves individually against predators.

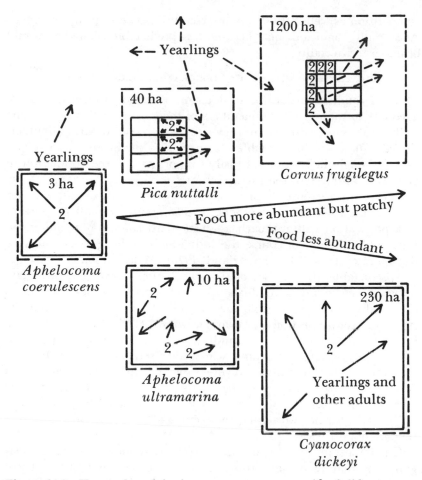

Figure 10.2 Types of exploitation systems among corvids. Solid squares = territory boundaries; broken squares = home range; solid arrows = territorial defense and range of food procurement; dashed arrows = food procurement outside territory; numbers inside squares = breeding pairs. (Redrawn from Verbeek, 1973.)

The fact that some animals change from territorial to hierarchical behavior when resources deteriorate (Hinde, 1952; Krebs, 1971) suggests that hierarchies provide more efficient methods of resource exploitation under stringent conditions than do territories. By abandoning territoriality an individual frees itself of some heavy demands for energy and time, allowing it to put proportionately more into maintenance. However, resource use may become less efficient than it was when the territory was maintained, owing to increased exploitation competition as well as to direct interference between group members. As a result, conditions may become less predictable than they would be under territoriality.

THE CONSEQUENCES OF A DETERIORATING RESOURCE BASE

If conditions continue to deteriorate resources may become inadequate to support all the members of the group. At this point some individuals must disperse if they are to survive; if all individuals shared equally in the available resources they might approach starvation, although the area might easily support part of the population in good condition. Here the hierarchy attains its greatest ecological significance, since it ensures that some individuals will obtain a larger share of the resources than others. Typically only a small proportion of a population will approach a substandard condition at any given time, and these are of course the individuals under greatest pressure to emigrate (Lack, 1954; Swingland, 1975). Where resource shortages are predictable (that is, seasonal), social interactions may stimulate dispersal before any individuals approach starvation. Such a process is probably often responsible for juvenile dispersal; it need not depend on group selection as Wynne-Edwards (1962) has maintained, because emigration may be an individually advantageous response to unfavorable social interactions, which themselves arise directly from competition for resources. Wynne-Edwards suggested that unless group selection is occurring the losers will fight to maintain their access to resources. But fighting may lead to repeated loss for certain individuals, further worsening their energy-balance problems, even if they are not otherwise hurt.

It is important to recognize that a hierarchy may not be the social guillotine that Wynne-Edwards imagined it to be. "Accepting" a low position at some point may maximize an individual's eventual fitness. To avoid fighting where one has a high probability of losing is to reduce net energy loss, which for some individuals at some times may mean the difference between starvation and survival. If dominant individuals do not have to fight, their use of the remaining resources will not be as heavy as they would be otherwise, thus further conserving the resource supply. There is an unfortunate shortage of data relevant to this hypothesis, but it is well known that energy expenditures associated with rigorous activity are several times larger than those associated with maintenance.

Emigrating individuals are not necessarily doomed, as Wynne-Edwards argued, but four important factors act against them. First, they are often not in as good physical condition as the individuals left behind. Second, emigration itself may be energetically expensive.

Third, difficulties of procuring food in unfamiliar areas, very likely against stiff competition, should cause high levels of mortality for such individuals. Fourth, predation rates on dispersers may be high (Errington, 1946), probably because they are not familiar with their new areas. Selection will favor dispersal only if some individuals return to breed in the area of origin. Until recently there were no good data on the fate of dispersers. Existing data (Newton, 1970) indicate that many dispersers do indeed survive, and that some may even rejoin their populations of origin. Discounting this possibility led Wynne-Edwards to conclude that emigration could be favored only through group selection.

When dispersal involves movement into less preferred habitats adjacent to the preferred one, the results may be less severe than for emigrants that quit a local area entirely. Fretwell (1969b) found that, generally, smaller juncos left the preferred open weedy fields and entered the forest during winter. Although these birds were typically subordinate to the larger ones and took apparently suboptimal habitats, they had the advantage, by virtue of being small, of having smaller absolute food requirements than the larger birds.

Whether an animal will fight or disperse should differ greatly depending on its dispersal capabilities, the degree to which odds in a fight may be assessed, the condition of the resources locally, and their "expected" condition elsewhere. Some evidence suggests that under deteriorating (or poor) conditions aggression decreases; other information suggests that it increases. Pulliam et al. (1974) reported that flocks of juncos (*Junco hyemalis* and *J. phaenotus*) show less aggression under deteriorating experimental conditions (fig. 10.9). This was accomplished by reducing their probability of encounter (that is, by avoiding one another), rather than by reducing the probability per encounter of aggression. Fighting did not appear to increase during inclement periods in studies on free-ranging flocks of titmice and kinglets (Morse, 1970b). Barash (1973b, 1974b) hypothesized that the high degree of social tolerance demonstrated by yellow-bellied marmots (*Marmota flaviventris*) living at extremely high altitudes is an outcome of the poor energy supply that they experience relative to those at lower altitudes. Such tolerance should be associated with a strong ability to disperse under poor conditions.

If severe weather limits the number of feeding sites, the frequency of aggression may increase as a consequence of crowding (Ketterson, 1978). The increased frequency of fighting for food reported in tits and corvids under inclement conditions (Gibb, 1954; Lockie, 1956) may result from this factor, as well as from an increased tendency for subordinates to initiate encounters. Both Sabine (1959) and Ketterson

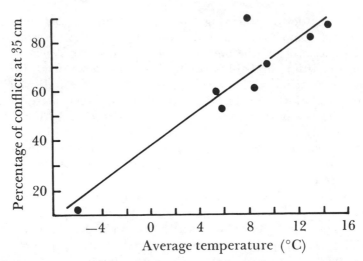

Figure 10.3 The relationship between aggression and temperature in juncos. The percentage of times a conflict resulted from simultaneous use of feeders placed 35 cm apart is shown in relation to the average environmental temperature. (Redrawn from Pulliam et al., 1974, with permission of the *Ibis*, journal of the British Ornithologists' Union.)

(1978) found that dominance reversals in juncos usually occurred during cold, snowy weather.

Conditions in which territories or hierarchies prevail

Brown (1964) considered holding a territory to be the best means of using a resource when an area can be defended. A model can be constructed based on two variables, resource patterning (clumped to spaced) and resource predictability (high to low). Where food is predictable and thinly spread, defending a type A territory (one with all resources included) should be feasible, but where food is highly clumped or unpredictable, defense may not be possible. Another concern is whether the resource can be defended under any circumstance (Brown and Orians, 1970; Orians, 1971). For example, oceanic feeding grounds considerable distances from a seabird's nesting area may be economically indefensible because of the high cost of reaching them and the impossibility of defending them while commuting or attending to nesting activities elsewhere.

Patterning and predictability have been investigated in natural systems. Crook (1965) found that birds spending considerable time searching for food are likely to be territorial. These birds are mostly carnivores (including insectivores) and show a pattern similar to that occurring in many small mammals (Lockie, 1966). Where resources are clumped, there is a tendency to feed in aggregations. Horn (1968) maintained that under some circumstances aggregations may have positive adaptive significance. For example, group feeding may increase the efficiency of food gathering where observational learning increases the rate at which food is found (Horn, 1968; Krebs, Mac-Roberts, and Cullen, 1972). This probably occurs most often where clumps of food are so large that one individual cannot fully exploit them (Ward, 1965a; Krebs et al., 1972). Alternatively, groups may occur where cooperation is required to exploit prey, as in killing large game or rounding up elusive prey. Using pack-hunting strategies some canids have been able to obtain prey far larger than they could if hunting in a solitary manner (Kleiman and Eisenberg, 1973). Some waterbirds such as cormorants (*Phalacrocorax* spp.) and mergansers (*Mergus* spp.) drive fish into constricted areas (against the shore), permitting efficient feeding by members of the group (Bartholomew, 1942; Emlen and Ambrose, 1970). Predators may also exert strong pressures that tend to concentrate their prey in groups. This will tend to reduce the probability that individual territories can be maintained. However, I know of no data on the role of predation in determining where territories will or will not be maintained.

AN ASIDE CONCERNING DOMINANCE AND COLONIZATION

Some dispersing individuals return to their original areas, but the majority of survivors probably breed in the areas in which they settle. Christian (1970) has emphasized the role of these individuals as the usual founders of new populations. If emigrants tend to lose in the competition for resources, they may not be the best potential colonizers, and this may contribute to the low success rate of colonization (see MacArthur and Wilson, 1967).

Synthesis

Aggression related to space is an important feature of the lives of many animals, especially where individuals are in some way resource-limited. The limiting resource is believed to be most often food, but food may only be the ultimate factor, with conventional competition for space occurring where food (or some other resource) is defendable. The nature of the aggressive response does not appear to differ qualitatively with the resource in question, although different displays may be required to defend a large space as opposed to a point source. On the other hand, the intensity or persistence of the aggressive behavior may differ markedly as a function of the prize in question. This may be seen in the constant and not infrequently debilitating aggression that takes place among harem-forming male mammals, such as some deer.

The relationship between aggressive behavior and resource limitation could be illuminated by comparing aggressive patterns of animals in populations that often experience substantial resource limitation with those of animals in populations that exist much of the time at a density well below carrying capacity. This low density may be a consequence of heavy predator pressure or other environmental factors. To the best of my knowledge no concerted effort has been made to do this, probably for a variety of reasons. First, forms generally believed to be under severe environmental or predatory constraints have not been subject to intensive studies of aggressive behavior. Appropriate groups include intertidal invertebrates (predator limitation) and many insects (environmental limitation). These animals probably exhibit behavioral patterns quite different from those of animals whose aggressive behavior has been intensively studied, if the former group exhibit aggressive behavior at all. In some cases chemical communication may prevail, and this is difficult to monitor. Thus, definite problems exist in attempting to make the appropriate comparisons in some taxa. For this reason the tactic of comparing selected phylogenetically allied animals may provide the most workable basis for comparison. However, any such analysis will require accurate information about aggressive behavior, resource abundance, and resource exploitation patterns, information that is all too rare in studies thus far published.

11 *Competition between Species*

MANY OF THE DIFFERENCES in resource exploitation observed among species are caused by interspecific competition. It is often argued that such competition has taken place mainly in the evolutionary past, that the mutual adjustments are a fait accompli, and that competition between the now "ecologically isolated" species therefore no longer exists in any meaningful sense (see Lack, 1965). This line of reasoning has led some workers (Andrewartha and Birch, 1954) to doubt whether the hypothesis of competitive exclusion is even testable. I will not join this debate here; references can be found in Ehrlich and Birch (1967) and Slobodkin, Smith, and Hairston (1967). Recent studies make it seem likely that a tension zone often exists between pairs or groups of species, and that it is occasionally or even regularly stretched (see MacArthur, 1972). Such tension zones may impinge on a species from more than one direction (the "species packing" of MacArthur, 1969, 1970).

Interspecific interactions

DIRECT INTERACTIONS

Interspecific interactions appear to be qualitatively similar to intraspecific interactions, although they are generally not as frequent or as overt (Morse, 1970b, 1976b; Myrberg and Thresher, 1974). Since conspecifics are typically more alike than are members of different species, competition for resources is usually more intense at the intraspecific than at the interspecific level.

Encounters between species more frequently take the form of supplanting actions, and less frequently the form of fights and chases, than do encounters within species. One might conclude from this that intraspecific encounters are of relatively greater intensity; however, because the relative intensity of these different patterns of interaction has not been established, this conclusion must remain tentative (see Hinde, 1970). Here I consider examples that fit this expectation as well as ones that do not, the latter typically occurring in pairs of extremely similar species.

As expected, in mixed-species foraging flocks of small birds interspecific encounters are less frequent than intraspecific encounters; the former more often involve supplanting; the latter, fights or chases (Morse, 1970b) (fig. 11.1). The high relative abundance of a few species (such as chickadees, titmice, and kinglets) in these flocks and the existence of niche partitioning both imply that a large proportion of interactions should, by chance alone, be intraspecific. But observed encounter frequencies are well in excess of those predicted. For example, Carolina chickadees (*Parus carolinensis*) make up 42 percent of the individuals in an average winter flock in Maryland, yet 93 percent of their interactions are intraspecific. Given that their level of niche overlap with other members of these flocks ranges from fifty percent to 80 percent, the frequency of intraspecific interactions is considerably higher than predicted by chance alone. Data from most other pairs of species in these flocks also appear to fit this pattern (Morse, 1970b).

On the other hand, the frequencies of hostile interactions between certain pairs of extremely similar species exceed the corresponding intraspecific frequencies. Ruby-crowned and golden-crowned kinglets (*Regulus calendula* and *R. satrapa*) and black-capped and boreal chickadees (*Parus atricapillus* and *P. hudsonicus*) serve as good examples (Morse, 1970b). Significantly, these species are largely allopatric or allotopic. Interspecific interactions between mountain (*P. gambelli*) and black-capped chickadees are also unexpectedly frequent (Minock, 1972), and a similar explanation probably applies. These cases suggest a parallel with the phenomenon of aggressive neglect described for some pairs of bird species (Ripley, 1959, 1961; Hutchinson and MacArthur, 1959), in which individuals spend much of their time attacking certain other ecologically similar species, at some point presumably to their detriment. In the situation described by Ripley, overlap in habitat between pairs of species appeared to be considerably greater than that among the members of mixed-species flocks; it seems unlikely that any of these cases is stable.

In certain situations the frequency of intraspecific interactions among spruce-woods warblers (*Dendroica* spp.) was greater than the

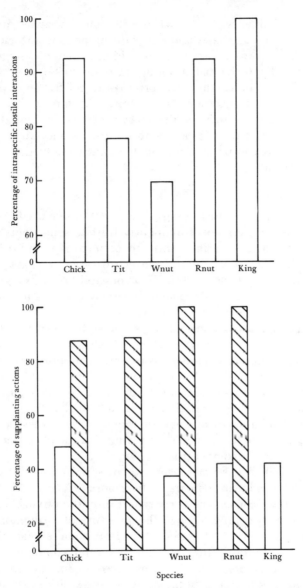

Figure 11.1 Hostile interactions in mixed-species foraging flocks. *Top,* percentage of intraspecific interactions in mixed flocks (versus interspecific interactions). *Bottom,* percentage of supplanting actions (versus fights and chases) in flocks. White bars = intraspecific interactions; hatched bars = interspecific interactions. Chick = Carolina chickadee; Tit = tufted titmouse; Wnut = white-breasted nuthatch; Rnut = red-breasted nuthatch; King = golden-crowned kinglet. (Modified from Morse, 1970b, copyright 1970 by the Ecological Society of America.)

interspecific frequency, but no clear differences occurred in the proportions of fights or chases and supplanting actions (Morse, 1976b). These were territorial breeding individuals spatially segregated to some degree by niche partitioning (Morse, 1968b). Myrberg and Thresher (1974) noted that territorial coral-reef fish initiated aggressive encounters with conspecifics at longer distances than they did with other species, which for the most part were potential competitors. Thus, in most cases responses to nonconspecifics qualitatively resemble those to conspecifics, although quantitatively they appear to be less important.

Displays

Displays, including songs, may act interspecifically. Male spruce-woods warblers not infrequently initiate song bouts with congeners as well as conspecifics, and the pattern of interspecific initiation fits with the corresponding pattern of attacks, chases, and supplanting actions (Morse, 1976b). The possibility that songs typically directed toward conspecific males have an important interspecific effect as well clearly deserves further study.

Two other species of wood warblers, the yellow warbler (*Dendroica petechia*) and chestnut-sided warbler (*D. pensylvanica*), sing somewhat similar songs in aggressive contexts, both interspecific and intraspecific (Morse, 1966). These species sometimes hold adjacent, exclusive territories. Thus songs may act as interspecific interference mechanisms, if only secondarily. Lein (1978) has argued that the birds should be able to distinguish songs of the two species. Their ability to do so would not necessarily diminish the effectiveness of the songs in interspecific competition.

Marler (1960) has reported convergent songs in several coexisting species. These may have interspecific competition functions, or, alternatively, their convergence may reflect only the physical constraints involved in broadcasting through the structure of a common environment (Chappuis, 1971; Morton, 1975); such an explanation is especially likely in birds that are not potential competitors. Similar environmental factors probably do not account for the similarities in the songs of yellow and chestnut-sided warblers, since interspecific interactions typically occur at distinct vegetational gradients that differ from the areas where intraspecific interactions are most frequent.

Spatial patterning

Interspecific competition very often leads to spatial segregation of the competing species, and such segregation tends to be more pronounced the more similar the species.

ECOLOGICAL ISOLATION

It is commonly believed that two ecologically similar, inter-acting populations are almost certain to diverge in a way that in-creases their ecological isolation (Lack, 1971). This divergence may be morphological, or it may involve changes in habitat selection or foraging area. Lack (1971) found that congeneric species of birds are segregated more often by feeding differences than by any other char-acteristic. Common sources of variation include differences in the size of body and beak and in diet; two or more such differences often occur simultaneously.

It now seems apparent that species are continually adjusting to each other; competition is not something that occurs only for a short period of time when two species first come into contact (Miller, 1967; Morse, 1974). Species may eventually become so different that inter-specific competition ceases to be an issue, but we need to investigate interactions that lead to such results, and to determine the degree to which sympatric species continue to interact even after considerable divergence.

Lack (1971) has offered voluminous evidence in support of eco-logical isolation. The bases for this isolation may differ markedly, de-pending on the nature of the resource at stake. When pairs of similar species first come into contact, their levels of overlap may be so great that coexistence is impossible, as in the kinglets and chickadees dis-cussed above, or interspecific territoriality may occur.

In some cases major shifts in habitat selection may provide isolation (see Hildén, 1965). For example, Johnson (1963, 1966) pointed out that two similar species of flycatchers (*Empidonax hammondi* and *E. wrightii*) do not use each other's habitats even when the oppor-tunity presents itself; they also use markedly different foraging pat-terns in spite of their morphological similarity. It is unclear why a second means of segregation is used, but it could be a secondary adaptation to a different type of habitat. It would be of interest to study the relationship between these two species or other pairs of morphologically similar species in several populations along their line of sympatry. Interspecific relationships may differ from place to place, presumably in response to both ecological and evolutionary variables that have not yet been carefully studied.

INTERSPECIFIC TERRITORIALITY

Interspecific territoriality has been recorded in several species pairs, mostly birds (reviews by Simmons, 1951; Orians and Will-son, 1964; Murray, 1971) and fish (Hartman, 1965; Myrberg and

Thresher, 1974). The functional significance and permanence of these relationships are matters of debate (see Cody, 1969, 1973, 1974; Cody and Brown, 1970). In birds the phenomenon has been recorded most often where low vertical relief occurs, with a correspondingly high relative frequency of reports from grassland and marsh birds. In such situations partitioning is two-dimensional. Interspecific defense also occurs among salmonids of shallow streams (Hartman, 1965) and fish on coral reefs (Myrberg and Thresher, 1974). Three-dimensional partitioning would be technically more difficult to document, which may account for its rarity in the literature. Interspecific territoriality may be a temporary solution to incompatibility where species have recently become sympatric. It is frequently argued that in time these species will achieve ecological isolation (see Orians and Willson, 1964).

Cody (1969, 1973) and Cody and Brown (1970) argued that character convergence has even occurred in some cases and that it facilitates interspecific territoriality by increasing the efficiency of resource partitioning. Character convergence as envisioned by Cody (1973, 1974) may take a variety of forms; color, vocalizations, and even morphology may be involved. Although it ought to be advantageous to the individuals involved to maximize efficiency, it is not clear exactly how selection would work to perpetuate such a system in most of the examples discussed by Cody. It certainly ought to be advantageous, to either species, to eliminate the other from the area, unless some undescribed mutual benefit occurs. Alternatively, convergence could represent parallel evolutionary responses to a common habitat. Cody's proposal has been attacked from both theoretical and empirical viewpoints (Gill, 1971; Kilham, 1972; Murray, 1976). More evidence is needed to establish the existence of such a phenomenon among potentially competitive species.

Perhaps the most plausible of Cody's examples is that in which convergence in signaling occurs between two species on opposite sides of a habitat gradient. Assume that species A is competitively superior in habitat 1 and species B in habitat 2, and that both species can survive in either habitat and will readily invade in the absence of the other. Vocalizations should succeed in decreasing the frequency of overt interactions, at least subsequent to the establishment of territories. Although species-specific vocalizations might provide additional information (that a particular species is involved), a song convergent with that of the congener might convey the essential message as effectively as a totally different song and would have the advantage of greater simplicity. Since repertoires of birds are distinctly limited (Mulligan, 1966; Moynihan, 1970), the economy involved may be substantial. Mexican sparrows studied by Cody and Brown (1970) may fit this model, although there appear to be alternative explanations (Gill, 1971).

Murray (1971) has advanced the argument that interspecific territoriality is not adaptive but is instead the result of mistaken identification. Three pairs of species are critical to Murray's argument: redwinged and tricolored blackbirds (*Agelaius phoeniceus* and *A. tricolor*) (Orians and Collier, 1963), sedge and reed warblers (*Acrocephalus schoenobaenus* and *A. scirpaceus*) (Brown and Davies, 1949; Catchpole, 1972, 1973), and Leconte's and sharp-tailed sparrows (*Passerherbulus caudatus* and *Ammospiza caudacuta*) (Murray, 1969). In these cases the persistently aggressive species is sometimes displaced from its territory, in spite of its actions. Murray believed that territory loss would inevitably occur or at least with an extremely high probability. However, the data are insufficient to allow a rigorous test of the argument in the case of the sparrows, and simpler explanations are possible for the blackbirds and warblers. Evidence is accumulating in support of the view that song discrimination among males is often quite precise (see Hinde, 1969), and that at the intraspecific level individual recognition clearly exists in some species (Stenger and Falls, 1959; Falls, 1969). Each of the three pairs of species has vocalizations clearly different from its counterpart. Although the members of each pair show considerable morphological similarity, distinct differences in coloration exist at least in the first two pairs.

Although Murray (1971) assumed that the aggressive behavior of the displaced species was not adaptive, redwinged blackbirds are not invariably usurped when tricolored blackbirds are present in small numbers (Orians and Collier, 1963). Catchpole (1972, 1973) brings up some serious questions about the function of the territory in reed and sedge warblers that must be answered before it is appropriate to consider the responses of these species to be nonadaptive in Murray's sense. Although reed warblers often usurp some space from previously established sedge warbler territories, Catchpole (1973) was unable to demonstrate that this reduced the success of nesting sedge warblers. The two species foraged in the same areas away from the nesting territories without strife. Although the habitats of these two species overlapped somewhat, the reed warbler mainly occupied low areas; the sedge warbler, uplands. Catchpole suggested that a function of territoriality in these species may be to ensure the widest possible spacing of nests, which would reduce predation pressure. If so, the interactions in question have to be evaluated in light of the nesting dispersion of the two species.

Murray's argument (1971) about the importance of mistaken identification may be valid where members of species pairs have only recently become sympatric (see, for example, Orians, 1961, on blackbirds). However, it seems less likely where habitat segregation usually occurs. Catchpole (1973) argued convincingly that reed-sedge warbler interactions are not a new phenomenon. Similarly, Murray indicated that contact among his sparrows is not new. Lyon, Crandall, and McKone (1977) found that male blue-throated hummingbirds (*Lampornis clemenciae*) responded differently to male and female conspecifics and other species of hummingbirds as a function of their location within a feeding territory (a set of feeders containing a 20 percent sucrose solution). When a feeding territory was small, all but conspecific females were ejected, but as the territory was progressively enlarged by adding more feeders, first black-chinned hummingbirds (*Archilochus alexandri*), then Rivoli's hummingbirds (*Eugenes fulgens*), then conspecific males were permitted to forage at the expanding outer perimeters. Although this is not a natural example, relations among hummingbirds observed in natural situations (Feinsinger, 1976) are consistent with it.

Abundant information from the literature on fish also indicates that the strong interspecific territoriality exhibited by many coral-reef fish is not the result of misidentification. Low (1971) found that the damselfish *Pomacentrus flavicauda* excluded only other substrate-feeding herbivorous fish from its territories, suggesting that the territories served to maintain an exclusive food source. Only carnivorous forms were not attacked by territory holders, which is consistent with this argument. Myrberg and Thresher (1974) demonstrated convincingly that interspecific territoriality in the threespot damselfish (*Eupomacentrus planifrons*) does not result from misidentification, although the more ecologically similar the antagonists, the stronger the responses (fig. 11.2); this pattern is similar to that seen in the hummingbirds studied by Leon and coworkers. Moran and Sale (1977) reported that the pomacentrid *Parma microlepis* responds most strongly to other species that use its living space in a way similar to itself and to egg predators such as wrasses. Although there is no correlation between levels of aggression on the part of *P. microlepis* and the similarity of diet, this species does respond to other species that might interfere with its breeding success. The series of graded responses is of particular interest in its own right, and as Moran and Sale emphasized, it indicates that territory holders have information about the relative likelihoods that different species will attack them. Moran and Sale assumed that this information is learned. If so, it is a most impressive result. Low (1971) reported that his damselfish responded to thirty-eight species and ignored sixteen. Thresher (1976)

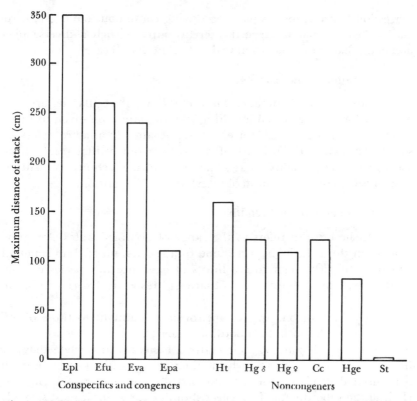

Figure 11.2   Maximum distances of attack by territorial damselfish (*Eupomacentrus planifrons*) against intruding male conspecifics, male congeners, and several species of intruding noncongeners. Data plotted are from January. Epl = *Eupomacentrus planifrons;* Efu = *E. fuscus;* Eva = *E. variabilis;* Epa = *E. partitus;* Ht = *Holacanthus tricolor;* Hg♂ = *Halichoeres garnoti* male; Hg♀ = *Halichoeres garnoti* female; Cc = *Chromis cyanea;* Hge = *Hypoplectrus gemma;* St = *Serranus tigrinus.* (Modified from Myrberg and Thresher, 1974.)

argued that interspecific recognition in threespot damselfish is based on characteristics of form, but that intraspecific recognition makes use of color also. The commonness of this behavior among coral-reef fish in undisturbed habitats suggests that interspecific territoriality is permanent, not a transient phase in the development of ecological isolation, as is suggested for some pairs of bird species by Orians and Willson (1964) and Rowlett (in Hertz, Remsen, and Zones, 1976). Although Murray indicated that his hypothesis is the first to account for a number of observed phenomena (the coupling of widely sympatric aggressive and nonaggressive species pairs, mutual territorial aggression between species pairs occupying contiguous habitats), it remains possible that interspecific territoriality is such a diverse phenomenon that one explanation will not suffice for all cases.

Dominance hierarchies

Interspecific dominance hierarchies are of regular occurrence, but they have not received anything like the attention given to intraspecific hierarchies (see Morse, 1974). Again, dominance refers to social dominance: the priority of access to resources that results from successful attacks, fights, chases, or supplanting actions, present or past. It is thus one mechanism of interference competition.

Changes in niche breadth

Niche breadth refers to the range of occupied states along one dimension through the hypervolume of a species' niche. Fundamental and realized niche refer to the limits of existence for species in noncompetitive and competitive situations, respectively (Hutchinson, 1957).

If individuals of one species are socially dominant to those of another species, the socially subordinate species will experience a relatively greater reduction of their realized niche breadth when they are in contact. The extent to which the dominant's realized niche breadth is reduced depends on the amount of time and energy it spends excluding the subordinate. This effect should express itself most strongly where dominants are rare relative to subordinates (Stoecker, 1972). In principle the subordinate species could avoid reducing its realized niche breadth when in contact with a dominant by shifting its exploitation patterns, but in most cases reductions rather than shifts occur. For example, certain members of mixed-species flocks of insectivorous birds forage more diversely when with dominants than when away from them (Morse, 1970b; R. Greenberg, personal communication). These species appear to be moving from a few preferred sites to a wider range of less preferred sites.

When members of one species are dominant in certain parts of an overlapping niche parameter and those of a second species are dominant in other parts of it, both species should reduce their realized niche breadths when together. Hartman (1965) described a case in which the dominance of two salmonids varied with the habitat in which they interacted; the result was an ecologically defined form of interspecific territoriality. If certain individuals of a first species are dominant to those of a second species and vice-versa, the effect will be to reduce the overlap between them and perhaps their niche breadth as well. Examples of this sort occur where size differences associated with indeterminate growth are involved, as in fish (Nilsson, 1955; Kalleberg, 1958) and lizards (Jenssen, 1973).

Hypothetically, niche dimensions might not change or shift in spite of hierarchical interactions among species. Calhoun (1963) suggested that the interspecific dominance hierarchies he reported among small mammals were of little relevance to the concept of competition, but he did not explore in detail any of the critical niche parameters. Subsequently, Myton (1974) argued that Calhoun's results may have been an artifact of the methods that he employed in gathering the data. I have found no case in which niche dimensions are reported not to change in the presence of interacting species. This is expected on the assumption that these hierarchical relationships have an adaptive basis for at least one of the species involved.

Consequences of Social Dominance for Community Structure

If competition consisted solely of interference, a socially subordinate species whose niche completely overlapped that of a socially dominant species in a resource-limited environment could coexist with that species indefinitely only if its fundamental niche breadth were larger than that of the dominant. Similar relationships might be predicted wherever overlap was high. MacArthur (1969, 1970, 1972), May and MacArthur (1972), and others have extensively studied the question of how much overlap is permissible under these circumstances (limiting similarity). It appears that guilds, groups of species using the same class of environmental resources in a similar way (Root, 1967), should be comprised of species with fundamental and realized niches of different sizes. Niche relationships within the spruce-woods warblers roughly fit the expected pattern (Morse, 1971b).

Given the postulated relationship between niche breadth and dominance, an inverse correlation should exist between plasticity and social dominance. This occurs in certain mixed-species groups of birds (Willis, 1966b; Morse, 1970b). For example, tufted titmice (*Parus bicolor*), the dominant species in these flocks, show little tendency to shift their patterns in response to the other species present. Golden-crowned kinglets (*Regulus satrapa*), the lowest ranking species, regularly shift in response to the titmice and to Carolina chickadees. The chickadees, which occupy an intermediate rank, regularly shift in response to titmice but not to kinglets.

CHARACTERISTICS OF DOMINANTS AND SUBORDINANTS

Dominant species are larger than subordinate species in most cases (Morse, 1974), but several lines of evidence suggest that large animals often have greater niche breadths than small ones. They often use a greater range of food sizes than small animals (Schoener, 1969a, 1971; D. S. Wilson, 1975), they often use larger areas (territories or home ranges) (McNab, 1963), and they often use a greater variety of habitats (Hutchinson, 1959). The fact that the dominance relationships may more than balance these tendencies indicates their importance. As is the case at an intraspecific level, reproductive condition may also influence rank at the interspecific level (Bailey and Batt, 1974).

Social subordinates may minimize the impact of dominants by exploiting resources before the latter can get to them. Field experiments using artificial baits have demonstrated that certain species of stingless bees (Johnson and Hubbell, 1974) and ants (Levins, cited in Wilson, 1971b) typically arrive at these resources before species dominant to them do. In this sense they are acting as fugitive species (Hutchinson, 1951), and if this is all that there is to it, they are dependent on this for their coexistence. A subordinate species not overlapping completely with a dominant might do rather well as a result of this advantage, together with its refuge in the area of nonoverlap. Willis and Oniki (1978) noted another tactic used by subordinates. *Pithys albifrons,* a small Amazonian antbird (Formicariidae), frequently darts into the center of the multispecific swarms of birds that habitually follow some species of army ants. Generally it retreats with a flushed insect before the larger, dominant species of antbirds frequenting the center of the throng can attack it. Werner and Hall (1977) suggested that the ecological displacement of bluegill sunfish (*Lepomis macrochirus*) by green sunfish (*L. cyanellus*) in areas where they coexist results partly from the more aggressive tendencies of the green sunfish. They proposed that this is a consequence of the sites typically inhabited by green sunfish, which are shallow and highly structured and, consequently, may be effectively defended. The bluegill, on the other hand, is typically found in more open water, where resources are widely scattered and the environment less structured, so that defense is not practical. The dominance relationship seen thus appears to be a consequence of the adaptations of the two species for these different, but adjacent, habitats.

The existence of a subordinate is uncertain if dominants preempt large parts of its fundamental niche. In this sense behavioral dominance may contribute to environmental predictability in the sense of Levins (1968). Large niches and a high level of plasticity are frequent outcomes of living in unpredictable environments. The first trait, large niches, is characteristic of most subordinates; the second, high plasticity, has only been documented occasionally to date (Morse, 1974).

RESTRAINTS ON DOMINANTS

If the niche breadths of social dominants are not set by their own competitors (at least not by direct interference), what factors do limit them? It may be that the very characteristics required for dominance can be maintained only at the expense of the ability to excel in other ways. Physical factors (temperature, substrate, moisture, oxygen concentration) have been clearly shown to limit dominants' niche breadths in at least one direction in, for example, chipmunks (Heller, 1971), voles (Stoecker, 1972), and crayfish (Bovbjerg, 1970). In other cases no evidence exists to suggest that physical factors directly limit niche breadth. Biological factors other than interference could be responsible, although few relevant data exist on this subject.

If resources are exploited similarly by more than one species, large (and hence dominant) species may be unable to exploit resources effectively when those resources are in short supply, although smaller subordinate species may still be able to use them. Gill and Wolf (1978) observed that when the average density of nectar rewards is low at some species of flowers, it may take large species of African sunbirds (*Nectarinia* spp.) twice as long to gather adequate amounts of this food as it does small species. Thus, *N. venusta* (7.5 g) may be able to maintain an energetic gain at patches of flowers of half the density required by *N. reichenowi* (15.5 g) (fig. 11.3). Size facilitates priority of access through interference, but it precludes options that are available to the smaller subordinate.

Competition (either interference or exploitation) of a diffuse nature may also occur. Diffuse competition is defined as the total competitive effect of all interspecific competitors (MacArthur, 1972), and the term is usually applied specifically to forms that do not overlap greatly. However, the effects of several such impinging species may be substantial (see Yeaton and Cody, 1974; Morse, 1977a). Numerous suggestions of diffuse competition appear in the literature, even between such distantly related (but ecologically convergent) forms as lizards and birds (Pianka, 1971), bats and birds (Morse, 1975a), reptiles and mammals (Kiester, 1971), hummingbirds and insects (Snow and Snow, 1972), and fish and squids (Packard, 1972). Although in-

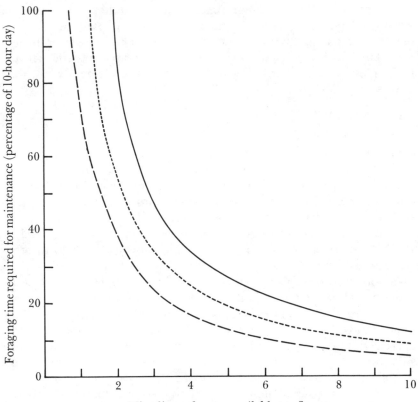

Figure 11.3 Calculated foraging time budgets required by sunbirds feeding at flowers of the mint, *Leonotis nepetifolia,* to maintain balanced daily energy budgets. The curves for the three species reflect the interactions of differences in average times per flower, average amounts of nectar removed from a flower, and total daily energy requirements. Solid line = *Nectarinia venusta* (7.5 g); short dashes = *N. famosa* (13.8 g); long dashes = *N. reichenowi* (15.5 g). (Redrawn from Gill and Wolf, 1978.)

terference and exploitation are typically discussed as alternatives, it is highly unlikely that one will operate to the total exclusion of the other (Miller, 1967). Brown (1971b) found that although one species of chipmunk could exclude a second from open forests, it was unable to do so in dense forests, because the socially subordinate species could easily escape by virtue of its greater agility in the trees. Such relationships could have marked effects where the subordinate species is much commoner than the dominant.

Alternatively, predators could exert a limiting effect on niche breadth by capturing a disproportionately high number of individuals that used resources in vulnerable areas or by requiring so much surveillance on the part of the potential prey species that they could not profitably exploit these resources (see Smythe, 1970b; Murton, Isaacson, and Westwood, 1971).

AVOIDANCE OF DOMINANTS

Overt interactions may not give a completely accurate impression of the importance of a given species in determining the niche dimensions of a subordinate. The mere presence of a dominant may often produce changes in the patterns of niche use of a subordinate. Interactions among members of mixed-species foraging flocks of small insectivorous birds are particularly intense in the fall when flocks are forming, but then decline greatly, presumably because members of the group avoid one another subsequent to forming a hierarchy (Morse, 1970b). Similarly, interactions among thrushes (*Hylocichla* spp.) are very frequent at the time that individuals first come in contact with one another on their breeding grounds, but then decline in frequency (Morse, 1971c). Cody (1973) argued that avoidance is a regular phenomenon in communities. For example, avoidance is the major response of a subordinate species of bumblebee (*Bombus ternarius*) when in close contact with a dominant species (*B. terricola*) (fig. 11.4), if individual food rewards are small. Foraging rates of the subordinate are also about 8.5 percent slower than when it is alone (Morse, 1977b, c).

DISTRIBUTION OF DOMINANCE-MEDIATED RELATIONSHIPS

Dominance-mediated interspecific relationships of the sort discussed in this chapter have been reported for mammals, birds, lizards (Jenssen, 1973), salamanders (Thurow, 1976), fish, starfish (Menge and Menge, 1974), insects, crustaceans, spiders (Enders, 1974), and limpets (Branch, 1976). The frequency of examples in groups that have been studied intensively, plus the recent reports for lizards, salamanders, starfish, spiders, and limpets, suggest that the phenomenon may be more widespread than currently appreciated. The existence of interactions bearing some characteristics of dominance-mediated niche partitioning in groups as widely disparate as corals (Lang, 1971; Jackson and Buss, 1975) and sponges (Jackson and Buss, 1975) suggests that this phenomenon may occur in other groups as well. However, it seems unlikely that mobile forms without intra-

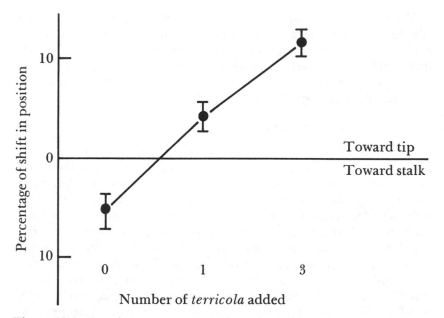

Figure 11.4 Foraging patterns on goldenrod (*Solidago juncea*) of a small species of bumblebee, *Bombus ternarius,* in the absence and presence of a large bumblebee, *B. terricola.* Individual flower heads of goldenrod are borne along several branches, each of which may reach a few centimeters in length. When foraging by themselves, *ternarius* tend to visit flower heads on the proximal, medial, and distal parts of these branches with similar frequency if the inflorescences are at their flowering peak; large *terricola* workers visit distal flower heads with much lower frequency than the other parts. Individual *ternarius* workers were placed in a 1-cu-m screened arena and allowed to acclimate. The location (proximal, medial, distal) of the next fifty flower heads visited was then recorded. Subsequently, 0 (control), 1, or 3 large *terricola* workers were introduced, and all bees present were again allowed to acclimate. A second similar set of data was then gathered. The data sets (approximately ten replicates for each) for each treatment were compared. 0 position equals no change between initial runs and treatments. Cumulative data points are shown, ±1 SE. *Ternarius* workers forage progressively more distally on goldenrod inflorescences as increasing numbers of *terricola* workers are added. No overt interactions were observed between the two species; therefore, the results may be attributed to avoidance of *terricola* by *ternarius.* (Modified from Morse, 1977c.)

specific aggressive behavioral patterns will demonstrate such behavior (see Miller, 1967) because interspecific patterns in forms studied to date resemble intraspecific patterns.

Interspecific dominance relationships also occur in a wide range of social situations, including both strict territorial situations and ones in which no stationary area is defended. Many interspecific groupings, including social groups, exhibit interspecific dominance relationships. Mixed-species social groups of this sort have been reported in birds (Moynihan, 1962; Morse, 1970b; Buskirk, 1972; Krebs, 1973), mammals (Lamprey, 1963; Keast, 1965b; Gartlan and Struhsaker, 1972), and fish (Breder, 1959; Hobson, 1968; Karst, 1968). They are discussed in chapter 12.

Synthesis

If interspecific dominance relationships involve ecologically important species, they could exert marked effects on the overall forms of communities. Intraspecific interactions should increase the breadth of resource use by a population, pushing it toward the dimensions of the fundamental niche even if niche dimensions of individual members decrease. Interspecific competitive interactions, on the other hand, should reduce the niche dimensions of the populations as a whole, at least of those species that lose in such encounters, and should have a similar effect on the patterns of use of individuals as well (see Svärdson, 1949). Thus the population effects of the two types of pressures are opposite, although the effects at the level of the individual may be similar. In an evolutionary sense the outcome of the two pressures will differ. The interspecific interactions will drive the two or more populations onto distinctly different adaptive peaks (regions of specialization) or different sides of the same peak, whereas intraspecific pressures will result in selection for either generalist individuals or a variety of specialists that occupy a continuum of different peaks effectively forming a ridge. Thus, intraspecifically a continuous divergence will occur, whereas a discontinuity or at least a sharp break will exist at the interspecific level.

The available evidence suggests that interspecific behavioral patterns used in competition for resources typically resemble intraspecific patterns. Thus, either the intraspecific behavioral patterns work well in interspecific situations, or interspecific interactions are relatively unimportant. There is also the possibility that interspecific interactions are the consequence of mistaken identifications. But even assuming that animals have powers of discrimination adequate to distinguish between similar and strikingly different species, mistaken identifications may not be maladaptive, because they involve individuals similar enough to the initiator to be likely competitors. Considerable evidence suggests that in many instances mistaken identification is not involved, but regardless of the basis for an attack, it may have essentially the same effect.

Even if interspecific interactions are less important to an individual than intraspecific interactions, a point generally agreed on by most workers since Darwin (1859), they may nonetheless be in large part responsible for the relative abundance and distribution of animals in many communities. Such interactions have pervasive long-term effects and may profoundly affect the evolution of a species' characteristics and the roles it plays in a community.

12 *Social Groups*

MANY ANIMALS spend much or all their time in close contact with others. Reviews of the factors associated with social and solitary existences appear in Crook (1965) and Lack (1968) (birds); Eisenberg (1966), Crook (1970a), Eisenberg, Muckenhirn, and Rudran (1972), and Jarman (1974) (mammals); and Morse (1977d) (general). Here I will concentrate on the advantages of living in social groups.

Types of groupings

At least three major factors could account for concentrations of animals. First, individuals may be brought together by physical factors beyond their control, as in planktonic animals, such as moon jellies (*Aurelia aurita*), collected by tidal movements. Second, some parts of an environment may be more satisfactory than others. Concentrations of animals about water holes and fruiting trees may be no more than the consequences of such environmental patchiness, as may the concentrations of birds feeding on insects flushed by army ants (Willis and Oniki, 1978). These two types of groupings may be referred to as aggregations, differing primarily in that one involves passive and the other active movement. Although aggregations pose interesting questions in their own right, I am mainly concerned with social groups, which reflect an intrinsic gregariousness. Intrinsic gregariousness has evolved to provide such concrete advantages (Moynihan, 1962) as the procurement of food through pack hunting, which would be unavailable to lone individuals. Flocks, herds, troops, packs, and schools are specific types of social groups in the sense that I use the term. In practice it may not be easy to distinguish social groups (in which individuals derive benefits by virtue of their presence with others) from aggregations. For example, social groups may use patchy resources. However, members of a social group should remain together when they are not at a resource concentration.

Most studies of the ecological benefits of sociality have been carried out on birds, mammals, and fish. I will therefore discuss mainly these groups, and in particular the birds with which I have worked extensively, but I will attempt to relate them to mammals and fish wherever appropriate. There are some striking similarities between the organization of fish schools and the organization of avian and mammalian groups (Radakov, 1972). Although schools are usually monospecific, multispecific groupings are known (Breder, 1959). Some species such as the jack mackerel (*Trachurus symmetricus*) (Hunter, 1968), appear to remain permanently schooled, at least above certain light thresholds; in other species schooling is clearly facultative. The members of a school tend to be similar in size, although this is by no means always the case (Radakov, 1972). Schools typically become more dense when attacked and tend to break up when individuals are feeding. They generally disperse at night and form again in the morning. Most, but not all (see Sale, 1972; Itzkowitz, 1974), schools lack hierarchical structure, in contrast to most flocks and herds. Radakov (1972) believed that the absence of hierarchies is a unique characteristic of schools, but some extremely large groupings such as herds of wildebeest (*Connochaetes taurinus*) on the move (Estes, 1969) or groups of New World blackbirds (Icteridae) or queleas (*Quelea quelea*) may be similar in this regard. For example, when a certain flock size is exceeded in domestic fowl (*Gallus domesticus*), an effective hierarchy breaks down (Guhl, 1962). Schools of squid show most of the behavioral patterns that characterize fish schools (Hurley, 1978).

Foraging groups

Foraging groups are common in birds, mammals, and fish. Often their members have somewhat similar diets (Morse, 1970b, 1977d). Many such groups remain continually on the move. Foraging groups are often only loosely clumped, with substantial individual distances. Many have social dominance hierarchies, which may markedly affect where individuals forage (Morse, 1967b, 1970b, 1978).

Groups vary in size, permanence, patterns of activity, and probably in function. Both single-species and multi-species groups occur commonly. Until recently few quantitative data have been available concerning their adaptive significance. Even fewer experimental data exist, and because of the difficulty of approximating natural conditions in experimental work, their relevance to natural groupings remains somewhat in doubt. Apart from possible reproductive advantages, which are discussed in chapter 7, two major advantages have been proposed: (1) obtaining maximum amounts of resources (typically food), and (2) gaining protection from predators. In many or even most cases, social groups probably deliver more than a single benefit, and attempts to attribute only one advantage to them are open to criticism (see Shaw, 1970; Lazarus, 1972; Radakov, 1972).

ENERGETIC ADVANTAGES

There are several possible energetic advantages to foraging in groups, and some of these may complement each other.

Beating. Swynnerton (1915) and others have suggested that flocks of small insectivorous birds benefit their members by flushing insects as they move. However, it has never been demonstrated that an individual flushes or catches more insects when in a flock than when alone. Unless the total number of insects captured increases in a flock, flocking per se would not bestow this advantage. If some individuals capture insects flushed by others, this would clearly be advantageous to them; but unless the other individuals obtain some compensating advantage there is no reason why they should participate in groups. In the mixed-species assemblages studied by Swynnerton (1915) and Winterbottom (1943, 1945), flycatching birds such as drongos (*Dicrurus* spp.), which never led the groups, appeared to obtain advantages through beating. Probably the association of wood pewees (*Contopus virens*), a flycatching species, with late-summer flocks in eastern North America has a similar basis (Morse, 1977d). The relationship of the drongos and pewees probably is that of commensals to the other flock members. Drongos may inadvertently reciprocate to some degree, in that they are extremely aggressive and do not hesitate to chase hawks or other potential avian predators. However, Swynnerton's and Winterbottom's descriptions make it quite clear that the drongos follow the flocks, and not vice versa, suggesting that the other species are not actively seeking out the drongos in order to benefit from any protection they may provide.

Even if some or all members of a flock obtain benefits from beating, this advantage will not exist when insects are inactive, as is usually the case at high latitudes during winter, when flocking is most pronounced. Clearly it is necessary to look beyond this hypothesis to account for the flocking tendencies of most species, and probably of all groups of insectivorous birds.

The group foraging behavior of herons, in which individuals advance through the vegetation several abreast, chasing and capturing the insects they flush, probably does genuinely qualify as cooperative beating, rather than merely as a commensal relationship. Wiese and Crawford (1974) reported cattle egrets (*Bubulcus ibis*) and a few other herons performing this behavior in the absence of cattle, which the egrets often tend for the insects that they flush. Wiese and Crawford provided no data on the relative success rates of group and individual foraging, but Dinsmore (1973) showed that cattle egrets are more than three times as efficient in capturing prey when attending cattle as they are when feeding alone (table 12.1), which strongly suggests that a similar advantage is obtained when the birds beat in a group for themselves. Beating flocks of cormorants (*Phalacrocorax auritus*) (Bartholomew, 1942) and mergansers (*Mergus* spp.) (Emlen and Ambrose, 1970) have also been noted; they drive fish and appear to obtain a feeding advantage.

Group hunting. Group hunting is highly developed in some mammals, particularly in certain canids (Kleiman and Eisenberg, 1973), permitting them to obtain prey that they could not obtain alone. Wolves (*Canis lupus*) regularly prey on animals as large as moose (*Alces alces*) (Mech, 1966), and African hunting dogs (*Lycaon pictus*) take wildebeest and zebra (*Equus burchelli*) (Estes and Goddard, 1967; van Lawick and van Lawick-Goodall, 1971). In particular, cooperative hunting techniques may allow animals to run down prey that as individuals they could not even catch, much less kill; this aspect of group hunting is highly developed in African hunting dogs (van Lawick and van Lawick-Goodall, 1971). Groups of porpoises may herd and attack schools of fish in a similar manner (Fink, 1959). The level of coordination seen in all these groups greatly surpasses that seen in any bird flock. Lions (*Panthera leo*) belonging to a pride may share captured prey (if sometimes reluctantly), although they do not hunt in the cooperative manner characteristic of canids (Schaller, 1972).

Some fish also profit from pack hunting in a way analogous to that

Table 12.1. Foraging efficiency of cattle egrets with cows and alone as determined by paired, two-minute observation periods. (From Dinsmore, 1973.)

Cattle Egrets Foraging	Minutes Watched	Captures	Steps Taken	Step Per Capture	Captures Per Minute	Efficiency Ratio
With cows	144	374	5,279	14.1	2.6	5.4
Alone	144	229	7,144	13.2	1.6	19.6

of wolves. Because schooling usually confers a large degree of protection from predators (Breder, 1959; Neill and Cullen, 1974), it is ironic that some predators respond by hunting in groups. Pack hunters usually attempt to prevent the schools of prey species from reaching safety, or restrict them in other ways. Prey are often driven into a confined space, such as a cove (Eibl-Eibesfeldt, 1962), or to the surface (Eibl-Eibesfeldt, 1962; Radakov, 1972). While the group is being herded, stragglers are picked off, although sometimes predators may rush into the midst of schools (Eibl-Eibesfeldt, 1962). Jacks (*Caranx ignobilis*) often form opportunistic groups when attacking schools of anchovies (*Stolephorus purpureus*) (Major, 1978). The groups appear to form when several individuals follow a leading individual that initiates an attack, perhaps on a somewhat isolated anchovy. The attacks break up the schools of prey, permitting improved success by all. When the prey are confronted by a single jack they typically break up, parting in front of the predator to reassemble immediately behind it. Additional predators prevent the prey from performing this maneuver smoothly; the stream of prey is broken up before it can reassemble into a school, forcing it to repeat the evasive maneuver a second time and greatly increasing the probability that individuals will be separated from the main body of the school.

Certain predatory fish such as swordfish and sharks (Bullis, 1961) may attack a school, maiming many fish on which they subsequently feed. In some instances a substantial fraction of the school may be killed. Obviously, schooling would be disadvantageous if such attacks were frequent. There is no clear evidence that predators actively cooperate in these school hunts, but neither is there such evidence for most flocks or herds.

Another advantage can be seen in both single-species and multi-species schools of coral-reef fish, which may intrude en masse into the territories of highly territorial lavender tangs (*Acanthurus nigrofuscus*) (Barlow, 1974b) or damselfish (*Pomacentridae*) (Itzkowitz, 1974; Alevizon, 1976; Robertson et al., 1976). The territory holder attempts to drive off the intruders, but some schools are so large that while one individual is being driven off, others enter to feed (usually on benthic algae) within the territory.

Minimizing duplication of effort. Several authors have suggested that foraging in flocks may minimize duplication of effort, in that individuals will tend not to feed where others are foraging or have just foraged (Short, 1961; Cody, 1971). Not only would individuals be able to see where others had foraged, but by maintaining a customary individual distance they could reduce overlap. Complications may occur when food is strongly clumped or the quality of food items differs greatly. Where food is clumped the mere presence of an individual may stimulate others to gather at the source, but if food items are small and widely spaced, gathering would not be profitable. Unfortunately the field data are inadequate to test critically whether birds in flocks avoid areas in which others have fed; it would be difficult to monitor substantial numbers of foraging sites during the time a flock moved through an area to determine whether they were used more than once. Many flocks (for example, tit flocks) forage over the same general area day after day, compounding the problems of measurement. Maintaining spatial patterns that minimize duplication of effort might be advantageous to flocks that typically forage in an area only once, such as flocks of nomadic species. It may also be efficient where resources are replenished rapidly.

A few relevant data do exist. Flocks often move forward on a rather broad front, thus achieving considerably more lateral spacing than they would if they moved in single file. Tit flocks are often roughly elliptical in shape; however, their length usually considerably exceeds their breadth (Morse, 1970b). If nonduplication is of major importance in this case, the ellipse should have its axes reversed. However, followers may be able to see which sites individuals at the front of the flock are using and thus avoid these areas, if the distance between the front and rear of a flock is not excessive. Problems would increase rapidly with larger or more diffuse flocks. There does appear to be a maximum size to tit flocks, and this differs with the area studied (Morse, 1970b). Seldom does a flock reach a size of more than forty-five individuals, and the largest flocks are often more compact than the smaller ones (Morse, 1970b) (fig. 12.1). It is not known whether large and small flocks differ markedly in shape. Although these data agree with the nonduplication hypothesis, much larger flocks, such as those of nomadic wintering finches, seem unlikely to do so.

Coatis (*Nasua nasua*) that feed in bands often forage in lines moving along a front (Kaufmann, 1962) in a way that should automatically reduce their duplication of effort. The foraging of peccaries (*Tayassu* spp.) (see Smythe, 1970b) might accomplish a similar result. Such species probably provide better opportunities for assessing the duplication-of-effort hypothesis than the bird flocks for which it was originally proposed.

Cody (1971) suggested that his desert finch flocks avoided duplication by observing evidence of their own previous activity. There is some suggestive evidence, but the possibility is untested. Gibb (1958) showed that tits preying on larvae of the lepidopteran *Ernamoria conicolana*, which dwell in pine cones, leave marks that are conspicu-

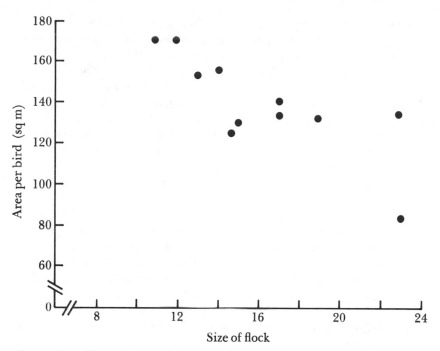

Figure 12.1 Mean area available to a flock member at a given instant as a function of flock size. The data were obtained from mixed-species foraging flocks composed primarily of chickadees, titmice, and kinglets during midwinter in a deciduous forest in Maryland. (Modified from Morse, 1970b, copyright 1970 by the Ecological Society of America.)

ous to a human observer. Despite the ease with which humans note these marks, it is not known whether birds recognize them in the field. The exploitation of most food items seems unlikely to leave such apparently obvious reminders as does the predation on *E. conicolana*. If abundant clues are available, participating in a flock is of little advantage in this regard. It seems most likely that flocking is important as a means of reducing duplication in cases where clues are *not* left.

Some tropical flocks appear to retrace their routes regularly (even daily) (Belt, 1874; Stanford, 1947), whereas arctic flocks seldom do so during winter. However, it is not known whether individuals use the same foraging stations on subsequent trips. Although many reports imply habitual movement patterns through a homogeneous habitat, MacDonald and Henderson (1977) found that the regular flock movements they noted in Kashmir clearly resulted from preferences for certain types of vegetation, a characteristic also noted in North American flocks by Morse (1970b). In both cases flocks favor dense brushy growth bordering streams. In a relatively homogeneous tropical forest, Buskirk (1972) found no tendency for flocks to retrace their routes. More fragmentary data suggest that tit flocks in the arctic forage over the same area only a few times during a winter; this conclusion is based on observations of individuals within a given area no more frequently than once every several days (Meinertzhagen, 1938; Gruzdev, 1952; Snow, 1952). Insect food in tropical areas should be constantly renewing itself, but it should be essentially nonrenewing to predators in high-latitude areas except at times when storms knock bark or limbs off trees. If these fragmentary reports are representative, they do seem to imply that one advantage of flocking is that it reduces duplication of effort.

Cody (1971) hypothesized that the movement patterns of his finch flocks minimized the frequency with which previous paths were crossed or retraced. He assumed that his flocks had a "reflecting boundary," that is, that when they reached the edge of their range their next movement was backward from the boundary, rather than across it or along its edge. The birds would retrack their approaching route only if the approach happened to be perpendicular to the boundary. The actual paths taken by these birds closely approximated the model. This model would probably work better in a situation where the boundary is set by some physical characteristic than it would where a flock's area bounded that of another flock (chapter 3). When groups meet there may be frequent movements along the periphery. Chickadee and titmouse flocks patrol their boundaries regularly even in the absence of contesting parties (Morse, 1970b).

The lack of critical field data remains a primary difficulty for the avoidance-of-duplication hypothesis. In the laboratory, Smith and Dawkins (1971) and Smith and Sweatman (1974) showed that great tits (*Parus major*) can concentrate their foraging in areas of high food density, which might improve the ability of individuals to avoid areas of past feeding.

Facilitation of food finding. Under some conditions individuals may enhance their feeding rates by observing the foraging of others and joining them at the sites at which they find food or by seeking similar sites (Murton, 1971a; Krebs, MacRoberts, and Cullen, 1972; Krebs, 1973). Under laboratory conditions, where hidden clumps of food were available, great tits in groups of four obtained more food than single birds or groups of two, and groups of two gathered somewhat more food than single birds (Krebs et al., 1972) (fig. 12.2). Individuals showed a strong tendency to join others who had found a food item and then to forage in that area or in a similar area nearby.

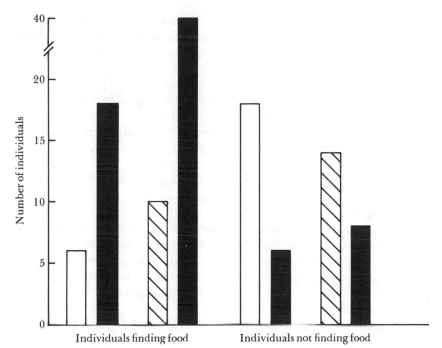

Figure 12.2 Success of great tits in finding food over a fifteen-minute period in groups of four versus single birds and in groups of four versus groups of two. White bars = single birds; hatched bars = groups of two; black bars = groups of four. (Modified from Krebs et al., 1972, with permission of the *Ibis,* journal of the British Ornithologists' Union.)

This process will work to the advantage of individual birds only if food is clumped, and a single patch must regularly contain enough for more than one individual. Subordinates may be parasitized to some degree under these conditions (see Baker, 1978). Even a socially dominant joiner would find this habit unrewarding if the food item were largely or totally eaten by the time it arrived. Information on the spacing of insect food under natural conditions is very scanty, but there are indications that insects are usually clumped (Gibb, 1960; Taylor, 1961). There is relatively good information on the spatial patterning of seed crops.

One might ask why some groups continually move even when food remains available in an area. The items taken by such groups are for the most part cryptic; finding several items at one site probably predicts subsequent success, but there may be advantages to taking only the most easily found or profitable items. This argument is consistent with those of Krebs, Ryan, and Charnov (1974) and Charnov (1976b), who argued that when an individual has not found a new item for a particular length of time it ought to move on. The constant motion of these groups may also serve a patrolling function in relatively productive and at least partially defensible areas.

English blue tits (*Parus caeruleus*) maintain extremely small individual distances, and they show a strong tendency to bunch together on a single foraging site. In addition to permitting small individual distances, they frequently forage simultaneously at similar sites. Blue tits often exhibit a rather conspicuous "peering" behavior, in which they appear to be observing other individuals several meters away, following which they frequently join these individuals (Morse, 1978c). This behavior accords very well with the observational learning hypothesis advanced by Krebs et al. (1972).

North American chickadees, titmice, and kinglets show relatively large individual distances and a relatively low frequency of hostile behavior (Morse, 1970b); this suggests a rather small role for the observational learning system hypothesized by Krebs and coworkers. Krebs (1973) did, however, demonstrate this phenomenon in two North American chickadees in an experimental setup similar to the one used for great tits. Thompson, Vertinsky, and Krebs (1974) suggested that rather than maximizing the average amount of food gathered, flocking may minimize the probability that an individual will go without food for a significantly long period (fig. 12.3). Where metabolic demands are high and food rather scarce, as at high latitudes during the winter, such an advantage might be great. Most small species of birds do flock at this season; under severe conditions their energy reserves may last no more than a few hours. Among some taxonomic groups (for example, woodpeckers), the tendency to participate in flocks is inversely related to body size (Morse, 1970b). On a given day when climatic conditions are of intermediate severity, individuals of large species of a taxonomic group show a lower tendency to participate in flocks than do small species in the same taxon— for example, tufted titmouse (*Parus bicolor*) versus Carolina chickadee (*Parus carolinensis*); great tit versus blue tit (Morse, 1970b, 1978).

There is some evidence that members of schools may increase their feeding efficiency by observing the feeding activities of other school members. Radakov (1958) has demonstrated experimentally that members of pollock (*Polachia virens*) schools are attracted to other fish that have found food, even when they clearly are unable to see this food themselves. The observational learning hypothesis seems most attractive where a highly contagious prey distribution exists and where resources are shifting in time and space, as they would be during the winter at high latitudes. However, more information is needed on

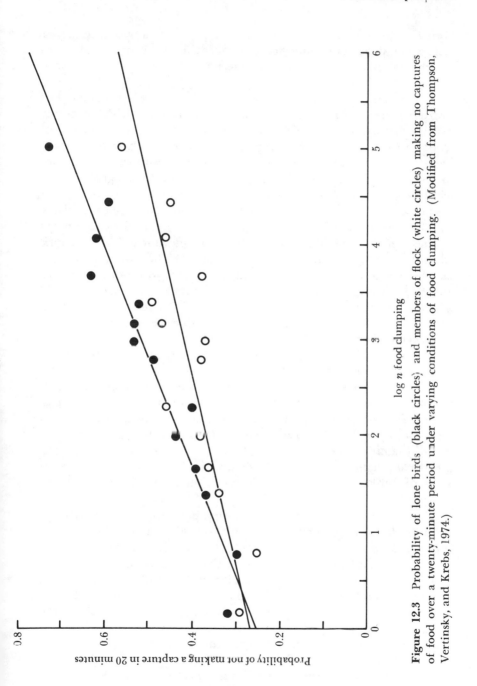

Figure 12.3 Probability of lone birds (black circles) and members of flock (white circles) making no captures of food over a twenty-minute period under varying conditions of food clumping. (Modified from Thompson, Vertinsky, and Krebs, 1974.)

both the distribution of food supplies and on their mobility. These factors could be responsible for differences in the behavior of flocks in England and in eastern North America.

Breaking the way. Some workers have stressed the marked hydrodynamic advantages that may be associated with schooling (Weihs, 1973; Breder, 1976), which could allow the members of schools to save considerable energy (see chapter 5). Water has a much higher viscosity than air, and therefore offers considerable resistance to objects moving through it. Many schools adopt a wedge-shaped or teardrop-shaped form when in motion, which appears to be an efficient design. When it slows down to feed, the school will often lose shape (Tikhonov, in Radakov, 1972). Schools also tend to be composed of like-sized individuals, and this leads to additional hydrodynamic advantages (Breder, 1965; Radakov, 1972). The size relationship may be related to swimming speed in menhaden (*Brevoortia tyrannus* and *B. patronus*) because injured or parasitized individuals typically school with smaller conspecifics (Guthrie and Kroger, 1974). Hergenrader and Hasler (1968) noted a considerable difference in rates of movement of single and schooling yellow perch (*Perca flavescens*) in extremely cold water (which is maximally dense).

Schools may have many other functions, however, and some of them may be of primary importance in certain cases. For example, schools of horse mackerel (*Trachurus trachurus*) swimming at different speeds, resting, feeding, or being attacked by predators have different shapes (Tikhonov, in Radakov, 1972), which do not always maximize their hydrodynamic advantage. Although individuals of similar sizes frequently occur in schools, this is not invariably the case. Schools composed of similar individuals may simply be members of the same hatch.

Additional evidence. Species often participate most frequently in winter flocks when their energy demands are highest, such as during times of heavy snow or extreme cold (Johnston, 1942; Morse, 1970b; Cody, 1971). In late winter individual tits may drop out of flocks and occupy their breeding territories during mild weather, but a return to wintery conditions will find them back in their groups (Hinde, 1952; Morse, 1970b). Because ambient temperature greatly affects the energy demand placed on individuals, the increase in flocking suggests that energetic considerations are important.

In studies on tropical flocks, both Fogden (1972) and Croxall (1976) found that during the periods when insect availability was lowest, flocking reached the highest frequency of the year. Davies

(1976b) was able to demonstrate that pied wagtails (*Motacilla alba*) in the winter shifted freely between flocks and territories, depending on the feeding conditions in the territories (fig. 12.4).

Flocking may decrease markedly when food is superabundant. Brown-headed nuthatches (*Sitta pusilla*) reduced their participation in flocks at the same time that a bumper crop of pine seeds became available, and their participation in flocks increased to its former level when this temporarily superabundant resource was exhausted (Morse, 1967b) (fig. 12.5). Highly concentrated resources may attract some normally gregarious species such as tits into even larger groups. Gibb (1954) noted that during years of heavy beech mast production great tits may be found in groups up to one hundred or more. But these

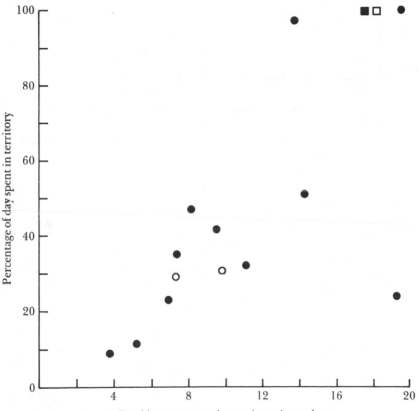

Figure 12.4 Relationship between rate of feeding and proportion of day spent in territory by pied wagtails during midwinter. Different symbols represent different individuals. (Modified from Davies, 1976b.)

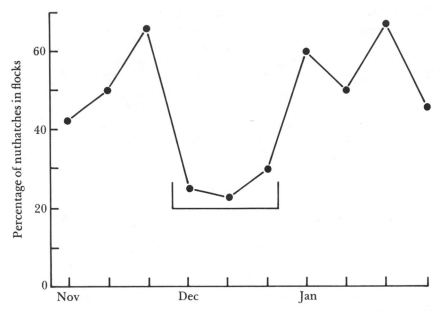

Figure 12.5 Proportion of brown-headed nuthatches participating in mixed-species foraging flocks before, during, and after a superabundant source of longleaf pine seeds was available. (Modified from Morse, 1967b, copyright 1967 by the Ecological Society of America.)

groups qualify more as aggregations drawn together by the configuration of the resource than as flocks.

DEFENSE AGAINST PREDATORS

Many groups exhibit strong responses to predators, so it is not surprising that there have been several attempts to account for flocking as an antipredation mechanism. Unfortunately, few quantitative data exist on predation rates in groups.

Group awareness. The presence of many eyes should improve the efficiency of predator detection by groups. Alarm responses, either visual or auditory, may then communicate a warning (see Marler, 1955; Morse, 1970b). Goss-Custard (1970a, b) believed this to be the basis for flocking in redshanks (*Tringa totanus*) feeding on mudflats; he argued that the groupings themselves probably hampered the efficiency of feeding. However, several workers have reported that birds or mammals in groups individually spend less time in surveillance than they do when working alone (Dimond and Lazarus, 1974; G. V. N. Powell, 1974; Estes, 1974) (fig. 12.6). Siegfried and Underhill (1975)

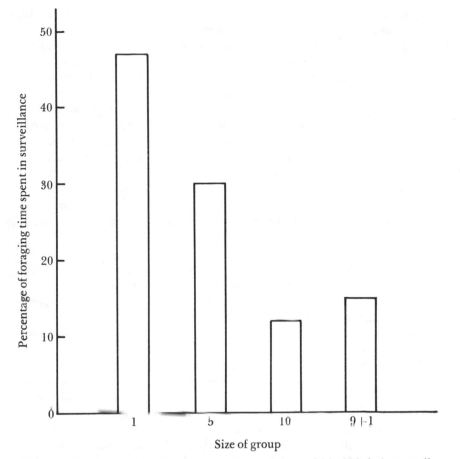

Figure 12.6 Proportion of time spent by starlings and blackbirds in surveillance as a function of their number. The three left-hand bars refer to starlings; the right-hand bar refers to nine starlings and one tricolored blackbird (*Agelaius tricolor*). (Modified from G. V. N. Powell, 1974.)

demonstrated that laughing doves (*Streptopelia senegalensis*) in large flocks were more likely to spot a flying model of a hawk than were small groups.

One might expect individuals in large, dense schools to be extremely vulnerable to blind-side attacks. However, one may readily elicit waves of agitation in schools that apparently result from the response of members to other individuals in the group (Breder, 1959; Radakov, 1972), which may minimize this possible disadvantage. However, Lutz (in Kenward, 1978) reported that goshawks (*Accipiter gentilis*) are more successful at large flocks of crows than at small ones, and Kenward suggested that the confusion resulting when the first crows take flight may conceal the hawk from other individuals.

It is sometimes argued that the conspicuousness of flocks causes individuals to be found more readily by predators than they would be if alone (Odum, 1942; Tinbergen, 1946) ; the advantages of protection associated with improved awareness could be offset (partially or totally) by increased conspicuousness, according to this line of reasoning. There seem to be no data adequate to determine conclusively whether individuals in flocks are more easily found than lone individuals under natural conditions. But a variety of conspicuous vocalizations are given primarily or only in flocks (see Morse, 1970b). These calls appear to be associated with maintaining the cohesion of the flocks and make it likely that individuals are more conspicuous (in the sense that they are more likely to be seen at all) than they would be if dispersed.

Confusing or discouraging predators. Mass responses to predators may permit individuals to reduce their vulnerability when in groups by presenting the predator with a confusing welter of targets from which it is unable to select just one before all have reached cover. This idea has been confirmed experimentally with fish (Neill and Cullen, 1974).

Many species of flocking birds give vocalizations after being warned of a predator. Examples include bushtits (*Psaltriparus minimus*) (Miller, 1922), chaffinches (*Fringilla coelobs*) (Marler, 1956), and chickadees (*Parus* spp.) (Morse, 1970b). These choruses are believed to confuse predators, but they are not found in all species (Buskirk, 1972). The calls given by these species about predators usually are considered particularly difficult for predators to localize (Marler, 1955), although Shalter and Schleidt (1977) and Shalter (1978a) have questioned this argument (see chapter 5).

Page and Whitacre (1975) have provided one of the few substantial sets of data on predation of flocking species when individuals are alone and in flocks of varying size (fig. 12.7). This work, done on wintering sandpipers, demonstrates that members of flocks are less vulnerable to hawks, primarily merlins (*Falco columbarius*), than are lone individuals, and that this advantage continues up to the largest flock size (fifty or more individuals), at which point predation rates on the sandpipers go up. This increase in predation rate in the largest groups may be the result of individuals getting in one another's way when responding to a hawk. Page and Whitacre's data do not permit one to distinguish between advantages of increased awareness and of the confusion effect. Both elements are probably important. In over

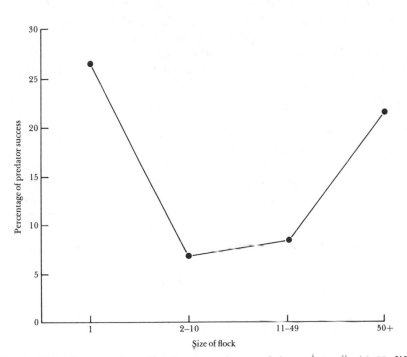

Figure 12.7 Success of merlins in capturing sandpipers from flocks of differing size. (Modified from Page and Whitacre, 1975.)

eighty attempts, the hawks regularly attempted to surprise the sandpipers but never caught one in the air.

It is possible that the mere presence of prey in groups discourages predators from attacking (see Moynihan, 1962). The tendency of sparrowhawks (*Accipter nisus*) to locate many more flocks than they actually attack (Morse, 1973b) is consistent with this idea. There is some experimental support from fish. Radakov (1958) presented young pollock to cod (*Gadus morrhua*) singly and in groups and found that the cod usually took stragglers rather than members of the integral school, a phenomenon similar to that observed in predation on bird flocks. Radakov obtained similar results in subsequent studies using atherinids (*Atherina mochon*) as prey and horse mackerel as predator. In fact, in one experiment the horse mackerel virtually gave up chasing the schooled prey but attacked vigorously (and successfully) when they were introduced one at a time immediately afterward (Radakov, 1972).

Neill and Cullen (1974) found that capture rates by predatory fish— pike (*Esox lucius*) and perch (*Perca fluviatilis*) —and cephalopods (*Loligo vulgaris, Sepia officinalis*) were lower when the prey (various species of small fishes) were in schools than when prey were alone (fig. 12.8). Prey taken from schools required longer capture times than did lone individuals. Under natural circumstances this increased capture time might permit prey an opportunity to find cover. Neill and Cullen also hypothesized that lone predators in natural situations specialize on stray individuals, given this difficulty in attacking schools. Major's studies (1978) on jacks and anchovies produced results generally similar to those of Neill and Cullen.

Forming groupings that deter predators. Crook (1970a, b) and others have attempted to relate differences in group structure to ecological variables, especially with reference to protection against predators. It is now clear that the structure of primate groupings differs with ecological conditions. For example, multi-male groups are typical in

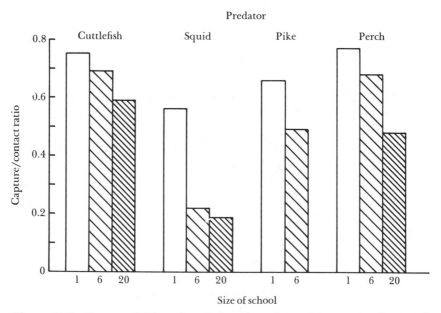

Figure 12.8 Success of fish and cephalopod predators in capturing lone and schooled prey fishes. (Modified from Neill and Cullen, 1974, with permission of the Zoological Society of London.)

baboons of relatively stable savanna-woodland, but one-male groups are typical in dry savannalike and forested areas. Moderate environmental fluctuation occurs in the savanna-woodlands, but with richer food supplies than in the dry savannas; conditions there consequently support both high predator and prey populations. The high predator pressure is generally believed to be responsible for multi-male groupings, the males serving an important protective function (Crook, 1970a). However, these multi-male groups can be supported only because of the relatively high productivity of the savanna-woodlands. In the poor savanna-type situation, which experiences considerable environmental fluctuation (long dry seasons), individuals are subject to a poor and variable food supply but also to relatively few predators.

Eisenberg et al. (1972) have noted that one-male and multi-male groups are really end points on a continuum, with a high proportion of the so-called multi-male systems really forming age-graded male systems. They have also stressed that social structure may to some degree reflect phylogeny, as well as current environmental conditions. Thus, Madagascan lemuroids have a tendency to form multi-male groups with more males than females, and South American primates show a trend toward male parental care and pair bonding. Although vulnerability to predators and characteristics of the food supply may be major determinants of social systems, factors such as size of groups, age of males, and sex ratios may be affected by a variety of other ecological variables (Eisenberg et al., 1972; Clutton-Brock and Harvey, 1977). Jarman (1974) has shown striking parallels to the primate patterns in the structure of ungulate groups in savanna and forest habitats.

Some social mammals take up defensive formations when challenged by a predator. Large adults form the outer perimeter, with young and sometimes females behind. This type of defense has been reported for elephants, baboons (*Papio* spp.), eland (*Taurotragus* spp.), and musk oxen (*Ovibos moschatus*), with the latter's defense against wolves perhaps being the best known. When threatened the oxen adopt a collective frontal stance to the predator, which usually succeeds in deterring it (Tener, 1965).

Flocks of starlings (*Sturnus vulgaris*) under attack when in flight tighten their ranks considerably (Tinbergen, 1953b; Gersdorf, 1966). It would be dangerous for a hawk to strike such an "object" (Tinbergen, 1953b). Gersdorf (1966) recounted several cases in which flocks of starlings in flight turned on pursuing sparrowhawks, on occasion driving them into the ground or water. Smith and Holland (1974) reported a similar attack by redwinged blackbirds (*Agelaius phoeniceus*) on an American kestrel (*Falco sparverius*), in which the falcon was driven into the water. Goss-Custard (1970b) proposed that this protective advantage exists for redshanks, which also tighten their formation when attacked.

Cover seeking. Neill and Cullen (1974) noted that individuals in large dense schools might gain considerable protection by seeking positions near the center of the school. This advantage of herding had been proposed earlier by Williams (1964) and Hamilton (1971). Schools of verkhovka (*Leucaspius delineatus*) and other fish studied by Radakov (1972) close ranks when confronted by predators, but individuals retain the same relative positions, suggesting that not all individuals are attempting to minimize their vulnerability by seeking cover inside the school. In principle, the advantage to some individuals could be very high, if they are usually able to obtain a favorable position. But position may be determined by dominance rank in many types of groups, and in any event schooling should not persist unless even the most vulnerable individuals gain some degree of protection (Pulliam, 1973). Milinski (1977) showed that this does occur for *Daphnia magna* preyed on by sticklebacks (*Gasterosteus aculeatus*); individuals at the center of a swarm of prey are least vulnerable of all, but peripheral members of the swarm are also less vulnerable than stragglers (fig. 12.9).

Several workers have discussed the intriguing possibility that some schools engage in collective mimicry (Cott, 1940; Springer, 1957; Breder, 1959). The idea is that group members might behave so as to make the group as a whole resemble some formidable object that the predator will avoid (for example, catfish young collectively resembling a large fish). As Shaw (1970) and Radakov (1972) have pointed out, there is not yet any conclusive evidence for this practice. It seems most likely to be occurring in fish that markedly increase the densities of their schools when predators appear. On the other hand, the mere tightening of such groups probably deters predators for purely physical reasons.

Mathematical advantages of clumping. Several workers have demonstrated mathematically that members of schools should run lower risks of being found by predators than they would if solitary (Breder, 1959; Orfeev, 1959, 1963, cited in Radakov, 1972; Brock and Riffenburgh, 1960; Olson, 1963. These arguments deal specifically with fish, but the similar spatial nature of other groups should make the calculations relevant to them as well. These models are based on random collisions and are therefore highly unrealistic (Shaw, 1970;

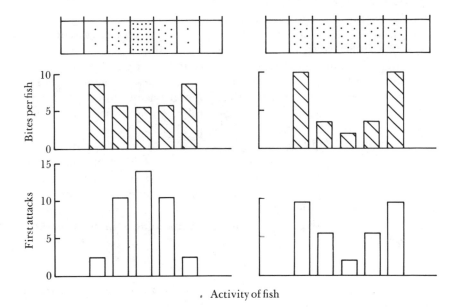

Figure 12.9 Attention paid to the various parts of daphnia swarms by sticklebacks. Upper blocks with dots indicate the density of individuals through a cross section of the swarm. *Left,* a swarm with a high central density; *right,* a swarm with an equal overall density. (Modified from Milinski, 1977.)

Radakov, 1972; Treisman, 1975) ; they seem unlikely to provide a complete explanation. Predators that succeed in preying on school members may actually follow the schools or frequent areas in which they are to be found, thus more than offsetting this theoretical advantage.

Additional evidence. Both Tinbergen (1946) and Rudebeck (1950 51) found that predation was particularly high on individuals that behaved differently from the majority of a flock. According to Tinbergen, house sparrows that flew in the opposite direction from other members of their group were most likely to be captured by hawks, and Rudebeck mentioned additional similar examples involving several species of hawks and flocking prey.

Barlow (1974b) found that surgeonfishes (Acanthuridae) that fed in the open by sifting sand remained in schools at such times, but nearby reef-grazing congeners were much less social. Fraser (1973) also noted a stronger tendency of killifishes (*Fundulus* spp.) to school in deep water, where there are predators and relatively little cover and where they do little of their feeding. This pattern suggests that these groupings have mainly defensive functions and that schooling per se may interfere with some patterns of feeding.

Although he could not ascertain the exact nature of the advantage obtained (confusion effect, increased watchfulness), Seghers (1974) made the interesting observation that populations of guppies (*Poecilia reticulata*) in areas where predatory fishes were present showed a marked propensity to school, but those in areas with few or no predators did not. In addition, the difference had a clear genetic basis. Differences were found even on opposite sides of waterfalls that had halted the predators, but not their prey. The schooling tendency appeared to be only a part of a coadapted complex of antipredator adaptations, which also included increased reaction distances to predators, reduced alarm thresholds, and differences in microhabitat selection.

COMPARISON OF ADVANTAGES

It is difficult to characterize flocks, herds, or schools from a single ecological viewpoint. The data and arguments do not permit a clear decision as to whether energetic advantages or predator avoidance form the major advantages of groups, although some authors have taken strong stands on the issue (see Moynihan, 1962; Lack, 1968; Murton, 1971b; Zahavi, 1971). The tendency to disperse when food is superabundant and to flock when conditions become severe is widespread and is consistent with major feeding benefits, as is the observation that flocks of more than one species usually contain only members that overlap one another in foraging to some degree (Morse, 1970b). Countering this is the possibility that these changes occur because under poor resource conditions an individual cannot by itself afford the luxury of spending more than a minimal amount of time watching out for predators or defending itself from them.

Lack and Lack (1972), Willis (1973a, b), and Post (1978) have observed that cohesive flocks are not formed on certain islands or parts of islands that have no aerial predators (Jamaica, Hawaii, parts of Puerto Rico). This is consistent with the idea that the primary function of flocking is to secure protection against predators. It is probably unwise to ask the either-or question, because of the possible interactions between these factors (Lazarus, 1972). In fact, G. V. N. Powell (1974) obtained experimental evidence that starlings reduce their surveillance as groups get larger, and that their foraging rate goes up simultaneously.

The data from mammals closely resemble those from birds. Size and composition differ depending on ecological variables, namely the nature of the food supply and the extent of predator pressure. But mammalian groupings have become considerably more complex socially than bird flocks. Schooling appears to be very largely a response to predators, and energetic advantages, particularly feeding advantages, seem to be of lesser importance to fish than they are to birds and mammals.

Interspecific groups

Mixed-species social groups have been reported in the following taxa: birds (Moynihan, 1962; Morse, 1970b, 1977d; Moriarty, 1976), large hoofed mammals (Lamprey, 1963; Keast, 1965b), primates (Gartlan and Struhsaker, 1972; Klein and Klein, 1973; Dunbar and Dunbar, 1974), and fish (Breder, 1959; Hobson, 1968; Karst, 1968; Fishelson, Popper and Avidor, 1974; Itzkowitz, 1974). Except in birds they have received much less attention than comparable intraspecific groups, and they have sometimes been explained away as chance aggregations. But they display gregariousness similar to that seen in single-species groups. Several advantages have been hypothesized for mixed-species groups, most being similar to those discussed for single-species groups. Here I will concentrate on aspects that are of particular significance at the interspecific level.

Different species in mixed flocks usually forage in the parts of the habitat that they appear to be adapted to use with maximum efficiency (Morse 1970b). For example, golden-crowned kinglets (*Regulus satrapa*) in flocks typically feed in the tips of the vegetation, whereas most woodpeckers feed on trunks and the inner parts of large limbs. Since differences in resource use among species are generally greater than those within species, common demands upon any particular resource should be of lower intensity among individuals of different species than among individuals of a single species. With few exceptions, interspecific competition is much less intense than intraspecific competition in mixed flocks, as judged by frequencies of interaction (Morse, 1970b). Mixed-species flocks should be less efficient at avoiding duplication of effort than single-species flocks, but some advantages may nonetheless be gained, since some overlap does occur between species in mixed-species groups. Similarly, the overall advantage of observational learning should on the average be lower in mixed groups than in single-species groups. Resources used by an individual of one species are less likely to be used by individuals of another species than by a conspecific, although again the overlap in exploitation of resources should enable some such advantage to be gained. On the other hand, since different species hunt for food somewhat differently or in different places, a relatively greater range of foods might be discovered for subsequent common exploration by mixed-species flocks.

Krebs (1973) demonstrated interspecific social facilitation in laboratory experiments with black-capped chickadees and chestnut-backed chickadees (*Parus rufescens*) (fig. 12.10), using the design of Krebs et al. (1972). He used naive, hand-reared birds and a distribution of food probably more clumped than that found in nature. Because the laboratory setups were artificial and the birds were not accustomed to natural conditions, it is difficult to compare this study to observations made under natural conditions, especially since the two species segregate strongly by habitat and by foraging station where they are sympatric (Sturman, 1968). On the other hand, B. Shephard (personal communication) found that wild-caught great tits and blue tits segregated when placed together in laboratory experiments. The food sources within Shephard's experimental setup were somewhat more dispersed than in the studies of Krebs et al. (1972) and Krebs (1973). Krebs and coworkers used hand-reared birds exclusively, and Shephard used wild-caught adults, suggesting that early experience is of importance in partitioning. The degree of clumping of food provided by Krebs (1973) in and prior to the experiments may also be an important variable. This degree of clumping may not occur typically in the field, although information on this point is not adequate. It is interesting that in some cases where food appeared clumped in the field, blue tits showed a strong tendency to feed in a way consistent with the pattern of social facilitation described by Krebs (Morse, 1978).

Interspecific dominance hierarchies tend to appear in mixed-species flocks, and some studies have reported that individuals change their patterns of foraging when participating in them (Morse, 1970b; Hogstad, 1978b). Socially subordinate species use foraging stations favored by the dominant species less frequently when the dominant is present than when it is absent (Morse, 1970b). Where an apparently equivocal dominance relationship exists, both members of a species pair reduce their use of areas of overlap (Morse, 1967; Austin and Smith, 1972). These shifts are similar to those seen in interspecific competitive interactions taking place outside of social groupings (Morse, 1974; also see chapter 11).

Participating in single-species groups or remaining solitary might again seem to be more practical than participating in mixed flocks. However, if one can extrapolate from Barash's observations (1974a) on single-species flocks of black-capped chickadees, hostile interactions may be less frequent among mixed flock members than among individuals not continually in contact. If so, proximity in flocks may not be

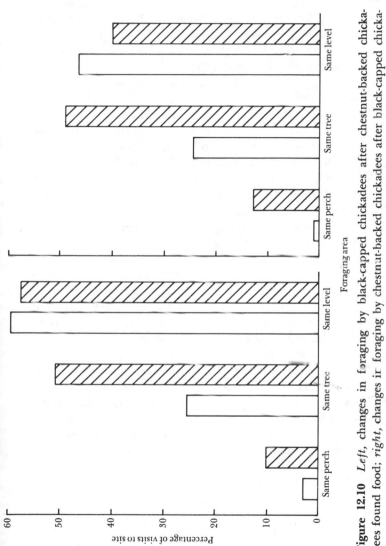

Figure 12.10 *Left,* changes in foraging by black-capped chickadees after chestnut-backed chickadees found food; *right,* changes in foraging by chestnut-backed chickadees after black-capped chickadees found food. White bars = percentage of visits to a site before observing the opposite species obtain food there; hatched bars = percentage of visits to that site after observing a food find by the other species. (Modified from Krebs, 1973; reproduced by permission of the National Research Council of Canada from the *Canadian Journal of Zoology* 51, pp. 1275–1288, 1973.)

as much of a disadvantage for the subordinate as it seems. For the dominant, the disadvantage should be minimal, since it may obtain any possible advantage of flocking without significantly reducing its niche breadth (see Morse, 1970b). The tendency of subordinates to increase their foraging activities in the niche space of species dominant to them when the latter are absent would reduce their net feeding disadvantages. If most or all individuals foraged in flocks, subordinates would suffer reduced risks of being attacked.

If either minimizing duplication of effort or maximizing social facilitation of feeding were the only goals, single-species flocks should be more prevalent than mixed flocks under natural conditions. Yet mixed-species flocks typically contain relatively small numbers of individuals representing several species, and in some cases this distribution appears to result from the exclusion of additional individuals by conspecifics (Moynihan, 1962). Although the energetic gains realized by participating in these groups may be lower than they are in one-species flocks, losses from competition may also be lower. But the fact remains that individuals of different species band together in a way that does not, on balance, appear to maximize their feeding advantages. This suggests that the principal adaptive basis for mixed-species flocks, as opposed to single-species flocks, should be looked for elsewhere.

Moynihan (1962) and Buskirk (1972) noted a strong tendency for few individuals of many species to be present in the tropical flocks they studied. Large flocks are thought to offer better protection from predators than small ones; this has been demonstrated in shorebird flocks under attack by merlins (Page and Whitacre, 1975) and in fish schools (Radakov, 1972). For a given intensity of competition, there could be more eyes to watch for predators in an interspecific group than in an intraspecific group. If the different species collectively feed in a broader range of locations than the members of intraspecific groups, the warning system might work even better, as it also would if the sensory capabilities of the different species varied considerably. Some species of shorebirds that occur in mixed-species groups are particularly wary (Jourdain, in Bent, 1927).

Several authors have reported mixed groups of fish and mixed groups of mammals, but there has been little treatment of their significance. An exception is Itzkowitz's study (1974) on Jamaican reef fishes, which shows that some members of schools change their foraging behavior in these groups. Fishelson (1977) noted that members of coral-reef schools in the Red Sea tend to feed similarly, and that they respond to one another's foraging. These observations suggest similarities to bird flocks, although Itzkowitz (personal communication) cautioned against equating mixed-species schools with flocks of birds.

Barnard (1979) has raised the point that minority individuals in mixed-species groups might be subject to disproportionately high predation, being perceived as odd by predators. This clearly could affect the composition of groups. He suggested that it may explain why mixed-species group members tend to resemble one another more closely than would be predicted by chance (social mimicry). There is no evidence suggesting that members of temperate zone mixed-species flocks of woodland birds obey these guidelines (see Morse, 1970b, 1978). In such diffuse groupings as these, oddity may not be an important cue used by predators. The proposition is more attractive when applied to dense groupings that predators could view synoptically. Under these conditions a minority individual might be perceived quickly and become subject to attack. Hobson (1968) has reported selective predation on minority species of fish in schools; in other situations one is often impressed by the homogeneity of school members.

Group size

Energetic and predator avoidance factors should largely determine not only whether animals will be grouped or solitary but also group size. If reproduction is involved, as in permanently pair-bonded individuals or family groups, additional complications may result. Unfortunately, there are few data on the factors affecting group size. However, several authors have attempted to link group size to ecological variables (reviewed by Caraco and Wolf, 1975).

Parid flocks tend not to exceed forty-five individuals, and flock size may vary with the habitat. The flock size is inversely related to the apparent insect productivity of the habitat and positively related to the size of the area used (Morse, 1970b, 1978). Hoffman and Braun (1977) found that after summers of high reproductive success, white-tailed ptarmigans (*Lagopus leucurus*) formed larger numbers of wintering flocks, rather than larger flocks. On the other hand, Beck (1977) reported that sage grouse (*Centrocercus urophasianus*) broke into a large number of small flocks during severe winters, rather than a few large ones. A similar pattern apparently occurs in black grouse (*Lyurus tetrix*) and capercaille (*Tetrao urogallus*) (Koskimies, 1957).

The cost of watching for predators may be so large for individuals or small groups that it favors a certain minimum group size. For example, Berger (1978) found that bighorn sheep (*Ovis canadensis*) in groups of five or less spent so much time scanning the surroundings (presumably searching for predators) that their foraging efficiency was considerably lower than that of individuals in larger groups. In species where groups are obligate or nearly so, the danger of falling below a critical group size may be more serious than that of exceeding the optimal size (Bertram, 1978), with the consequence that optimal group size is often exceeded.

Considerable size constancy may occur in permanent groups of related individuals, as reported for several species including superb blue wrens (*Malurus cyaneus*), choughs, babblers, jays, and bushtits (reviewed by Ervin, 1977). However, this size may differ with the characteristics of the habitat, as shown by Brown and Balda (1977) in a study of Hall's babblers (*Pomatostomus halli*) (fig. 7.6).

Caraco and Wolf (1975) used Schaller's data (1972) on lions to look for correlations between group size and ecological variables (fig. 12.11). No specific foraging advantages accrue to increasing group size if the prey to be captured are small. For example, when specializing on Thomson's gazelles (*Gazella thomsoni*), there are no advantages for groups of greater than two individuals, and groups larger than this are unlikely to meet their minimum requirements. Individual rewards to members of larger groups that specialize on zebra or wildebeest are great enough to support larger prides, although individual food rewards are greatest for groups of two or three. Survival of young is greatest in large prides, apparently because the closely related females protect and feed one another's cubs. Such prides provoke intense competition among male lions, because possession of one guarantees high reproductive success. Groups of males are much more successful in securing control of a pride than are lone males, and this in turn selects for increased numbers of males operating as cooperative units, with their maximum number, and that of the females, being controlled ultimately by the resource base. Thus, a variety of interacting factors appears to determine group size in lions.

Using a variety of references that provided information on pack sizes in wolves preying on white-tailed deer (*Odocoileus virginianus*), caribou (*Rangifer tarandus*), American elk (*Cervus elaphus*), and moose, Nudds (1978) obtained results strikingly similar to those of Caraco and Wolf, which suggests that this relationship between group size and type of prey taken may characterize most social animals feeding on formidable prey.

Zimen (1976) reported on behavioral mechanisms that regulate the pack size of wolves. Curtailment of food, resulting from a reduction of prey size or prey biomass, or from an increase in pack size, caused an increase in the frequency and intensity of aggressive behavior. As a result, pack size was reduced directly through increased mortality, especially of pups, and indirectly through the emigration of low-ranking adults and juveniles, sometimes through active ejection.

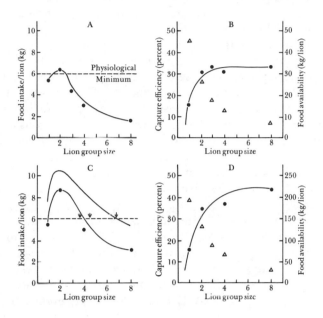

Figure 12.11 Hunting behavior of lions as a function of group size. *A,* mean edible prey (Thomson's gazelle) biomass (kg) per lion per three chases (predicted daily index). Dashed line represents daily minimum for an individual lion, males and females averaged. *B,* capture efficiency (circles) with food availability (triangles) per lion. Solid line (efficiency of capture of Thomson's gazelle as a function of lion group size) has been visually fitted to observed data. Triangles indicate mean available prey biomass (kg) per lion per three captures (corrected for multiple kills) for each group size. Three chases approximate daily prey availability during dry season along woodlands-plains border. Three chases is predicted minimum for group of any size to sustain itself physiologically on Thomson's gazelle. *C,* mean edible prey (wildebeest) biomass (kg) per lion per chase per twenty-four hours. Upper curve shows hypothesized wet-season intake in woodlands-plains border, where vegetation cover increases capture efficiencies. Lower line gives same calculations for eastern plains and western woodlands, when wildebeest are available prey. Dashed line represents daily physiological minimum. Observed mean lion group sizes are given for eastern plains (left arrow), western woodlands (middle arrow), and border region (right arrow). *D,* capture efficiency (circles) and food availability (triangles) per lion. Solid line (efficiency of capture of wildebeest as a function of lion group size) has been visually fitted to observed data. Triangles indicate mean available prey biomass (kg) per lion per wildebeest capture (corrected for multiple kills) for each group size. (Slightly modified from Caraco and Wolf, 1975, copyright 1975 by The University of Chicago.)

Antelope species demonstrate nicely how certain factors impinge on group size and other characteristics. Jarman (1974) found a close relationship between group size and body size, which appears to be a consequence of demands for quality food by small species. Since demands per unit body weight are higher in smaller than in larger species, smaller species must select higher-quality food than larger species in order to obtain their requirements. They are often too small to enjoy appreciable group defense, and they are also often unable to outrun predators. Under these constraints, inconspicuousness would seem to be favored. As antelope increase in size, their selectivity tends to decrease along one or more parameters, the largest species being quite unselective and taking substantial amounts of food of low nutritional value. Larger species give warning signals, and the largest may attack would-be predators.

A certain minimum flock size may be required in order to defend an area. MacRoberts and MacRoberts (1976) found that in interactions between groups of acorn woodpeckers (*Melanerpes formicivorous*), large groups always prevailed over smaller ones and were sometimes able to usurp space from the small groups. But the space required to support progressively larger flocks may eventually become so great as to be indefensible, unless some division of labor is employed, which would break up the flock. No evidence exists for such a division of labor (see Hamilton and Watt, 1970), and perhaps the relative constancy of upper group sizes reflects this constraint. Clutton-Brock and Harvey (1978) reached a similar conclusion for mammalian groups. The tendency for large tit flocks to move faster than small ones (Morse, 1970b) may permit them to patrol large areas, up to a certain size limit.

Roosts and colonies

Roosts and colonies may be defined as resting and breeding groups, respectively. Certain thermoregulatory and reproductive aspects of these groups are discussed in chapters 5 and 7. Sometimes only limited numbers of sites provide adequate protection for these activities. Some night roosts formed by blackbirds (Icteridae) and starlings appear to result from a shortage of safe areas. Presumably this accounts for the formation of some gigantic roosts, from which birds may fly as much as 80 km to their feeding grounds (Hamilton and Gilbert, 1969; Hamilton and Watt, 1970).

Individuals may be subject to some predation even in such roosts, however, and the social structure of the groups may impose risks on their members. Fuchs (1977) showed that sandwich terns (*Sterna sandvicensis*) along the periphery of a colony were subject to far higher levels of egg predation than were those in the center. The center of a colony is occupied typically by the senior breeders, who probably maintain access to these areas as a consequence of their dominance. Similarly, predation may be concentrated on the outer fringes of the roost, or, if ground predators are involved, the lower parts of the roost as well. Swingland (1977) noted that dominant rooks (*Corvus frugilegus*) typically perched high in roosts, despite the fact that this might subject them to greater thermal stress than would roosting at lower elevations. In particularly severe weather the dominant birds moved to lower elevations, displacing younger subordinate birds to the less favorable sites. Presumably the tendency to roost high was associated with minimizing predation, with compromises being made only when microclimatic conditions became so severe that their cost exceeded that of remaining in a position of minimal vulnerability.

Although protection may be a primary advantage for such groupings, other advantages may be obtained as well. Ward (1965a) and Zahavi (1971) proposed that information about feeding areas is transmitted in such communally roosting species as the pied wagtail, a species that disperses considerably in its feeding. Ward and Zahavi (1973) argued that roosts in general have evolved primarily in response to energetic considerations (the need to find extremely patchy food efficiently), rather than as an antipredation mechanism. Ward (1965a) noted that queleas that left their roosts quickly in the morning consisted of two subgroups, those that dispersed directly in groups and others that initially dispersed only short distances. The later typically flew up from the ground and joined large groups as they flew overhead. This would permit indecisive individuals to join others that were already oriented toward satisfactory feeding locations. Siegfried (1971) observed similar behavior in cattle egrets and also noted that juveniles were typically not the first to disperse from roosts. The proposals of Ward and Zahavi (1973) are consistent with the data they present, but these data do not rule out the possibility that roosts have important antipredator functions. Roosts are usually situated where they are maximally free from predation. Perhaps the strongest evidence against an antipredator argument is provided by Ffrench (1967) who found that predation on dickcissels (*Spiza americana*) on their wintering roosts in Trinidad was extremely heavy over long periods.

Krebs (1974) presented what is perhaps the most direct evidence in support of the information-center hypothesis in his study of great blue heron (*Ardea herodias*) breeding colonies. He demonstrated that individuals tend to leave the colony on hunting forays in groups (fig. 12.12) and that synchrony is most marked between near neighbors. The areas visited change continually over time. At the feeding grounds,

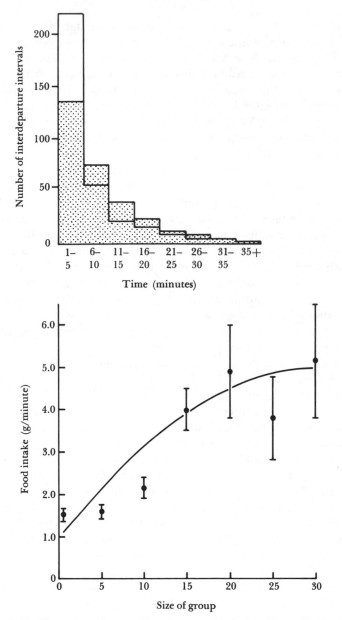

Figure 12.12 Foraging of great blue herons. *Top,* the observed distribution of intervals between successive departures from the colony by adults making foraging trips (white histogram). The shaded histogram represents the expected distribution according to a negative exponential distribution. *Bottom,* rate of food intake (±1 SE) of adults as a function of flock size. (Redrawn from Krebs, 1974.)

individuals in groups are conspicuously more successful in capturing food than are lone individuals, largely or entirely as a result of aggregation at places where fishing is good. The information-center hypothesis is of particular interest for cases in which predator avoidance or thermoregulation (chapter 6) does not appear to provide a likely explanation, but as yet it has no experimental verification.

Other groupings

The relationship among some members of groups appears to be commensal. This probably accounts for the relationship of wood pewees to the flocks they follow, presumably for insects flushed by other flock members. It is difficult to draw a clear line between commensal relationships and mutualistic ones. Other associates of bird flocks probably are obtaining commensal advantages similar or analogous to those enjoyed by the pewees. Juncos (*Junco hyemalis*), bluebirds (*Sialia sialis*), and chipping sparrows (*Spizella passerina*) follow flocks composed primarily of titmice and chickadees but share few if any feeding similarities with them. They typically forage on the ground, while the conventional flock members forage in the trees above. Most likely these birds are taking advantage of the warning systems of the flock members located in the trees above, who enjoy a vantage not available to their ground-feeding associates (Morse, 1970b). Other workers have reported tree shrews (Stresemann, 1917) and squirrels (Chapin, 1932; Moynihan, 1962) following these flocks, presumably for some food-gathering or predator-avoidance advantage, which they would seem unlikely to provide in return to the flock members.

Synthesis

It is difficult to characterize flocks, schools, and herds from a single ecological viewpoint because of their wide range of attributes. Numerous apparent cases of both feeding and antipredator advantages may be found in groups of birds, fish, and mammals. They range from situations in which one function appears to be of far greater importance than the other to ones in which both aspects are clearly important. A strong defensive element, even to the detriment of feeding efficiency, seems more characteristic of fish than of mammalian and avian groupings, and a stronger defensive element is more apparent in mammalian than in avian groupings. Schooling often exists in spite of disadvantages in feeding, with the result that individuals frequently disperse considerably when feeding. The absence of dominance hierarchies in most schools that have been investigated (Radakov, 1972) is consistent with this interpretation. Differential antipredator advantages might be secured if dominants monopolize particularly invulnerable positions within a group (cover seeking), but few data currently support this possibility.

Most workers who have written about groups, especially mixed ones, have felt it necessary to speculate about their evolutionary origins. Whether energetic or antipredator advantages arose first in any given system, development of one advantage should stimulate rapid development of the other. Sorting out primary and secondary elements would appear to be difficult, if not futile. Continued speculation about evolutionary primary and secondary functions seems unlikely to provide new insight, and time might be spent more profitably attempting to formulate new questions.

Mixed social groups have great potential utility for the testing of hypotheses in areas such as niche theory or ecological release. They typically consist of species with overlapping resource exploitation patterns, which are subject to change in predictable ways depending on whether they are in contact or apart. Thus, one can readily achieve natural experimental conditions by exploiting the different and changing compositions of the groups (Morse, 1970b). Although manipulations may be difficult in the field, they may not pose the problems associated with nonsocial species. Because the species in question are regularly in close contact, observations under relatively confined conditions, as in the laboratory, may be more valid in social than in solitary species.

13 *Future Directions*

PROGRESS IN THESE AREAS points in several exciting new directions, from feeding strategies to group cohesion. In general the rigor of current work is improving, with manipulative studies increasing in the laboratory, and even more importantly, in the field. This movement lags in some areas, however. Not all subjects are equally amenable to rigorous investigation, even though it may be prerequisite to further progress.

Foraging

Studies of foraging are currently proceeding along several more or less independent pathways. Integration between such areas as optimal foraging theory and switching theory is needed. With a few conspicuous exceptions, no effort has been made to integrate foraging theory and niche theory, despite the fact that the latter is intimately connected with resource exploitation. This is not too surprising because, until the late 1970s, there was little integration even of such a well-defined topic as optimal foraging; as many as nine independent derivations of optimal diet theory had been published by 1977 (Pyke et al., 1977). Greater attention to the importance of factors other than energy gain will also increase our understanding of differences in the behavior of different species. For example, there is no reason to suspect that nutrient constraints will be of equal importance at the herbivore level and at the levels above it. An adequate technology exists for the study of this subject, since the nutrition of domesticated animals is rather well understood. Yet little precise information is available about the needs of wild animals, and few ecologists have training in nutrition.

Most work on foraging theory in ecology has developed independently of learning theory in psychology, but there are definite connections between the problems of the two fields. Ecologists might profit greatly from exploring the extensive literature of learning theory (Gill and Wolf, 1977), taking what is relevent to their needs and perhaps adapting experimental techniques to their own purposes.

In ecological studies, data for different individuals in a population are usually lumped together. Thus although the overall patterns of exploitation may be known, individual differences are not revealed, at least not beyond some summary measure of variability. To the animals in question, individual differences may be critical, and of course selection usually operates at the level of individuals. Further attention must be paid to this matter if we hope to gain a sophisticated understanding of the forces that determine the overall exploitation patterns of populations.

Short-term studies may be seriously misleading unless they are conducted in extremely constant environments. In particular, they may miss critical stressful periods in the lives of the animals in question. For this reason there is no substitute for studies long enough to follow a cohort through its life span; unfortunately, our present system of awarding grant support does not encourage this approach.

Most of our knowledge about many species derives from studies conducted on single populations. Even if truly representative data are gathered at one site, they may not resemble those obtained elsewhere. This could occur for a variety of reasons. Competition, predation, and environmental factors—all may cause differences in population size or density. We will not obtain an accurate understanding of the relative importance of these factors until a variety of different populations within a species are studied using standard techniques. Many ecologists have tended to think along implicitly typological lines, assuming that the pattern observed in one population will be the same as that seen in others. That this is not always so is evident from those few studies in which interpopulation variability has been looked for. This seems to be one of the most profitable directions that foraging studies might take.

Work on foraging and predator avoidance has been heavily concentrated in birds and mammals, who are surely not representative of all animals. The ability to remain active in a variety of conditions conveys unique benefits, but the exorbitant energy costs that result lead to compromises that set endotherms apart from all other animals. Thus current concepts of foraging may be biased in a number of ways. Greater attention to other groups would help balance our perspective.

Interactions between predators and prey

In spite of the paucity of data on behavioral interactions between predators and prey, there are hints that the tactics involved may be very complex; this can be appreciated merely by surveying some of the remarkable adaptations promoting crypticity and aposematism. But these are only a fraction of the antipredator adaptations found in animals. If predator pressure is apt to fluctuate strongly in time and space we might expect adaptations that provide their possessors with appropriate flexibility. Quantitative data for animals that hunt difficult-to-capture prey are very hard to obtain. Nevertheless, these data are essential. They could be accumulated at relatively small cost in the course of other major studies if workers remained alert for them. The interactions between sparrowhawks (*Accipiter nisus*) and tits described in chapter 5 (Morse, 1973b) serve as a good case in point. They appear to involve an initial response by the prey, a counter-response by the predator, and a subsequent response to the counter-response by the prey. There is no reason to assume that patterns of comparable complexity are unusual in predator-prey interactions, yet information is not adequate to evaluate their frequency. I discovered this pattern quite by chance in the process of conducting other studies on tits. Had I not been fortuitously located with regard to both the flock and the hawk on two occasions I would not have noticed the pattern.

Complex interactions between predator and prey are of interest not merely because they offer insight into the mechanics of a fundamental link in the food chain. They also almost certainly reduce the abilities of both predator and prey populations to gather resources. Current practice would subsume these interactions under the heading of foraging costs, but they may be important qualitative determinants of foraging repertoires.

Selfish behavior and kin selection

The importance of kin selection to breeding behavior is currently a subject of intense interest. Detailed studies have focused on the role of nonbreeding individuals in rearing young. The weight of the evidence suggests that nonbreeding individuals do provide some assistance, but almost without exception this evidence is correlational and is subject to other interpretations. Several workers hold that the presence of nonbreeding individuals can be accounted for without using inclusive fitness arguments. To resolve this question, experimental manipulations are needed in vertebrate systems. Systems with nonbreeding helpers may have some interesting ecological implications, because nonbreeders (whether related or not) may make reproduction possible under conditions that would not otherwise permit successful fledging. Foraging patterns that individually fail to provide adequate energetic margins may be usable where several individuals share the load, or the added protection gained through numbers may make reproduction possible where it otherwise would not be. In either case the consequence may be colonization of otherwise inaccessible areas.

It is now becoming apparent that demography is of fundamental importance to the social systems of many animals (Brown, 1978). The effects of competition, predation, and environmental factors on demographic patterns deserve much more detailed investigation. Of particular interest is the question whether particular demographic patterns are consequences of these external pressures in the sense that they are inherently better buffered against these pressures than other demographic patterns would be.

The consequences of parent-offspring conflict and sibling rivalry are another important side of the kin selection question. Trivers' papers on these subjects have elicited great excitement, but more studies explicitly testing the ideas are needed; in many cases the tests will not be easy, as Howe (1979) has noted, but they are certainly warranted. Conflicts within family groups are likely to be molded in interesting ways by ecological factors, and some insight into these might be gained through experimental approaches. Such studies might in turn illuminate the demographic characteristics exhibited by family groups and by whole populations.

Sexual selection

Sexual selection has produced some of the best illustrations of the connection between behavior and ecology. It is now clear that the limits to sexual selection may be set by ecological factors, but most of the relevant work has concerned size dimorphism, which is only one of the many dimorphisms sexual selection may produce (chapter 8). If differences of size are for some reason precluded, sexual selection may act on a host of other characters, such as coloration and vocalization. The ecological correlates and consequences of these different manifestations of sexual selection are important subjects and deserve much more study than they have received.

Sexual selection may have surprising implications. It can go to extremes only where sufficient niche space exists. Considering the great success of mammals, their diversity seems surprisingly low (about four thousand species), and that of many important groups within the class (ungulates, carnivores) is quite modest, despite radiations such as those of the African bovids. Is this low diversity a correlate of marked size dimorphism? If so, is it a direct consequence of dimorphism?

Aggressive behavior

The losers of aggressive encounters may be driven from an area or may leave it voluntarily, but there seem to be contradictions in the available data on frequencies of aggression as a function of environmental conditions. If aggression does not become more frequent and more overt as conditions worsen, why not? We need much more information on levels of aggression under carefully monitored environmental conditions. Is there an inverse relationship between the tendency to disperse and to increase the level of aggression under severe conditions? Dispersal may be the key variable, but it is difficult to study under natural conditions. Also needed is information on survival rates of individuals that do and do not disperse. Unfortunately, few data exist on the survival of individuals that disperse more widely than most members of their species. This is not surprising, since it is very difficult to track such individuals. But it would be profitable to look for systems in which such information might be gathered.

Interspecific interactions

It would be worthwhile to investigate the impact of social dominance relationships on the way species pack themselves into communities. Data compiled by Morse (1974) suggest that included or nearly included niches occur in communities more frequently than would be predicted by chance. This pattern, which often has a strongly spatial basis, may allow more species to fit into a community than could otherwise do so. However, the few cases studied so far do not constitute a test of this idea. Is the pattern commoner in some places than in others? How do such cases differ from those in which interspecific territoriality occurs, and do they suggest why interspecific territoriality is not more common than it is? Why is it relatively common in coral-reef fish?

Sometimes it is unclear why a particular species is socially dominant. Usually the larger species plays the dominant role, but in cases involving species of similar size, one often shows clear dominance, and occasionally it is the smaller. Is the dominant usually more abundant for some independent reason, and thus adapted to higher frequencies of intraspecific interaction? Although this may sometimes be the case, it clearly is not so in others—for example, voles (Stoecker, 1972). What other variables might determine this relationship?

If dominant species are so successful, why do they frequently occupy smaller niches than their subordinates? Levins (1968) suggested that niche specialization is achieved only at the expense of niche breadth (chapter 2) ; but why need this always hold, given the advantage of dominance? The data strongly suggest that it pays to be a specialist wherever this is possible. Specialists probably often have relatively small resource bases, and they should therefore be relatively vulnerable to changing conditions. If this is true, it implies that the most success-ful species may have a shorter expected evolutionary lifetime than some of the least successful. Do socially dominant species share other characteristics (such as small or disjoint ranges) that make them candidates for early extinction?

Dominance, at both the intraspecific and interspecific level, appears to be important in the organization of populations and communities. Overt aggressive behavior is much less frequent in such systems than would be predicted by chance. The discrepancy can be attributed to avoidance reactions on the part of subordinates and in some cases on the part of dominants as well. But there is very little detailed informa-tion on the extent and mechanics of avoidance. Although not easy to quantify, studies that evaluate the costs and benefits of avoidance are very much needed.

The state of the art

We still lack fundamental data on several basic phenomena, such as the detailed characteristics of predator-prey interactions. Ken-ward's innovative work (1978) on predator avoidance shows how some of these data might be obtained. Kenward flew goshawks (*Accip-iter gentilis*) trained for falconery at woodpigeons (*Columba palum-bus*) and was able to assess directly the utility of antipredator responses by prey and choice of prey by the predator. The literature to date suggests that it would be virtually impossible to gather similar data under completely natural conditions. Although studies such as Ken-ward's are difficult to perform, they fill a major gap and are clearly well worth the effort.

Although the studies reviewed in this book clearly reveal a growing emphasis on experimentation, some readers may be wondering why more experimental work has not already been done. It certainly does not suffice to rest content merely because the bulk of available evidence is consistent with a certain argument. I regard current attempts to test optimal foraging theory as an encouraging step, in spite of some reserva-tions (chapter 3) . The lack of experimental work on some subjects is surprising. For example, only one experiment (so far unpublished) has been run on the effect of helpers at the nest (cited by Brown, 1978) . Similarly, although the literature is replete with speculation on whether flocking is primarily important as an aid to foraging or to the avoidance of predators, almost no experiments have been performed under field conditions. The critical experiments are extremely difficult, but they are needed to get us beyond the already sufficient descriptive and comparative baselines, and they will yield very worthwhile results. There is little point to additional observational studies of foraging relationships, especially ones not making use of standardized techniques that might allow better comparison of the results of different studies.

For many questions there is no need to use birds or mammals as experimental subjects, although they have certainly provided a great deal of necessary information at the descriptive and comparative levels. In my own work on foraging I have shifted largely from insectivorous birds to the far more tractable bumblebees. Numbers of individuals, their food supplies, and their predators can all be readily manipulated. Members of the population can be individually marked with ease. The colonial nesting and breeding system of bumblebees allows one to control the relatedness of individuals, if this is important. Such manipulations are difficult or impossible to perform on insectivorous birds.

It is highly desirable to relate the presumably evolved variable under study to direct measures of Darwinian fitness. Studies of foraging behavior frequently conclude that the patterns observed are efficient, in the sense that they permit the forager to obtain high net energy intakes or other comparable advantages. It is assumed that these measures of success correlate with fitness, that is, with success in rearing offspring that contribute genetically to the next generation. Such arguments would be considerably strengthened if groups of individuals were followed closely enough to determine the number of their offspring and the fates of those offspring. It would be reassuring to know beyond reasonable doubt that a correlation does exist between foraging behavior and fitness. It would also be valuable to know whether the correlation varies systematically as a function of various environmental conditions. Studies attempting to make the connection should be encouraged, in spite of the fact that if they merely confirm what was assumed, the perceived payoff may be rather small. One often hears the argument that the adaptations under consideration would never have evolved in the first place unless they conferred high levels of fitness. But this is potentially a circular argument and, strictly speaking, depends on the very data we do not have.

We know that patterns of behavior do not always seem to be optimal. This could result from variable conditions that necessitate compromise strategies or from situations in which the organism in question is actually out of evolutionary equilibrium with its environment. The latter seems particularly likely where human activities have dramatically changed environmental conditions in a short period of time. Intensive, long-term studies that incorporate direct measures of fitness might enable us to assess the relative frequencies of (1) effectively optimal fits, (2) compromise strategies, and (3) real disequilibriums. Although it is clear that instances of all three types exist, their relative frequencies are still unknown.

References

Index

References

ALCOCK, J. 1969. Observational learning in birds. *Ibis* 111:308–321.

———. 1972. The evolution of the use of tools by feeding animals. *Evolution* 26:464–473.

———. 1973. Cues used in searching for food by red-winged blackbirds (*Agelaius phoeniceus*). *Behaviour* 46:174 188.

———. 1975. Territorial behaviour by males of *Philanthus multimaculatus* (Hymenoptera: Sphecidae) with a review of territoriality in male sphecids. *Animal Behaviour* 23:889–895.

ALDRICH, J. W., and A. J. DUVALL. 1958. Distribution and migration of races of the mourning dove. *Condor* 60:108–128.

ALEVIZON, W. S. 1976. Mixed schooling and its possible significance in a tropical western Atlantic parrotfish and surgeonfish. *Copeia* 1976:796–798.

ALLAN, R. G. 1962. The Madeiran storm petrel *Oceanodroma castro*. *Ibis* 103b:274–295.

ALLEN, A. A. 1914. The red-winged blackbird: a study in the ecology of a cat-tail marsh. *Abstracts and Proceedings of the Linnaean Society of New York* 24–25:43–128.

ALLEN, J. A. 1976. Further evidence for apostatic selection by wild passerine birds.—9: 1 experiments. *Heredity* 36:173–180.

ALTMANN, M. 1952 Social behavior of elk, *Cervus canadensis nelsoni,* in the Jackson Hole area of Wyoming. *Behaviour* 4:116–143.

———. 1956. Patterns of herd behavior in free-ranging elk of Wyoming. *Zoologica* 41:65–71.

AMADON, D. 1967. Galapagos finches grooming marine iguanas. *Condor* 69:311.

ANDERSEN, D. C., K. B. ARMITAGE, and R. S. HOFFMANN. 1976. Socioecology of marmots: female reproductive strategies. *Ecology* 57:552–560.

ANDERSSON, M., and C. G. WIKLUND. 1978. Clumping versus spacing out: experiments on nest predation in fieldfares (*Turdus pilaris*). *Animal Behaviour* 26:1207–12.

ANDREW, R. J. 1972. Changes in search behaviour in male and female chicks, following different doses of testosterone. *Animal Behaviour* 20:741–750.

ANDREW, R. J., and L. J. ROGERS. 1972. Testosterone, search behaviour and persistence. *Nature* 237:343–346.

ANDREWARTHA, H. G., and L. C. BIRCH. 1954. *The distribution and abundance of animals*. Chicago: University of Chicago Press.

ANDREWS, R. M. 1971. Structural habitat and time budget of a tropical *Anolis* lizard. *Ecology* 52:262–270.

ARCHER, J. 1970. Effects of population density on behaviour in rodents. In J. H. Crook, ed., *Social behaviour in birds and mammals*. New York: Academic Press, pp. 169–210.

ARMITAGE, K. B. 1974. Male behaviour and territoriality in the yellow-bellied marmot. *Journal of Zoology, London* 172:233–265.

ARMSTRONG, E. A. 1947. *Bird display and behaviour*. London: Lindsay Drummond.

———. 1955. *The wren*. London: Collins.

———. 1963. *A study of bird song*. New York: Oxford University Press.

ARNOLD, G. W. 1964. Factors within plant associations affecting the behaviour and performance of grazing animals. *Symposium of the British Ecological Society* 4:133–154.

ASHMOLE, N. P. 1971. Sea bird ecology and the marine environment. In D. S. Farner and J. R. King, eds., *Avian biology*, vol. 1. New York: Academic Press, pp. 223–286.

ASHMOLE, N.P., and S. H. TOVAR. 1968. Prolonged parental care in royal terns and other birds. *Auk* 85:90–100.

ASSEM, J. VAN DEN. 1967. Territory in the three-spined stickleback *Gasterosteus aculeatus* L. *Behaviour, Supplement* 16:1–164.

AUSTIN, G. T. 1976. Sexual and seasonal differences in foraging of ladder-backed woodpeckers. *Condor* 78:317–323.

AUSTIN, G. T., and E. L. SMITH. 1972. Winter foraging ecology of mixed insectivorous bird flocks in oak woodland in southern Arizona. *Condor* 74:17–24.

AUSTIN, O. L. 1949. Site tenacity, a behavior trait in the common tern. *Bird-banding* 20:1–39.

AUSTIN, O. L., JR. 1968. Life histories of North American cardinals, grosbeaks, buntings, towhees, finches, sparrows, and allies. *Bulletin of the United States National Museum* 237:1–1889.

AVERY, R. 1973. External energy sources and the feeding strategy of lizards. *Bulletin of the British Ecological Society* 4 (3):25.

BACH, C., B. HAZLETT, and D. RITTSCHOF. 1976. Effects of interspecific competition on fitness of the hermit crab *Clibanarius tricolor*. *Ecology* 57:579–586.

BAERENDS, G. P., K. A. BRIL, and P. BULT. 1965. Versuche zur Analyse einer erlernten Reizsituation bei einem Schweinsaffen (*Macaca nemestrina*). *Zeitschrift für Tierpsychologie.* 22:394–411.

BAILEY, R. O., and B. D. J. BATT. 1974. Hierarchy of waterfowl feeding with whistling swans. *Auk* 91:488–493.

BAKER, A. J. 1974. Prey-specific feeding methods of New Zealand oystercatchers. *Notornis* 21:219–233.

BAKER, M. C. 1977. Shorebird food habits in the eastern Canadian arctic. *Condor* 79:56–62.

———. 1978. Flocking in the great tit *Parus major:* an important consideration. *American Naturalist* 112:779–781.

BAKER, M. C., and A. E. M. BAKER. 1973. Niche relationships among six species of shorebirds on their wintering and breeding ranges. *Ecological Monographs* 43:193–212.

BAKER, R. R. 1977. *The evolution of animal migration.* London: English Universities Press.

BALDA, R. P., and G. C. BATEMAN. 1971. Flocking and annual cycle of the piñon jay, *Gymnorhinus cyanocephalus. Condor* 73:287–302.

BALLARD, W. B., and R. J. ROBEL. 1974. Reproductive importance of dominant male greater prairie chicken. *Auk* 91:75–85.

BARASH, D. P. 1973a. Territorial and foraging behavior of pika (*Ochotona princeps*) in Montana. *American Midland Naturalist* 89:202–207.

———. 1973b. Social variety in the yellow-bellied marmot (*Marmota flaviventris*). *Animal Behaviour* 21:579–584.

———. 1974a. An advantage of winter flocking in the black-capped chickadee, *Parus atricapillus. Ecology* 55:674–676.

———. 1974b. The evolution of marmot societies: a general theory. *Science* 185:415–420.

———. 1976. Pre-hibernation behavior of free-living hoary marmots, *Marmota caligata. Journal of Mammalogy* 57:182–185.

BARDIN, A. V. 1975. (Behavior of tits and nuthatches at stores of seeds.) *Vestnik Leningrad University* 15:7–14. (In Russian)

BARGHUSEN, H. R. 1975. A review of fighting adaptations in dinocephalians (Reptilia, Therapsida). *Paleobiology* 1:295–311.

BARLOW, G. W. 1974a. Hexagonal territories. *Animal Behaviour* 22:876–878.

———. 1974b. Extraspecific imposition of social grouping among surgeonfishes (Pisces: Acanthuridae). *Journal of Zoology, London* 174:333–340.

BARNARD, C. J. 1979. Predation and the evolution of social mimicry in birds. *American Naturalist* 113:613–618.

BARRY, W. J. 1976. Environmental effects on food hoarding in deermice (*Peromyscus*). *Journal of Mammalogy* 57:731–746.

BARTHOLOMEW, G. A. 1942. The fishing activities of double-crested cormorants on San Francisco Bay. *Condor* 44:13–21.

———. 1970. A model for the evolution of pinniped polygyny. *Evolution* 24:546–559.

———. 1972. Aspects of timing and periodicity of heterothermy. In F. S. South, J. P. Hannon, J. R. Willis, E. T. Pengelley, and N. R. Alpert, eds.,

Hibernation and hypothermia: perspectives and challenges. Amsterdam: Elsevier, pp. 663–680.

BARTHOLOMEW, G. A., and T. M. CASEY. 1977. Endothermy during terrestrial activity in large beetles. *Science* 195:882–883.

BARTHOLOMEW, G. A., F. N. WHITE, and T. R. HOWELL. 1976. The thermal significance of the nest of the sociable weaver *Philetairus socius:* summer observations. *Ibis* 118:402–410.

BATEMAN, G. C., and R. P. BALDA. 1973. Growth, development, and food habits of young piñon jays. *Auk* 90:39–61.

BATTEN, L. A. 1971. Bird population changes on farmland and in woodland for the years 1968–69. *Bird Study* 18:1–8.

BATTEN, L. A., and J. H. MARCHANT. 1977. Bird population changes for the years 1974–75. *Bird Study* 24:55–61.

BEACH, F. 1950. The snark was a boojum. *American Psychologist* 5:115–124.

BECK, B. B. 1976. Tool use by captive pigtailed macaques. *Primates* 17:301–310.

BECK, T. D. I. 1977. Sage grouse flock characteristics and habitat selection in winter. *Journal of Wildlife Management* 41:18–26.

BEITINGER, T. L., and J. J. MAGNUSON. 1975. Influence of social rank and size on thermoselection behavior of bluegill (*Lepomis macrochirus*). *Journal of the Fisheries Research Board of Canada* 32:2133–36.

BELLAIRS, A. 1969. *The life of reptiles,* vols. 1 and 2. London: Weidenfeld and Nicolson.

BELT, T. 1874. *The naturalist in Nicaragua.* London: John Murray.

BENINDE, J. 1937. Naturgeschichte des Rothirsches. *Monographie Wildsäugetiere, IV.* Leipzig: P. Schöps.

BENT, A. C. 1927. Life histories of North American shore birds, pt. 1. *Bulletin of the United States National Museum* 142:1–420.

———. 1932. Life histories of North American gallinaceous birds. *Bulletin of the United States National Museum* 162:1–490.

———. 1937. Life histories of North American birds of prey, pt. 1. *Bulletin of the United States National Museum* 167:1–409.

———. 1938. Life histories of North American birds of prey, pt. 2. *Bulletin of the United States National Museum* 170:1–482.

———. 1948. Life histories of North American nuthatches, wrens, thrashers, and their allies. *Bulletin of the United States National Museum* 195:1–475.

———. 1953. Life histories of North American wood warblers. *Bulletin of the United States National Museum* 203:1–774.

———. 1958. Life histories of North American blackbirds, orioles, tanagers, and allies. *Bulletin of the United States National Museum* 211:1–549.

BERGER, J. 1978. Group size, foraging, and antipredator ploys: an analysis of bighorn sheep decisions. *Behavioral Ecology and Sociobiology* 4:91–99.

BERGERUND, A. T. 1974. The role of the environment in the aggregation, movement and disturbance behaviour of caribou. In V. Geist and F. Walther, eds., *The behaviour of ungulates and its relation to management.*

Morges, Switzerland: International Union for the Conservation of Nature, pp. 552–584.

BERLYNE, D. E. 1966. Curiosity and exploration. *Science* 153:25–33.

BERNSTEIN, I. S. 1976. Dominance, aggression and reproduction in primate societies. *Journal of Theoretical Biology* 60:459–472.

BERTRAM, B. C. R. 1975. The social system of lions. *Scientific American* 232 (5) :54–65.

———. 1978. Living in groups: predators and prey. In J. R. Krebs and N. B. Davies, eds., *Behavioural ecology: an evolutionary approach*. Sunderland, Mass.: Sinauer Associates, pp. 64–96.

BEUKEMA, J. J. 1968. Predation by the three-spined stickleback (*Gasterosteus acculeatus* L.) : the influence of hunger and experience. *Behaviour* 31:1–126.

BEUSEKOM, C. F. VAN. 1972. Ecological isolation with respect to food between sparrowhawk and goshawk. *Ardea* 60:72–96.

BISHOP, J. A. 1972. An experimental study of the cline of industrial melanism in *Biston betularia* (L.) (Lepidoptera) between urban Liverpool and rural North Wales. *Journal of Animal Ecology* 41:209–243.

BLACK, C. P. 1975. The ecology and bioenergetics of the northern black-throated blue warbler (*Dendroica caerulescens caerulescens*) . Ph.D. thesis, Dartmouth College.

BLEST, A. D. 1961. The concept of ritualization. In W. H. Thorpe and O. L. Zangwill, eds., *Current problems in animal behaviour*. Cambridge: Cambridge University Press, pp. 102–124.

BOCK, W. J. 1961. Salivary glands in the gray jays (*Perisoreus*) . *Auk* 78:355–365.

BOCK, W. J., R. P. BALDA, and S. B. VANDER WALL. 1973. Morphology of the sublingual pouch and tongue musculature in Clark's nutcracker. *Auk* 90:491–519.

BOER, M. H. DEN. 1971. A colour-polymorphism in caterpillars of *Bupalus piniarius* (L.) (Lepidoptera: Geometridae) . *Netherlands Journal of Zoology* 21:61–116.

BOGERT, C. M. 1949. Thermoregulation in reptiles: a factor in evolution. *Evolution* 3:195–211.

———. 1959. How reptiles regulate their body temperature. *Scientific American* 200 (4) :105–120.

BOVBJERG, R. 1970. Ecological isolation and competitive exclusion in two crayfish (*Orconectes virilis* and *Orconectes immunis*) . *Ecology* 51:225–236.

BOWMAN, R. I. 1961. Morphological differentiation and adaptation in the Galapagos finches. *University of California Publications in Zoology* 58:1–326.

BOWMAN, R. I., and S. L. BILLEB. 1965. Blood-eating in a Galapagos finch. *Living Bird* 4:29–44.

BOWMAN, R. I., and A. CARTER. 1971. Egg-pecking behavior in Galapagos mockingbirds. *Living Bird* 10:243–270.

BRAITHWAITE, R. W. 1974. Behavioural changes associated with the population cycle of *Antechinus stuartii* (Marsupialia) . *Australian Journal of Zoology* 22:45–62.

BRANCH, G. M. 1975. Mechanisms reducing intraspecific competition in *Patella* spp.: migration, differentiation and territorial behaviour. *Journal of Animal Ecology* 44:575–600.

———. 1976. Interspecific competition experienced by South African *Patella* species. *Journal of Animal Ecology* 45:507–529.

BRATTSTROM, B. H. 1963. A preliminary review of the thermal requirements of amphibians. *Ecology* 44:238–255.

BREDER, C. M., JR. 1959. Studies on social groupings in fishes. *Bulletin of the American Museum of Natural History* 117:393–482.

———. 1965. Vortices and fish schools. *Zoologica* 50:977–114.

———. 1976. Fish schools as operational structures. *Fisheries Bulletin* 74:471–502.

BRIAN, M. V. 1955. Food collection by a Scottish ant community. *Journal of Animal Ecology* 24:336–351.

———. 1956. The natural density of *Myrmica rubra* and associated ants in West Scotland. *Insectes Sociaux* 3:474–487.

BRIAN, M. V., and A. D. BRIAN. 1951. Insolation and ant population in the west of Scotland. *Transactions of the Royal Entomological Society of London* 102:303–330.

BROCK, V. E., and R. H. RIFFENBURGH. 1960. Fish schooling: a possible factor in reducing predation. *Journal du Conseil, Conseil Permanent International pour l'Exploration de la Mer* 25:307–317.

BRODIE, P. F. 1975. Cetacean energetics, an overview of intraspecific size variation. *Ecology* 56:152–161.

BROOKS, W. S. 1968. Comparative adaptations of the Alaskan redpolls to the Arctic environment. *Wilson Bulletin* 80:253–280.

BROSSET, A. 1973. Evolution des Accipiter forestiers de l'est du Gabon. *Alauda* 41:185–202.

———. 1974. La nidification des oiseaux en foret Gabonaise: architecture, situation des nids et predation. *La Terre et la Vie* 28:579–610.

BROWN, J. H. 1971a. The desert pupfish. *Scientific American* 225 (5) :104–110.

———. 1971b. Mechanisms of competitive exclusion between two species of chipmunks. *Ecology* 52:305–311.

BROWN, J. H., and C. R. FELDMETH. 1971. Evolution in constant and fluctuating environments: thermal tolerances of desert pupfish (*Cyprinodon*). *Evolution* 25:390–398.

BROWN, J. L. 1963a. Social organization and behavior of the Mexican jay. *Condor* 65:126–153.

———. 1963b. Aggressiveness, dominance and social organization in the Steller jay. *Condor* 65:460–484.

———. 1964. The evolution of diversity in avian territorial systems. *Wilson Bulletin* 76:160–169.

———. 1969. Territorial behavior and population regulation in birds. *Wilson Bulletin* 81:293–329.

———. 1970. Cooperative breeding and altruistic behavior in the Mexican jay, *Aphelocoma ultramarina*. *Animal Behaviour* 18:366–378.

————. 1972. Communal feeding of nestlings in the Mexican jay (*Aphelocoma ultramarina*): interflock comparisons. *Animal behaviour* 20:395–403.

————. 1975a. *The evolution of behavior.* New York: Norton.

————. 1975b. Helpers among Arabian babblers *Turdoides squamiceps*. *Ibis* 117:243–244.

————. 1978. Avian communal breeding systems. *Annual Review of Ecology and Systematics* 9:123–155.

BROWN, J. L., and R. P. BALDA. 1977. The relationships of habitat quality to group size in Hall's babbler (*Pomatostomus halli*). *Condor* 79:312–320.

BROWN, J. L., and G. H. ORIANS. 1970. Spacing patterns in mobile animals. *Annual Review of Ecology and Systematics* 1:239–262.

BROWN, L. E. 1966. Home range and movement of small mammals. *Symposium of the Zoological Society of London* 18:111–142.

BROWN, L., and D. AMADON, 1969. *Eagles, hawks, and falcons of the world.* Feltham, England: Country Life.

BROWN, P. E., and M. G. DAVIES. 1949. *Reed warblers.* East Molesey, Surrey: Foy Publishing Co.

BROWN, R. G. B. 1969. Seed selection by pigeons. *Behaviour* 34:115–131.

BRYAN, J. E., and P. A. LARKIN. 1972. Food specialization by individual trout. *Journal of the Fisheries Research Board of Canada* 29:1615–24.

BUCKLEY, F. G., and P. A. BUCKLEY. 1974. Comparative feeding ecology of wintering adult and juvenile royal terns (Aves: Laridae, Sterninae). *Ecology* 55:1053–63.

BUCKLEY, P. A., and F. G. BUCKLEY. 1977. Hexagonal packing of royal tern nests. *Auk* 94:36–43.

BULLER, W. L. 1888. *A history of the birds of New Zealand,* 2d ed. London: W. L. Buller.

BULLIS, H. R., JR. 1961. Observations on the feeding behavior of white-tip sharks on schooling fishes. *Ecology* 42:194–195.

BÜRKLI, W. 1973. Tannenmeisen lagern in Winter Vorräte von Zudkmücken an. *Ornithologische Beobachter* 70:135–136.

BURT, W. H. 1943. Territoriality and home range concepts as applied to mammals. *Journal of Mammalogy* 24:346–352.

BUSKIRK, W. H. 1972. Foraging ecology of bird flocks in a tropical forest. Ph.D. thesis, University of California, Davis.

BYGOTT, J. D. 1972. Cannibalism among wild chimpanzees. *Nature* 238:410–411.

CADE, T. J. 1967. Ecological and behavioral aspects of predation by the northern shrike. *Living Bird* 6:43–86.

CAIN, A. J., and P. M. SHEPPARD. 1954. Natural selection in *Cepaea*. *Genetics* 39:89–116.

CALDER, W. A. 1973. Microhabitat selection during nesting of hummingbirds in the Rocky Mountains. *Ecology* 54:127–134.

————. 1974. The thermal and radiant environment of a winter hummingbird nest. *Condor* 76: 268–273.

CALDER, W. A., and J. BOOSER. 1973. Hypothermia of broad-tailed hummingbirds during incubation in nature with ecological correlations. *Science* 180:751–753.

CALHOUN, J. B. 1963. The social use of space. In W. V. Mayer and R. C. van Gelder, eds., *Physiological mammalogy*, vol. 1. New York: Academic Press, pp. 1–187.

CAMPANELLA, P. J., and L. L. WOLF. 1974. Temporal leks as a mating system in a temperate zone dragonfly (Odonata: Anisoptera). I, *Plathemis lydia* (Drury). *Behaviour* 51:49–87.

CARACO, T., and L. L. WOLF. 1975. Ecological determinants ot group size of foraging lions. *American Naturalist* 109:343–352.

CARL, E. A. 1971. Population control in arctic ground squirrels. *Ecology* 52:395–413.

CARLQUIST, S. 1974. *Island biology.* New York: Columbia University Press.

CARPENTER, C. C. 1966. Comparative behavior of the Galapagos lava lizards (*Tropidurus*). In R. I Bowman, ed., *The Galapagos.* Berkeley and Los Angeles: University of California Press.

CARPENTER, C. R. 1958. Territoriality: a review of concepts and problems. In A. Roe and G. G. Simpson, eds., *Behavior and evolution.* New Haven: Yale University Press, pp. 224–250.

CARPENTER, F. L. 1974. Torpor in an Andean hummingbird: its ecological significance. *Science* 183:545–547.

———. 1978. A spectrum of nectar-eater communities. *American Zoologist* 18:809–819.

CARPENTER, F. L., and R. E. MACMILLEN. 1976. Threshold model of feeding territoriality and test with a Hawaiian honeycreeper. *Science* 194:639–642.

CASEY, T. M. 1976. Activity patterns, body temperature and thermal ecology in two desert caterpillars (Lepidoptera: Sphingidae). *Ecology* 57:485–497.

CATCHPOLE, C. K. 1972. A comparative study of territoriality in the reed warbler (*Acrocephalus scirpaceus*) and sedge warbler (*A. schoenbaenus*). *Journal of Zoology, London* 166:213–231.

———. 1973. Conditions of co-existence in sympatric breeding populations of *Acrocephalus* warblers. *Journal of Animal Ecology* 42:623–635.

———. 1974. Habitat selection and breeding success in the reed warbler (*Acrocephalus scirpaceus*). *Journal of Animal Ecology* 43:363–380.

CEDERHOLM, G., and J. EKMAN. 1976. A removal experiment on crested tit *Parus cristatus* and willow tit *P. montanus* in the breeding season. *Ornis Scandinavica* 7:207–213.

CHALMERS, N. R., and T. E. ROWELL. 1971. Behaviour and female reproductive cycles in a captive group of mangabeys. *Folia Primatologica* 14:1–14.

CHANCE, M. R. A., and W. M. S. RUSSELL. 1959. Protean displays: a form of allaesthetic behaviour. *Proceedings of the Zoological Society of London* 132:65–70.

CHAPIN, J. P. 1932. The birds of the Belgian Congo, pt. 1. *Bulletin of the American Museum of Natural History* 65:1–736.

———. 1939. The birds of the Belgian Congo, pt. 2. *Bulletin of the American Museum of Natural History* 75:1–632.

CHAPLIN, S. B. 1974 Daily energetics of the black-capped chickadee, *Parus atricapillus*, in winter. *Journal of Comparative Physiology* 89:321–330.

CHAPMAN, R. M., and N. LEVY. 1957. Hunger drive and reinforcing effect of novel stimuli. *Journal of Comparative and Physiological Psychology* 50: 233–238.

CHAPPUIS, C. 1971. Un example de l'influence du milieu sur les emissions vocales des oiseaux: l'evolution des chants en foret equatoriale. *La Terre et la Vie* 1971:183–202.

CHARNOV, E. L. 1976a. Optimal foraging: attack strategy of a mantid. *American Naturalist* 110:141–151.

———. 1976b. Optimal foraging: the marginal value theorem. *Theoretical Population Biology* 9:129–136.

CHARNOV, E. L., and J. R. KREBS. 1975. The evolution of alarm calls: altruism or manipulation? *American Naturalist* 109:107–112.

CHENEY, D. L. 1978. Interactions of immature male and female baboons with adult females. *Animal Behaviour* 26:389–408.

CHENEY, D. L., and R. M. SEYFARTH. 1977. Behaviour of adult and immature male baboons during inter-group encounters. *Nature* 269:404–406.

CHILDRESS, J. R. 1972. Behavioral ecology and fitness theory in a tropical hermit crab. *Ecology* 53:960–964.

CHRISTIAN, J. J. 1970. Social subordination, population density, and mammalian evolution. *Science* 168:84–90.

CLARKE, B. C. 1969. The evidence for apostatic selection. *Heredity* 24:347–352.

CLENCH, H. K. 1966. Behavioral thermoregulation in butterflies. *Ecology* 47:1021–34.

CLOUDSLEY-THOMPSON, J. L. 1972. Temperature regulation in desert reptiles. *Symposium of the Zoological Society of London* 31:39–59.

CLUTTON-BROCK, T. H., ed. 1977. *Primate ecology: studies of feeding and ranging behaviour in lemurs, monkeys and apes*. London: Academic Press.

CLUTTON-BROCK, T. H., and P. H. HARVEY. 1977. Primate ecology and social organization. *Journal of Zoology, London* 183:1–39.

———. 1978. Mammals, resources, and reproductive strategies. *Nature* 273: 191–195.

CODY, M L. 1969. Convergent characteristics in sympatric species: a possible relation to interspecific competition and aggression. *Condor* 71:222–239.

———. 1971. Finch flocks in the Mohave Desert. *Theoretical Population Biology* 2:142–158.

———. 1973. Character convergence. *Annual Review of Ecology and Systematics* 4:189–211.

———. 1974. Competition and the structure of bird communities. *Monographs in Population Biology* 7:1–318.

CODY, M. L., and J. H. Brown. 1970. Character convergence in Mexican finches. *Evolution* 24:304–310.

COLE, G. F. 1972. Grizzly bear-elk relationships in Yellowstone National Park. *Journal of Wildlife Management* 36:556–561.

COLE, L. C. 1946. A study of the Cryptozoa of an Illinois woodland. *Ecological Monographs* 16:49–86.

COLLIAS, N. E. 1944. Aggressive behavior among vertebrate animals. *Physiological Zoology* 17:83–123.

COLLOPY, M. W. 1977. Food caching by female American kestrels in winter. *Condor* 79:63–68.

CONDER, P. 1949. Individual distance. *Ibis* 91: 649–655.

CONNELL, J. 1975. Some mechanisms producing structure in natural communities: a model and evidence from field experiments. In M. L. Cody and J. M. Diamond, eds.. *Ecology and evolution of communities*. Cambridge: Harvard University Press, pp. 460–490.

COOK, D. C. 1978. Foraging behaviour and food of grey herons *Ardea cinerea* on the Ythan estuary. *Bird Study* 25:17–22.

COPPINGER, R. P. 1969. The effect of experience and novelty on avian feeding behavior with reference to the evolution of warning coloration in butterflies. I, Reactions of wild-caught adult blue jays to novel insects. *Behaviour* 35:45–60.

———. 1970. The effect of experience and novelty on avian feeding behavior with reference to the evolution of warning coloration in butterflies. II, Reactions of naive birds to novel insects. *American Naturalist* 104:323–335.

CORKHILL, P. 1973. Food and feeding ecology of puffins. *Bird Study* 20:207–220.

COTT, H. B. 1940. *Adaptive colouration in animals*. London: Methuen.

———. 1961. Scientific results of an enquiry into the ecology and economic status of the Nile crocodile (*Crocodilus niloticus*) in Uganda and Northern Rhodesia. *Transactions of the Zoological Society of London* 29:211–356.

COULSON, J. C. 1966. The influence of the pair-bond and age on the breeding biology of the kittiwake gull *Rissa tridactyla*. *Journal of Animal Ecology* 35:269–279.

COWIE, R. J. 1977. Optimal foraging in great tits (*Parus major*). *Nature* 268:137–139.

COX, G. W. 1961. The relation of energy requirements of tropical finches to distribution and migration. *Ecology* 42:253–266.

CRAIG, R. B. 1978. An analysis of the predatory behavior of the loggerhead shrike. *Auk* 95:221–234.

CRAIG, R. B., D. L. DEANGELIS, and K. R. DIXON. 1979. Long- and short-term dynamic optimization models with application to the feeding strategy of the loggerhead shrike. *American Naturalist* 113:31–51.

CRAIGHEAD, J. J., and F. C. CRAIGHEAD, JR. 1956. *Hawks, owls and wildlife*. Harrisburg, Pa.: Stackpole Publishing Co.

CRISP, D. J. 1967. Chemical factors inducing settlement in *Crassostrea virginica* (Gmelin). *Journal of Animal Ecology* 36:329–335.

CRISP, D. J., and P. S. MEADOWS. 1963. Adsorbed layers: the stimulus to settlement in barnacles. *Proceedings of the Royal Society B* 158:364–387.

CROOK, J. H. 1965. The adaptive significance of avian social organization. *Symposium of the Zoological Society of London* 14:181–218.

———. 1970a. Social organization and the environment: aspects of contemporary social ethology. *Animal Behaviour* 18:197–209.

———. 1970b. The socio-ecology of primates. In J. H. Crook, ed., *Social behaviour in birds and mammals*. New York: Academic Press, pp. 103–166.

CROWCROFT, P. 1955. Territoriality in wild house mice, *Mus musculus*. *Journal of Mammalogy* 36:299–301.

CROWELL. K. L. 1962. Reduced interspecific competition among the birds of Bermuda. *Ecology* 43:75–88.

CROXALL, J. P. 1976. The composition and behaviour of some mixed-species bird flocks in Sarawak. *Ibis* 118:333–346.

———. 1977. Feeding behaviour and ecology of New Guinea rainforest insectivorous passerines. *Ibis* 119:113–136.

CROZE, H. 1970. Searching image in carrion crows. *Zeitschrift für Tierpsychologie, Beiheft* 5:1–86.

CURIO, E. 1969. Funktionsweise und Stammesgeschichte des Flugfeinderkennens einiger Darwinfinken (Geospizinae). *Zeitschrift für Tierpsychologie* 26:394–487.

———. 1975. The functional organization of anti-predator behaviour in the pied flycatcher: a study of avian visual perception. *Animal Behaviour* 23: 1–115.

———. 1976. *The ethology of predation*. New York: Springer-Verlag.

———. 1978. The adaptive significance of avian mobbing. I, Teleonomic hypotheses and predictions. *Zeitschrift für Tierpsychologie* 48:175–183.

DARE, P. J. 1966. The breeding and wintering populations of the oystercatcher (*Haematopus ostralegus* Linnaeus) in the British Isles. *Fisheries Investigations*, ser. 2, 25 (5) :1–69.

DARE, P. J., and A. J. MERCER. 1973. Foods of the oystercatcher in Morecambe Bay, Lancashire. *Bird Study* 20:173–184.

DARLEY, J. A., D. M. SCOTT, and N. K. TAYLOR. 1977. Effects of age, sex, and breeding success on site fidelity of gray catbirds. *Bird-banding* 48:145–151.

DARLING, F. F. 1937. *A herd of red deer*. London: Oxford University Press.

———. 1938. *Bird flocks and the breeding cycle*. Cambridge: Cambridge University Press.

DARROW, R. W. 1947. General habits. In G. Bump, R. W. Darrow, F. C. Edminster, and W. F. Crissey, eds., *The ruffed grouse*. Albany: New York State Conservation Department, pp. 247–298.

DARWIN, C. 1859. *On the origin of species*. London: John Murray.

———. 1872. *The expression of emotions in man and animals*. London: John Murray.

DAVIES, N. B. 1976a. Parental care and the transition to independent feeding in the young spotted flycatcher (*Muscicapa striata*). *Behaviour* 59:280–295.

———. 1976b. Food, flocking and territorial behaviour of the pied wagtail (*Motacilla alba yarrellii* Gould) in winter. *Journal of Animal Ecology* 45:235–253.

DAVIES, N. B., and R. E. GREEN. 1976. The development and ecological significance of feeding techniques in the reed warbler (*Acrocephalus scirpaceus*). *Animal Behaviour* 24:213–229.

DAVIES, N. B., and T. R. HALLIDAY. 1978. Deep croaks and fighting assessment in toads *Bufo bufo. Nature* 274:683–685.

DAVIES, S. J. J. F. 1963. Aspects of the behaviour of the magpie goose *Anseranas semipalmata. Ibis* 105:76–98.

DAVIS, D. E. 1967. The annual rhythm of fat deposition in woodchucks (*Marmota monax*). *Physiological Zoology* 40:391–402.

DAWKINS, M. 1971. Perceptual changes in chicks: another look at the "searching image" concept. *Animal Behaviour* 19:566–574.

DAWKINS, R. 1976. *The selfish gene.* New York: Oxford University Press.

DAWKINS, R., and T. R. CARLISLE. 1976. Parental investment, mate desertion and a fallacy. *Nature* 262:131–132.

DAWKINS, R., and J. R. KREBS. 1978. Animal signals: information or manipulation? In J. R. Krebs and N. B. Davies, eds., *Behavioural ecology: an evolutionary approach.* Sunderland, Mass.: Sinauer Associates, pp. 282–309.

DEAG, J. M. 1977. Aggression and submission in monkey societies. *Animal Behaviour* 25:465–474.

DEBENEDICTIS, P. A. 1966. The bill-brace feeding behavior of the Galapagos finch *Geospiza conirostris. Condor* 68:206–208.

DEVORE, B. I. 1971. The evolution of human society. In J. F. Eisenberg and W. S. Dillon, eds., *Man and beast: comparative social behavior.* Washington, D.C.: Smithsonian Press, pp. 297–311.

DHONDT, A. A. 1971. Some factors influencing territory in the great tit, (*Parus major* L.). *Gerfaut* 61:125–135.

DIAMOND, A. W., and R. W. SMITH. 1973. Returns and survival of banded warblers wintering in Jamaica. *Bird-banding* 44:221–224.

DIAMOND, J. M. 1970. Ecological consequences of island colonization by southwest Pacific birds. II, The effect of species diversity on total population density. *Proceedings of the National Academy of Sciences, U.S.A.* 67:1715–1721.

———. 1978. Niche shifts and the rediscovery of interspecific competition. *American Scientist* 66:322–331.

DICE, L. R. 1947. Effectiveness of selection by owls of deer mice (*Peromyscus maniculatus*) which contrast in color with their background. *Contributions of the Laboratory of Vertebrate Biology, University of Michigan* 34:1–20.

DIMOND, S., and J. LAZARUS. 1974. The problem of vigilance in animal life. *Brain, Behaviour, and Evolution* 9:60–79.

DINSMORE, J. J. 1973. Foraging success of cattle egrets, *Bubulcus ibis. American Midland Naturalist* 89:242–246.

DITTUS, W. P. J. 1977. The social regulation of population density and age-sex distribution in the toque monkey. *Behaviour* 63:281–322.

DIXON, K. L. 1963. Some aspects of social organization in the Carolina chickadee. *Proceedings of the International Ornithological Congress* 13:240–258.

DORST, J. 1974. *The life of birds.* London: Weidenfeld and Nicolson.

DOW, D. D. 1965. The role of saliva in food storage by the gray jay. *Auk* 82:139–154.

————. 1977. Reproductive behavior of the noisy miner, a communally breeding honeyeater. *Living Bird* 16:163–185.

Dow, M. A., and F. von Schilcher. 1975. Aggression and mating success in *Drosophila melanogaster*. *Nature* 254:511–512.

Downhower, J. F., and K. B. Armitage. 1971. The yellow-bellied marmot and the evolution of polygamy. *American Naturalist* 105:355–370.

Drost, R. 1955. Wo verbleiben im Binnenland frei aufgezogene Nordsee-Silbermowen? *Vogelwarte* 18:85–93.

————. 1958. Über die Ansiedlung von jung ins Binnenland verfrachteten Silbermowen (*Larus argentatus*). *Vogelwarte* 19:169–173

Drury, W. H., Jr., and W. J. Smith. 1968. Defense of feeding areas by adult herring gulls and intrusion by young. *Evolution* 22:193–201.

Dunbar, R. I. M., and E. P. Dunbar. 1974. Ecological relations and niche separation between sympatric terrestrial primates in Ethiopia. *Folia Primatologica* 21:36–60.

Dunn, E. K. 1972. Effect of age on the ability of sandwich terns *Sterna sandvichensis*. *Ibis* 114:360–366.

————. 1976. Predation by weasels *Mustela nivalis* on tits. *Ibis* 118:467.

Eaton, G. G. 1976. The social order of Japanese macaques. *Scientific American* 235 (4) :96–106.

Eaton, R. L. 1974. *The cheetah*. New York: Nostrand-Reinhold.

Edman, J. D., and H. W. Kale II. 1971. Host behavior: its influence on the feeding success of mosquitoes. *Annals of the Entomological Society of America* 64:513–516.

Edmunds, M. 1974. *Defence in animals*. Harlow, Essex: Longman.

Ehrlich, P. R., and L. C. Birch. 1967. The "balance of nature" and "population control." *American Naturalist* 101:97–107.

Eibl-Eibesfeldt, I. 1962. Freiwasserbeobachtungen zur Deutung des Schwarmverhaltens verschiedener Fische. *Zeitschrift für Tierpsychologie* 19:165–182.

————. 1964. *Im Reich der tausend Atolle*. Munich: Piper.

————. 1970. *Ethology*. New York: Holt, Rinehart and Winston.

Eisenberg, J. F. 1966. The social organization of mammals. *Handbuch der Zoologie* 8 (39) :1–92.

Eisenberg, J. F., and D. G. Kleiman. 1972. Olfactory communication in mammals. *Annual Review of Ecology and Systematics* 31:1–32.

Eisenberg, J. F., N. A. Muckenhirn, and R. Rudran. 1972. The relation between ecology and social structure in primates. *Science* 176:863–874.

Elner, R. W., and E. N. Hughes. 1978. Energy maximization in the diet of the shore crab, *Carcinus maenas*. *Journal of Animal Ecology* 47:103–116.

Elton, C. 1927. *Animal ecology*. London: Sidgwick and Jackson.

Elton, C., and R. S. Miller. 1954. The ecological survey of classifying habitats by structural characteristics. *Journal of Ecology*. 42:460–496.

El-Wailly, A. J. 1966. Energy requirements for egg-laying and incubation in zebra finch *Taeniopygia castanotis*. *Condor* 68:582–594.

Ely, C. A. 1973. Returns of North American birds to their wintering grounds in southern Mexico. *Bird-banding* 44:228–229.

EMLEN, J. M. 1966. The role of time and energy in food preference. *American Naturalist* 100:611–617.

EMLEN, J. T. 1957. Defended area? A critique of the territory concept and of conventional thinking. *Ibis* 99:352.

EMLEN, S. T. 1978. The evolution of cooperative breeding in birds. In J. R. Krebs and N. B. Davies, eds., *Behavioural ecology: an evolutionary approach.* Sunderland, Mass.: Sinauer Associates, pp. 245–281.

EMLEN, S. T., and H. W. AMBROSE III. 1970. Feeding interactions of snowy egrets and red-breasted mergansers. *Auk* 87:164–165.

EMLEN, S. T., and L. W. ORING. 1977. Ecology, sexual selection, and the evolution of mating systems. *Science* 197:215–223.

ENDERS, F. 1974. Vertical stratification in orb-web spiders (Araneidae, Araneae) and a consideration of other methods of coexistence. *Ecology* 55:317–328.

ENDLER, J. A. 1978. Predators' view of animal color patterns. *Evolutionary Biology* 11:319–364.

ENEMAR, A., and B. SJOSTRAND. 1972. Effects of the introduction of pied flycatchers, *Ficedula hypoleuca,* on the composition of a passerine bird community. *Ornis Scandinavica* 3:79–89.

ERRINGTON, P. L. 1946. Predation and vertebrate populations. *Quarterly Review of Biology* 21:144–177, 221–245.

ERVIN, S. 1977. Flock size, composition, and behavior in a population of bushtits (*Psaltriparus minimus*). *Bird-banding* 48:97–109.

ESTABROOK, G. F., and A. E. DUNHAM. 1976. Optimal diet as a function of absolute abundance, relative abundance, and a relative value of available prey. *American Naturalist* 110:401–413.

ESTES, R. D. 1966. Behaviour and life history of the wildebeest (*Connochaetes taurinus* Burchell). *Nature* 212:999–1000.

———. 1969. Territorial behavior of the wildebeest (*Connochaetes taurinus,* Burchell 1823). *Zeitschrift für Tierpsychologie* 26:284–370.

———. 1974. Social organization of the African Bovidae. In V. Geist, F. Walther, eds., *The behaviour of ungulates and its relation to management.* Morges, Switzerland: International Union for the Conservation of Nature, pp. 166–205.

ESTES, R. D, and J. GODDARD. 1967. Prey selection and hunting behavior of the African wild dog. *Journal of Wildlife Management* 31:52–70.

ETKIN, W. 1964. Co-operation and competition in social behavior. In W. Etkin, ed., *Social behavior and organization among vertebrates.* Chicago: University of Chicago Press, pp. 1–34.

EVANS, P. R. 1969. Winter fat deposition and overnight survival of yellow buntings (*Emberiza citrinella* L.). *Journal of Animal Ecology* 38:415–423.

———. 1976. Energy balance and optimal foraging strategies in shorebirds: some implications for their distributions and movements in the non-breeding season. *Ardea* 64:117–139.

EWER, R. F. 1967. The behaviour of the African giant rat (*Cricetomys gambianus* Waterhouse). *Zeitschrift für Tierpsychologie* 24:6–79.

EWING, L. S., and A. W. EWING. 1973. Correlates of subordinate behaviour in the cockroach, *Nauphoeta cinerea*. *Animal Behaviour* 21:571–578.

FAGEN, R. 1974. Selective and evolutionary aspects of animal play. *American Naturalist* 108:850–858.

FALL, M. W. 1971. Seasonal variations in the food consumption of woodchucks (*Marmota monax*). *Journal of Mammalogy* 52:370–375.

FALLS, J. B. 1969. Functions of territorial song in the white-throated sparrow. In R. A. Hinde, ed., *Bird vocalizations*. Cambridge: Cambridge University Press, pp. 207–232.

FARENTINOS, H. C. 1972. Social dominance and mating activity in the tasseleared squirrel (*Sciurus aberti ferreus*). *Animal Behaviour* 20:316–326.

FEDYK, A. 1971. Social thermoregulation in *Apodemus flavicollis* (Melchior 1834). *Acta Theriologica* 16:221–229.

FEINSINGER, P. 1976. Organization of a tropical guild of nectarivorous birds. *Ecological Monographs* 46:257–291.

FFRENCH, R. P. 1967. The dickcissel on its wintering grounds in Trinidad. *Living Bird* 6:123–140.

FICKEN, M. S., and R. W. FICKEN. 1962. The comparative ethology of the wood warblers: a review. *Living Bird* 1:103–122.

———. 1967. Age-specific differences in the breeding behavior of the American redstart. *Wilson Bulletin* 79:188–199.

FICKEN, R. W., M. S. FICKEN, and J. P. HAILMEN. 1978. Differential aggression in genetically different morphs of the white-throated sparrow (*Zonotrichia albicollis*). *Zeitschrift für Tierpsychologie* 46:43–57.

FINK, B. D. 1959. Observations of porpoise predation on a school of Pacific sardines. *California Fish and Game* 45:210–217.

FISHELSON, L. 1977. Sociobiology of feeding behavior of coral fish along the coral reef of the Gulf of Elat (= Gulf of Aqaba), Red Sea. *Israel Journal of Zoology* 26:114–134.

FISHELSON, L., D. POPPER, and A. AVIDOR. 1974. Biosociology and ecology of pomacentrid fishes around the Sinai Peninsula (northern Red Sea). *Journal of Fish Biology* 6:119–133.

FISHER, H. I., and E. E. DATER. 1961. Esophageal diverticula in the redpoll, *Acanthis flammea*. *Auk* 78:528–531.

FISHER, J., and R. A. HINDE. 1949. The opening of milk bottles by birds. *British Birds* 42:347–357.

FISLER, G. F. 1969. Mammalian organizational systems. *Contributions to Science, Los Angeles County Museum of Natural History* 167:1–32.

———. 1976. Agonistic signals and hierarchy changes of antelope squirrels. *Journal of Mammalogy* 57:94–102.

FOGDEN, M. P. L. 1972. The seasonality and population dynamics of equatorial forest birds in Sarawak. *Ibis* 114:307–343.

FOSSEY, D. 1974. Observations on the home range of one group of mountain gorillas (*Gorilla gorilla beringei*). *Animal Behaviour* 22:568–581.

FOSTER, M. S. 1974a. A model to explain molt breeding overlap and clutch size in some tropical birds. *Evolution* 28:182–190.

————. 1974b. Rain, feeding behavior and clutch size in tropical birds. *Auk* 91:722–726.

————. 1975. The overlap of molting and breeding in some tropical birds. *Condor* 77:304–314.

————. 1976. Nesting biology of the long-tailed manakin. *Wilson Bulletin* 88:400–420.

FOTHERINGHAM, N. 1976. Population consequences of shell utilization by hermit crabs. *Ecology* 57:570–578.

FRANCIS, W. J. 1976. Micrometeorology of a blackbird roost. *Journal of Wildlife Management* 40:132–136.

FRANKIE, G. W. 1976. Pollination of widely dispersed trees by animals in Central America with an emphasis on bee pollination systems. In J. Burley and B. T. Styles, eds, *Tropical trees: variation, breeding, and conservation.* New York: Academic Press, pp. 151–159.

FRASER, A. B. 1973. Foraging strategies in three species of *Fundulus.* M.S. thesis, University of Maryland.

FRASER, D. F. 1976. Coexistence of salamanders in the genus *Plethodon:* a variation of the Santa Rosalia theme. *Ecology* 57:238–251.

FRAZIER, A., and V. NOLAN, JR. 1959. Communal roosting by the eastern bluebird in winter. *Bird-banding* 30:219–226.

FREE, J. B., and C. G. BUTLER. 1959. *Bumblebees.* London: Collins.

FRENCH, N. R., and R. W. HODGES. 1959. Torpidity in cave-roosting hummingbirds. *Condor* 61:223.

FRETWELL, S. D. 1969a. Ecotypic variation in the nonbreeding season in migratory populations: a study of tarsal length in some Fringillidae. *Evolution* 23:406–420.

————. 1969b. Dominance behavior and winter habitat distribution in juncos (*Junco hyemalis*). *Bird-banding* 40:1–25.

————. 1969c. On territorial behaviour and other factors influencing habitat distribution in birds. III, Breeding success in a local population of field sparrows (*Spizella pusilla* Wils.). *Acta Biotheoretica* 19:47–52.

————. 1972. Populations in a seasonal environment. *Monographs in Population Biology* 5:1–218.

FRETWELL, S. D., and J. S. CALVER. 1969. On territorial behaviour and other factors influencing habitat distribution in birds. II, Sex ratio variation in the dickcissel (*Spiza americana* Gmel.). *Acta Biotheoretica* 19:37–46.

FRETWELL, S. D, and H. L. LUCAS, JR. 1969. On territorial behaviour and other factors influencing habitat distribution in birds. I, Theoretical development. *Acta Biotheoretica* 19:16–36.

FRITH, H. J. 1959. Breeding of the mallee fowl, *Leipoa ocellata* Gould (Megapodiidae). *CSIRO Wildlife Research* 4:31–65.

FRY, C. H. 1977. The evolutionary significance of co-operative breeding in birds. In B. Stonehouse and C. Perrins, eds., *Evolutionary ecology.* Baltimore: University Park Press, pp. 127–135.

FRY, F E. J. 1964. Animals in aquatic environments: fishes. In D. B. Dill, ed., *Handbook of physiology: section 4, adaptation to the environment.* Washington, D.C.: American Physiological Society, pp. 715–728.

Fuchs, E. 1977. Predation and anti-predator behaviour in a mixed colony of terns *Sterna* spp. and black-headed gulls *Larus ridibundus* with special reference to the sandwich tern *Sterna sandvicensis*. *Ornis Scandinavica* 8:17–32.

Gaines, M. S., and C. J. Krebs. 1971. Genetic changes in fluctuating vole populations. *Evolution* 25:702–723.

Gartlan, S. J., and T. T. Struhsaker. 1972. Polyspecific associations and niche separation of rain forest anthropoids in Cameroon, West Africa. *Journal of Zoology, London* 168:221–266.

Gass, C. L., G. Angehr, and J. Centa. 1976. Regulation of food supply by feeding territoriality in the rufous hummingbird. *Canadian Journal of Zoology* 54:2046–54.

Gaston, A. J. 1976. Group territorial behaviour in long-tailed tits and jungle babblers. *Ibis* 118 (abstr.) :304.

————. 1977. Social behaviour within groups of jungle babblers (*Turdoides striatus*). *Animal Behaviour* 25:828–848.

————. 1978a. Ecology of the common babbler *Turdoides caudatus*. *Ibis* 120:415–432.

————. 1978b. The evolution of group territorial behavior and cooperative breeding. *American Naturalist* 112:1091–1100.

Gauthreaux, S. A., Jr. 1978. The ecological significance of behavioral dominance. *Perspectives in Ethology* 3:17–54.

Geer, T. A. 1978. Effects of nesting sparrowhawks on nesting tits. *Condor* 80:419–422.

Geist, V. 1965. On the rutting behavior of the mountain goat. *Journal of Mammalogy* 45:551–568.

————. 1966a. The evolution of horn-like organs. *Behaviour* 27:175–214.

————. 1966b. The evolutionary significance of mountain sheep horns. *Evolution* 20:558–566.

————. 1971. *Mountain Sheep*. Chicago: University of Chicago Press.

Gentry, R. L. 1973. Thermoregulatory behavior of eared seals. *Behaviour* 46:73–93.

Gersdorf, E. 1966. Beobachtungen über das Verhalten von Vogelschwärmen. *Zeitschrift für Tierpsychologie* 23:37–43.

Gibb, J. A. 1954. Feeding ecology of tits, with notes on the treecreeper and goldcrest. *Ibis* 96:513–543.

————. 1958. Predation by tits and squirrels on the eucosmid, *Ernarmonia conicolana* (Heyl). *Journal of Animal Ecology* 27:376–396.

————. 1960. Populations of tits and goldcrests and their food supply in pine plantations. *Ibis* 102:163–208.

Gill, F. B. 1971. Review of "Character convergence in Mexican finches" by M. L. Cody. *Bird-banding* 42:150.

Gill, F. B., and L. L. Wolf. 1975a. Foraging strategies and energetics of East African sunbirds at mistletoe flowers. *American Naturalist* 109:491–510.

————. 1975b. Economics of feeding territoriality in the golden-winged sunbird. *Ecology* 56:333–345.

————. 1977. Nonrandom foraging by sunbirds in a patchy environment. *Ecology* 58:1284–1296.

————. 1978. Comparative foraging efficiencies of some montane sunbirds in Kenya. *Condor* 80:391–400.

GILL, J. C., and W. THOMSON. 1956. Observations on the behaviour of suckling pigs. *British Journal of Animal Behaviour* 4:46–51.

GLASE, J. C. 1973. Ecology of social organization in the black-capped chickadee. *Living Bird* 12:235–267.

GODFREY, G., and P. CROWCROFT. 1960. *The life of the mole*. London: Museum Press.

GOODWIN, D. 1968. Some possible functions of sun-bathing in birds. *British Birds* 60:363–364.

GÖRANSSON, G., G. HÖGSTEDT, J. KARLSSON, H. KÄLLANDER, and S. ULFSTRAND. 1974. Sångens foll för revirhållandet hos náktergal *Luscinia luscinia:* några experiment med play-back-teknik. *Vår Fågelvärld* 33:201–209.

GÖRANSSON, G., J. KARLSSON, S. G. NILSSON, and S. ULFSTRAND. 1975. Predation on birds' nests in relation to antipredator aggression and nest density: an experimental study. *Oikos* 26:117–120.

Goss-Custard, J. D. 1970a. The responses of redshank *Tringa totanus* (L) to spatial variations in the density of their prey. *Journal of Animal Ecology* 39:91–113.

————. 1970b. Feeding dispersion in some overwintering wading birds. In J. H. Crook, ed., *Social behaviour in birds and mammals*. New York: Academic Press, p. 3–35.

————. 1977a. Optimal foraging and the size selection of worms by the redshank, *Tringa totanus,* in the field. *Animal Behaviour* 25:10–29.

————. 1977b. The energetics of prey selection by redshank, *Tringa totanus* (L) in relation to prey density. *Journal of Animal Ecology* 46:1–19.

GOTTIER, R. F. 1968. The dominance-submission hierarchy in the social behavior of the domestic chicken. *Journal of Genetical Psychology* 112:205–226.

GOULD, L. L., and F. HEPPNER. 1974. The vee formation of Canada geese. *Auk* 91:494–506.

GOULD, S. J. 1974. The origin and function of "bizarre" structures: antler size and skull size in the "Irish elk," *Megaloceros giganteus. Evolution* 28:191–220.

GRANT, P. R. 1968. Polyhedral territories of animals. *American Naturalist* 102:75–80.

————. 1969. Experimental studies of competitive interaction in a two-species system. I, *Microtus* and *Clethrionomys* species in enclosures. *Canadian Journal of Zoology* 47:1059–82.

————. 1971. The habitat preference of *Microtus pennsylvanicus,* and its relevance to the distribution of this species on islands. *Journal of Mammalogy* 52:351–361.

————. 1972. Convergent and divergent character displacement. *Biological Journal of the Linnaean Society* 4:39–68.

GRANT, P. R., B. R. GRANT, J. N. M. SMITH, I. J. ABBOTT, and L. K. ABBOTT. 1976. Darwin's finches: population variation and natural selection. *Proceedings of the National Academy of Science, U.S.A.* 73:257–261.

GRANT, W. C., JR., and K. M. ULMER. 1974. Shell selection and aggressive behavior in two sympatric species of hermit crabs. *Biological Bulletin* 145:32–43.

GREER, A. E. 1970. Evolutionary and systematic significance of crocodilian nesting habits. *Nature* 227:523–524.

GRISCOM, L. 1941. The recovery of birds from disaster. *Audubon Magazine* 43:191–196.

GROSSMAN, A. F., and G. C. WEST. 1977. Metabolic rate and temperature regulation of winter acclimatized black-capped chickadees *Parus atricapillus* of interior Alaska. *Ornis Scandinavica* 8:127–138.

GROVES, S. 1978. Age-related differences in ruddy turnstone foraging and aggressive behavior. *Auk* 95:95–103.

GRUBB, T. C., JR. 1973 Absence of "individual distance" in the tree swallow during adverse weather. *Auk* 90:432–433.

————. 1975. Weather-dependent foraging behavior of some birds wintering in a deciduous woodland. *Condor* 77:175–182.

————. 1977. Weather-dependent foraging behavior of some birds wintering in a deciduous woodland: horizontal adjustments. *Condor* 79:241–244.

GRUZDEV, V. V. 1952. (The importance of light in the distribution of insectivorous birds in the forests.) *Zoologeschii Zhurnal* 31:556–563. (In Russian)

GUHL, A. M. 1958. The development of social organization in the domestic chick. *Animal Behaviour* 6:92–111.

————. 1962. The social environment and behaviour. In E. S. E. Hafez, ed., *The behaviour of domestic animals*. Baltimore: Williams and Wilkins, pp. 96–108.

GUTHRIE, J. F., and R. L. KROGER. 1974. Schooling habits of injured and parasitized menhaden. *Ecology* 55:208–210.

GWINNER, E. 1965. Über den Einfluss des Hungers und anderen Faktoren auf die Verstuck-Aktivität des Kolkraben (*Corvus corax*). *Vogelwarte* 23:1–4.

HAARTMAN, L. VON. 1949. Die Trauerfliegenschnäpper. I, Ortstreue und Rassenbildung. *Acta Zoologica Fennica* 56:1–104.

————. 1973. Talmespopulationen på Lemsjöholm. *Lintumies* 1:7–9. (Finnish with English summary)

HAFTORN, S. 1956. Contribution to the food biology of tits, especially about storing of surplus food. IV, A comparative analysis of *Parus atricapillus* L., *P. cristatus* L. and *P. ater* L. *Det Kgl Norske Videnskabers Selskabs Skrifter* 1956 (4):1–54.

————. 1972. Hypothermia of tits in the Arctic winter. *Ornis Scandinavica* 3:153–166.

————. 1973. (A study of the Siberian tit *Parus cinctus* during the breeding season.) *Sterna* 12:91–155. (Norwegian with English summary)

————. 1974. Storage of surplus food by the boreal chickadee *Parus hudsonicus* in Alaska, with some records on the mountain chickadee *Parus gambeli* in Colorado. *Ornis Scandinavica* 5:145–161.

HAIGH, J., and J. MAYNARD SMITH. 1972. Can there be more predators than prey? *Theoretical Population Biology* 3:290–299.

HAINSWORTH, F. R., B. G. COLLINS, and L. L. WOLF. 1977. The function of torpor in hummingbirds. *Physiological Zoology* 50:215–222.

HAIRSTON, N. G., F. E. SMITH, and L. B. SLOBODKIN. 1960. Community structure, population control, and competition. *American Naturalist* 94:421–425.

HAMERSTROM, F. 1942. Dominance in winter flocks of chickadees. *Wilson Bulletin* 54:32–42.

HAMILTON, W. D. 1964. The genetical evolution of social behavior. *Journal of Theoretical Biology* 7:1–52.

———. 1971. Geometry for the selfish herd. *Journal of Theoretical Biology* 31:295–311.

———. 1972. Altruism and related phenomena, mainly in social insects. *Annual Review of Ecology and Systematics* 3:193–232.

HAMILTON, W. J., JR. 1939. *American mammals.* New York: McGraw-Hill.

HAMILTON, W. J., III. 1973. *Life's color code.* New York: McGraw-Hill.

HAMILTON, W. J., III, R. E. BUSKIRK, and W. H. BUSKIRK. 1975. Defensive stoning by baboons. *Nature* 256:488.

———. 1976. Defense of space and resources by chacma (*Papio ursinus*) baboon troops in an African desert and swamp. *Ecology* 57:1264–72.

HAMILTON, W. J., III, and C. D. BUSSE. 1978. Primate carnivory and its significance to human diets. *Bioscience* 28:761–766.

HAMILTON, W. J., III, and W. M. GILBERT. 1969. Starling dispersal from a winter roost. *Ecology* 50:886–898.

HAMILTON, W. J., III, W. M. Gilbert, F. H. Heppner, and R. Planck. 1967. Starling roost dispersal and a hypothetical mechanism regulating movement to and away from dispersal centers. *Ecology* 48:825–833.

HAMILTON, W. J., III, and F. H. HEPPNER. 1967. Radiant solar energy and the function of black homeotherm pigmentation: an hypothesis. *Science* 155:196–197.

HAMILTON, W. J., III, and K. E. F. WATT. 1970. Refuging. *Annual Review of Ecology and Systematics* 1:263–286.

HAMMEL, J. T. 1956. Infrared emissivities of some arctic fauna. *Journal of Mammalogy* 37:375–378.

HARRINGTON, B. A., and S. GROVES. 1977. Aggression in foraging migrant semipalmated sandpipers. *Wilson Bulletin* 89:336–338.

HARRIS, M. P. 1968. Egg-eating by Galapagos mockingbirds. *Condor* 70:269–270.

———. 1970. Territory limiting the size of the breeding population of the oystercatcher (*Haematopus ostralegus*): a removal experiment. *Journal of Animal Ecology* 39:707–713.

HARRISSON, T. H., and J. N. S. BUCHAN. 1934. A field study of the St. Kilda wren (*Troglodytes troglodytes hirtensis*), with especial reference to its numbers, territory and food habits. *Journal of Animal Ecology* 3:133–145.

HARTLEY, P. H. T. 1953. An ecological study of the feeding habits of the English titmice. *Journal of Animal Ecology* 22:261–288.

HARTMAN, G. F. 1965. The role of behavior in the ecology and interaction of underyearling coho salmon (*Oncorhynchus kisutch*) and steelhead trout (*Salmo gairdneri*). *Journal of the Fisheries Research Board of Canada* 22:1035–81.

HARTWICK, E. B. 1976. Foraging strategy of the black oyster catcher (*Haematopus bachmani* Audubon). *Canadian Journal of Zoology* 54:142–155.

HARTZLER, J. E. 1974. Predation and the daily timing of sage grouse leks. *Auk* 91:532–536.

HARVEY, P. H., and P. J. GREENWOOD. 1978. Anti-predator defence strategies: some evolutionary problems. In J. R. Krebs and N. B Davies, eds, *Behavioural Ecology: an evolutionary approach*. Sunderland, Mass.: Sinauer Associates, pp. 129–151.

HATCH, D. E. 1970. Energy conserving and heat dissipating mechanisms of the turkey vulture. *Auk* 87:111–124.

HATCH, J. J. 1965. Only one species of Galapagos mockingbird feeds on eggs. *Condor* 67:354–355.

HAUSER, D. C. 1957. Some observations on sun-bathing in birds. *Wilson Bulletin* 69:78–90.

HAUSFATER, G. 1975. Dominance and reproduction in baboons. *Contributions in Primatology* 7:1–150.

HAYS, H., E. DUNN, and A. POOLE. 1973. Common, arctic, roseate, and sandwich terns carrying multiple fish. *Wilson Bulletin* 85:233–236.

HAZLETT, B. A. 1970. Interspecific shell fighting in three sympatric species of hermit crabs in Hawaii. *Pacific Science* 24:472–482.

HEATH, J. E. 1962. Temperature fluctuation in the turkey vulture. *Condor* 64:234–235.

HEDIGER, H. 1955. *Studies of the psychology and behavior of captive animals in zoos and circuses*. New York: Criterion Books.

HEINRICH, B. 1974. Thermoregulation in endothermic insects. *Science* 185:747–756.

HELLER, H. C. 1971. Altitudinal zonation of chipmunks (*Eutamias*): interspecific aggression. *Ecology* 52:312–319.

HEPPLESTON, P. B. 1971. The feeding ecology of oystercatchers (*Haematopus ostralegus* L.) in winter in northern Scotland. *Journal of Animal Ecology* 40:651–672.

———. 1972. The comparative breeding ecology of oystercatchers (*Haematopus ostralegus* L.) in inland and coastal habitats. *Journal of Animal Ecology* 41:23–51.

HEPPNER, F. 1970. The metabolic significance of differential absorption of radiant energy by black and white birds. *Condor* 72:50–59.

HERGENRADER, G. L., and A. D. HASLER. 1968. Influence of changing seasons on schooling behavior of yellow perch. *Journal of the Fisheries Research Board of Canada* 25:711–716.

HERREID, C. F., II. 1967. Temperature regulation, temperature preference and tolerance, and metabolism of young and adult free-tailed bats. *Physiological Zoology* 40:1–22.

HERTER, W. R. 1940. Über das "Putten" einiger Meisen-Arten. *Ornithologische Monatsbericht* 48:105–109.

HERTZ, P. E., J. V. REMSEN, JR., and S. I. ZONES. 1976. Ecological complementarity of three sympatric parids in a California oak woodland. *Condor* 78:307–316.

HESPENHEIDE, H. A. 1966. The selection of seed size by finches. *Wilson Bulletin* 78:191–197.

HEWSON, R. 1976. Grazing by mountain hares *Lepus timidus* L., red deer *Cervus elaphus* L. and red grouse *Lagopus 1. scoticus* on heather moorland in north-east Scotland. *Journal of Applied Ecology* 13:657–666.

HICKEY, J. J., ed. 1969. *Peregrine falcon populations: their biology and decline*. Madison: University of Wisconsin Press.

HILDÉN, O. 1965. Habitat selection in birds. *Annales Zoologici Fennici* 2:53–75.

HINDE, R. A. 1952. The behaviour of the great tit (*Parus major*) and some other related species. *Behaviour, Supplement* 2:1–201.

————. 1954a. Factors governing the changes in strength of a partially inborne response, as shown by the mobbing behaviour of the chaffinch (*Fringilla coelobs*). I, The nature of the response, and an examination of its course. *Proceedings of the Royal Society B* 142:306–331.

————. 1954b. Factors governing the changes in strength of a partially inborne response, as shown by the mobbing behaviour of the chaffinch (*Fringilla coelobs*). II, The waning of the response. *Proceedings of the Royal Society B* 142:332–358.

————. 1956. The biological significance of the territories of birds. *Ibis* 98:340–369.

————. 1970. *Animal behaviour*, 2d ed. New York: McGraw-Hill.

————, ed. 1969. *Bird vocalizations*. Cambridge: Cambridge University Press.

HINDE, R. A., and J. FISHER. 1951. Further observations on the opening of milk bottles by birds. *British Birds* 44:393–396.

HOBSON, E. S. 1968. Predatory behavior of some shore fishes in the Gulf of California. *Research Report, Bureau of Sport Fisheries and Wildlife, U.S.A.* 73:1–92.

————. 1972. Activity of Hawaiian reef fishes during the evening and morning transitions between daylight and darkness. *Fisheries Bulletin* 70:715–740.

HOCHBAUM, H. N. 1955. *Travels and traditions of waterfowl*. Minneapolis: University of Minnesota Press.

HOFFMAN, D. L., and P. J. WELDON. 1978. Flight responses of two intertidal gastropods (Prosobranchia: Trochidae) to sympatric predatory gastropods from Barbados. *Veliger* 20:361–366.

HOFFMAN, R. W., and C. E. BRAUN. 1977. Characteristics of a wintering population of white-tailed ptarmigan in Colorado. *Wilson Bulletin* 89:107–115.

HOGLUND, N. H. 1964. Über die Ehrnahrung des Habichts (*Accipiter gentilis* Lin.) in Schweden. *Viltrevy* 2:271–328.

HOGSTAD, O. 1967. Seasonal fluctuation in bird populations within a forest area near Oslo (Southern Norway) in 1966–67. *Nytt Magasin Zoologie* 15:81–96.

———. 1975. Quantitative relations between hole-nesting and open-nesting species within a passerine breeding community. *Norwegian Journal of Zoology* 23:261–267.

———. 1978a. Sexual dimorphism in relation to winter foraging and territorial behaviour of the three-toed woodpecker *Picoides tridactylus* and three *Dendrocopos* species. *Ibis* 120:198–203.

———. 1978b. Differentiation of foraging niche among tits, *Parus* spp., in Norway during winter. *Ibis* 120:139–146.

HOLLING, C. S. 1959. The components of predation as revealed by a study of small-mammal predation of the European pine sawfly. *Canadian Entomologist* 91:293–320.

HOLM, C. H. 1973. Breeding sex ratios, territoriality, and reproductive success in the red-winged blackbird (*Agelaius phoeniceus*). *Ecology* 54:356–365.

HOLMES, R. T., T. W. SHERRY, and S. E. BENNETT. 1978. Diurnal and individual variability in the foraging behavior of American redstarts (*Setophaga ruticilla*). *Oecologia* 36:141–149.

HOLYOAK, D. T. 1969. Sex-differences in feeding behaviour and size in the carrion crow. *Ibis* 112:397–400.

HOPSON, J. A. 1975. The evolution of cranial display structures in hadrosaurian dinosaurs. *Paleobiology* 1:21–43.

HORN, H. S. 1968. The adaptive significance of colonial nesting in the Brewer's blackbird (*Euphagus cyanocephalus*). *Ecology* 49:682–694.

HORNOCKER, M. G. 1969. Winter territoriality in mountain lions. *Journal of Wildlife Management* 33:457–464.

———. 1970. An analysis of mountain lion predation upon mule deer and elk in the Idaho Primitive Area. *Wildlife Monographs* 21:1–39.

HOWARD, R. D. 1974. The influence of sexual selection and interspecific competition on mockingbird song (*Mimus polyglottos*). *Evolution* 28:428–438.

HOWE, H. F. 1976. Egg size, hatching asynchrony, sex, and brood reduction in the common grackle. *Ecology* 57:1195–1207.

———. 1979. Evolutionary aspects of parental care in the common grackle, *Quiscalus quiscula* L. *Evolution* 33:41–51.

HOWLAND, H. C. 1974. Optimal strategies for predator avoidance: the relative importance of speed and manoeuvrability. *Journal of Theoretical Biology* 47:333–350.

HUDSON, J. W. 1967. Variation in the pattern of torpidity of small homeotherms. In K. C. Fisher, A. R. Dawe, C. P. Lyman, E. Schonbaun, and F. E. South, eds., *Mammalian hibernation, III.* Edinburgh: Oliver and Boyd.

HUEY, R. B. 1974. Behavioral thermoregulation in lizards: importance of associated costs. *Science* 184:1001–3.

HUEY, R. B., and T. P. WEBSTER. 1975. Thermal biology of a solitary lizard: *Anolis marmoratus* of Guadeloupe, Lesser Antilles. *Ecology* 56:445–452.

HUGHES, R. N. 1979. Optimal diets under the energy maximization premise: the effects of recognition time and learning. *American Naturalist* 113:209–221.

HUMPHREYS, W. F. 1974. Behavioural thermoregulation in a wolf spider. *Nature* 251:502–503.

HUMPHRIES, D. A., and P. M. DRIVER. 1967. Erratic display as a device against predators. *Science* 156:1767–68.

———. 1970. Protean defence by prey animals. *Oecologia* 5:285–302.

HUNT, G. L., JR. 1972. Influence of food distribution and human disturbance on the reproductive success of herring gulls. *Ecology* 53:1051–61.

HUNTER, J. R. 1968. Effects of light on schooling and feeding of jack mackerel, *Trachurus symmetricus*. *Journal of the Fisheries Research Board of Canada* 25:393–407.

HURLEY, A. C. 1978. School structure of the squid *Loligo opalescens*. *Fisheries Bulletin* 76:433–442.

HUTCHINSON, G. E. 1951. Copepodology for the ornithologist. *Ecology* 32:571–577.

———. 1957. Concluding remarks. *Cold Spring Harbor Symposium of Quantitative Biology* 22:415–427.

———. 1959. Homage to Santa Rosalia *or* why are there so many kinds of animals? *American Naturalist* 93:145–159.

HUTCHINSON, G. E., and R. H. MACARTHUR. 1959. On the theoretical significance of aggressive neglect in interspecific competition. *American Naturalist* 93:133–134.

HUTCHINSON, J. C. D. 1954. Heat regulation in birds. In J. Hammond, ed., *Progress in the physiology of farm animals*, vol. 1. London: Butterworth, pp. 299–362.

IMMELMAN, K. 1963. Drought adaptations in Australian desert birds. *Proceedings of International Ornithological Congress* 13:649–657.

INGOLFSSON, A. 1969. Behaviour of gulls robbing eiders. *Bird Study* 16:45–52.

IRVING, L. 1972. *Arctic life of birds and mammals*. New York: Springer-Verlag.

ITÔ, Y. 1969. Groups and family bonds in animals in relation to their habitat. In L. L. Aronson et al., eds. *Development and evolution of behavior*. San Francisco: Freeman, pp. 389–415.

ITZKOWITZ, M. 1974. A behavioural reconnaissance of some Jamaican reef fishes. *Zoological Journal of the Linnaean Society* 55:87–118.

———. 1978. Group organization of a territorial damselfish, *Eupomacentrus planifrons*. *Behaviour* 65:125–137.

IVLEV, V. S. 1961. *Experimental ecology of the feeding of fishes*. New Haven: Yale University Press.

JACKSON, J. A. 1970. A quantitative study of the foraging ecology of downy woodpeckers. *Ecology* 51:318–323.

JACKSON, J. B. C., and L. BUSS. 1975. Allelopathy and spatial competition among coral reef invertebrates. *Proceedings of the National Academy of Sciences, U.S.A.* 72:5160–63.

JAEGER, R. G. 1974. Competitive exclusion: comments on survival and extinction of species. *Bioscience* 24:33–39.

JAMES, F. C. 1970. Geographic size variation in birds and the relationship to climate. *Ecology* 51:365–390.

JANES, S. W. 1976. The apparent use of rocks by a raven in nest defense. *Condor* 78:409.

JANIS, C. 1976. The evolutionary strategy of the Equidae and the origins of rumen and cecal digestion. *Evolution* 30:757–774.

JARMAN, P. J. 1974. The social organisation of antelope in relation to their ecology. *Behaviour* 58:215–267.

JEHL, J. R., JR. 1970. Sexual selection for size differences in two species of sandpipers. *Evolution* 24:311–319.

JENKINS, D., A. WATSON, and G. R. MILLER. 1967. Population fluctuations in the red grouse *Lagopus lagopus scoticus*. *Journal of Animal Ecology* 36: 97–122.

JENSSEN, T. A. 1973. Shift in the structural habitat of *Anolis opalinus* due to congeneric competition. *Ecology* 54:863–869.

JEWELL, P. A. 1966. The concept of home range in mammals. *Symposium of the Zoological Society of London* 18:85–109.

JOHNSON, L. K., and S. P. HUBBELL. 1974. Aggression and competition among stingless bees: field studies. *Ecology* 55:120–127.

JOHNSON, N. K. 1963. Biosystematics of sibling species of flycatchers in *Empidonax hammondi-oberholseri-wrightii* complex. *University of California Publications in Zoology* 66:79–238.

———. 1966. Bill size and the question of competition in allopatric and sympatric populations of dusky and gray flycatchers. *Systematic Zoology* 15:70–87.

JOHNSON, R. P. 1973. Scent marking in animals. *Animal Behaviour* 21:521–535.

JOHNSTON, V. M. 1942. Factors influencing local movements of woodland birds in winter. *Wilson Bulletin* 54:192–198.

JONES, P. J., and P. WARD. 1976. The level of reserve protein as the proximate factor controlling the timing of breeding and clutch-size in the red-billed quelea *Quelea quelea*. *Ibis* 118:547–574.

JUTRO, P. R. 1975. Territorial defense of people by laughing gulls. *Living Bird* 14:157–161.

KADLEC, J. A., and W. H. DRURY, JR. 1968. Structure of the New England herring gull population. *Ecology* 49:644–676.

KAHL, M. P., JR. 1963. Thermoregulation in the wood stork, with special reference to the role of the legs. *Physiological Zoology* 36:141–151.

KALLEBERG, H. 1958. Observations in a stream tank of territoriality and competition in juvenile salmon and trout (*Salmo salar* L. and *S. trutta* L.) . *Report of the Institute of Freshwater Research, Drottningholm* 39:55–98.

KARST, H. 1968. Unterwasserbeobachtungen an sozialen Gruppierungen von Süsserwasserfischen ausserhalf der Laichzeit. *Internationale Revue der gesamten Hydrobiologie und Hydrographie* 53:573–599.

KATZ, P. L. 1974. A long-term approach to foraging optimization. *American Naturalist* 108:758–782.

KAUFMAN, D. W. 1973a. Was oddity conspicuous in prey selection experiments? *Nature* 242:111–112.

———. 1973b. Shrike prey selection: color or conspicuousness? *Auk* 90:204–206.

———. 1974a. Differential owl predation on white and agouti *Mus musculus*. *Auk* 91:145–150.

———. 1974b. Adaptive coloration in *Peromyscus polionotus:* experimental selection by owls. *Journal of Mammalogy* 55:271–283.

KAUFMANN, J. H. 1962. Ecology and social behavior of the coati, *Nasua nasua* on Barro Colorado Island, Panama. *University of California Publications in Zoology* 60:95–222.

KAWABE, M. 1966. One observed case of hunting behaviour among wild chimpanzees living in the savanna woodland of western Tanzania. *Primates* 7:393–396.

KAWAI, M. 1965. Newly-acquired pre-cultural behavior of the natural troop of Japanese monkeys on Koshima Islet. *Primates* 6:1–30.

KEAR, J. 1962. Food selection in finches with special reference to interspecific differences. *Proceedings of the Zoological Society of London* 138:163–204.

———. 1970. The adaptive radiation of parental care in waterfowl. In J. H. Crook, ed., *Social behaviour in birds and mammals.* New York: Academic Press, pp. 357–392.

———. 1972. Feeding habits of birds. In R. N. T.-W. Fiennes, ed., *Biology of nutrition.* Oxford: Pergamon, pp. 471–503.

KEAST, A. 1965a. Resource subdivision amongst cohabiting fish species in a bay, Lake Opinicon, Ontario. *University of Michigan, Great Lakes Research Division Publication* 13:106–132.

———. 1965b. Interrelations of the two zebra species in an overlap zone. *Journal of Mammalogy* 46:53–66.

———. 1966. Trophic interrelationships in the fish fauna of a small stream. *University of Michigan, Great Lakes Research Division Publication* 15:51–79.

KEAST, A., and E. S. MORTON, eds. 1980. *Migrant birds in the American tropics.* Washington, D.C.: Smithsonian Institution Press.

KELTY, M. P., and S. I. Lustick. 1977. Energetics of the starling (*Sturnus vulgaris*) in a pine woods. *Ecology* 58:1181–85.

KENAGY, G. J. 1973. Daily and seasonal patterns of activity and energetics in a heteromyid rodent community. *Ecology* 54:1201–19.

KENDEIGH, S. C. 1941. Birds of a prairie community. *Condor* 43:165–174.

———. 1961. Energy of birds conserved by roosting in cavities. *Wilson Bulletin* 73:140–147.

———. 1969. Energy responses of birds to their thermal environments. *Wilson Bulletin* 81:441–449.

———. 1970. Energy requirements for existence in relation to size of bird. *Condor* 72:60–65.

KENNEDY, R. J. 1968. The role of sunbathing in birds. *British Birds* 61:320–322.

——. 1969. Sunbathing behavior of birds. *British Birds* 62:249–258.

KENWARD, R. E. 1978. Hawks and doves: factors affecting success and selection in goshawk attacks on woodpigeons. *Journal of Animal Ecology* 47:449–460.

KENWARD, R. E., and R. M. SIBLY. 1977. A woodpigeon (*Columba palumbus*) feeding preference explained by a digestive bottle-neck. *Journal of Applied Ecology* 14:815–826.

KESSEL, B. 1976. Winter activity patterns of black-capped chickadees in interior Alaska. *Wilson Bulletin* 88:36–61.

KETTERSON, E. D. 1978. Environmental influences upon aggressive behavior in wintering juncos. *Bird-banding* 49:313–320.

KETTERSON, E. D., and V. NOLAN, JR. 1976. Geographic variation and its climatic correlates in the sex ratio of eastern-wintering dark-eyed juncos (*Junco hyemalis hyemalis*). *Ecology* 57:679–693.

KETTLEWELL, H. B. D. 1955. Selection experiments on industrial melanism in the Lepidoptera. *Heredity* 9:323–342.

KIESEL, D. S. 1972. Foraging behavior of *Dendrocopos villosus* and *D. pubescens* in eastern New York State. *Condor* 74:393–398.

KIESTER, A. R. 1971. Species density of North American amphibians and reptiles. *Systematic Zoology* 20:127–137.

KILHAM, L. 1965. Differences in feeding behavior of male and female hairy woodpeckers. *Wilson Bulletin* 77:134–145.

——. 1971. Roosting habits of white-breasted nuthatches. *Condor* 73:113–114.

——. 1972. Habits of the crimson-crested woodpecker in Panama. *Wilson Bulletin* 84:28–47.

KING, J. A. 1955. Social behavior, social organization, and population dynamics in a black-tailed prairie dog town in the Black Hills of South Dakota. *Contributions of the Laboratory of Vertebrate Biology, University of Michigan* 67:1–123.

KING, J. R. 1973. Energetics of reproduction in birds. In D. S. Farner, ed., *Breeding biology of birds*. Washington, D.C.: National Academy of Sciences, pp. 78–107.

KLEIMAN, D. G. 1977. Monogamy in mammals. *Quarterly Review of Biology* 52:39–69.

KLEIMAN, D. G., and J. F. EISENBERG. 1973. Comparisons of canid and felid social systems from an evolutionary perspective. *Animal Behaviour* 21:637–659.

KLEIN, L. L., and D. J. KLEIN. 1973. Observations on two types of neotropical primate intertaxa associations. *American Journal of Physical Anthropology* 38:649–654.

KLOMP, H. 1972. Regulation of the size of bird populations by means of territorial behaviour. *Netherlands Journal of Zoology* 22:456–488.

KLOPFER, P. H. 1963. Behavioral aspects of habitat selection: the role of early experience. *Wilson Bulletin* 75:15–22.

————. 1965. Behavioral aspects of habitat selection: a preliminary report on stereotypy in foliage preferences of birds. *Wilson Bulletin* 77:376–381.

————. 1967. Behavioral stereotypy in birds. *Wilson Bulletin* 79:290–300.

————. 1969. *Habitats and territories.* New York: Basic Books.

KLOPFER, P. H., and J. P. HAILMAN. 1965. Habitat selection in birds. *Advances in the Study of Behavior* 1:279–303.

KLOPFER, P. H., and R. H. MACARTHUR. 1960. Niche size and faunal diversity. *American Naturalist* 94:293–300.

————. 1961. On the causes of tropical species diversity: niche overlap. *American Naturalist* 95:223–226.

KLUYVER, H. N. 1972. Regulation of numbers in populations of great tits (*Parus m. major*). In P. J. den Boer and G. R. Gradwell, eds., *Dynamics of populations*. Wageningen, the Netherlands: Center for Agricultural Publishing and Documention, pp. 507–523.

KNAPTON, R. W., and J. R. KREBS. 1974. Settlement patterns, territory size, and breeding density in the song sparrow (*Melospiza melodia*). *Canadian Journal of Zoology* 52:1413–20.

————. 1976. Dominance hierarchies in winter song sparrows. *Condor* 78: 567–569.

KOCH, R. F., A. E. COURCHESNE, and C T. COLLINS. 1970. Sexual differences in foraging behavior of white-headed woodpeckers. *Bulletin of the Southern California Academy of Science* 69:60–64.

KODRIC-BROWN, A., and J. H. BROWN. 1978. Influence of economics, interspecific competition, and sexual dimorphism on territoriality of migrant rufous hummingbirds. *Ecology* 59:285–296.

KOLATA, G. B. 1976. Primate behavior: sex and the dominant male. *Science* 191:55–56.

KOLENOSKY, G. 1972. Wolf predation on wintering deer in east-central Ontario. *Journal of Wildlife Management* 36:357–369.

KORSCHGEN, C. E. 1977. Breeding stress of female eiders in Maine. *Journal of Wildlife Management* 41:360–373.

KOSKIMIES, J. 1957. Flocking behaviour in capercaillie, *Tetrao urogallus* (L.), and the black grouse *Lyurus tetrix* in Finland. *Finnish Papers in Game Research* 18:1–32.

KREBS, C. J., and J. H. MYERS. 1974. Population cycles in small mammals. *Advnces in Ecological Research* 8:267–399.

KREBS, J. R. 1971. Territory and breeding density in the great tit, *Parus major* L. *Ecology* 52:2–22.

————. 1973. Experiments on the significance of mixed-species flocks of chickadees (*Parus* spp.). *Canadian Journal of Zoology* 51:1275–88.

————. 1974. Colonial nesting and social feeding as strategies for exploiting food resources in the great blue heron (*Ardea herodias*). *Behaviour* 51:99–134.

————. 1976. Habituation and song repertoires in the great tit. *Behavioral Ecology and Sociobiology* 1:215–227.

————. 1977a. Song and territory in the great tit *Parus major*. In B. Stonehouse and C. Perrins, eds., *Evolutionary ecology*. Baltimore: University Park Press, pp. 47–62.

———. 1977b. The significance of song repertoires: the Beau Geste hypothesis. *Animal Behaviour* 25:475–478.

———. 1978. Optimal foraging: decision rules for predators. In J. R. Krebs and N. B. Davies, eds., *Behavioural ecology: an evolutionary approach*. Sunderland, Mass.: Sinauer Associates, pp. 23–63.

KREBS, J. R., J. T. ERICHSEN, M. I. WEBBER, and E. L. CHARNOV. 1977. Optimal prey selection in the great tit *(Parus major)*. *Animal Behaviour* 25:30–38.

KREBS, J. R., M. H. MACROBERTS, and J. M. CULLEN. 1972. Flocking and feeding in the great tit *Parus major*—an experimental study. *Ibis* 114:507–530.

KREBS, J. R., J. RYAN, and E. L. CHARNOV. 1974. Hunting by expectation or optimal foraging? A study of patch use by chickadees. *Animal Behaviour* 22:953–964.

KRISCHIK, V. A. 1977. Choice of odd and conspicuous prey by largemouth bass, *Micropterus salmoides*. M.S. thesis, University of Maryland.

KRUUK, H. 1964. Predators and anti-predator behaviour of the black-headed gull *(Larus ridibundus L.)*. *Behaviour, Supplement* 11:1–130.

———. 1972a. Surplus killing by carnivores. *Journal of Zoology, London* 166:233–244.

———. 1972b. *The spotted hyaena*. Chicago: University of Chicago Press.

KUMMER, H. 1971. *Primate societies*. Chicago: Aldine-Atherton.

KUNZ, T. H. 1974. Feeding ecology of a temperate insectivorous bat *(Myotis velifer)*. *Ecology* 55:693–711.

KUSHLAN, J. A. 1973. Spread-wing posturing in cathartid vultures. *Auk* 90:889–890.

———. 1978. Nonrigorous foraging by robbing egrets. *Ecology* 59:649–653.

KYTE, M. A., and G. W. COURTNEY. 1977. A field observation of aggressive behavior between two North Pacific octopus, *Octopus dofleini martini*. *Veliger* 19:427–428.

LACK, D. 1937. The psychological factor in bird distribution. *British Birds* 31:130–136.

———. 1947. *Darwin's finches*. Cambridge: Cambridge University Press.

———. 1954. *The natural regulation of animal numbers*. Oxford: Clarendon Press.

———. 1956. *Swifts in a tower*. London: Methuen.

———. 1965. Evolutionary ecology. *Journal of Animal Ecology* 34:223–231.

———. 1966. *Population studies of birds*. Oxford: Clarendon Press.

———. 1968. *Ecological adaptations for breeding in birds*. London: Methuen.

———. 1971. *Ecological isolation in birds*. Oxford: Blackwell.

LACK, D., and P. LACK. 1972. Wintering warblers in Jamaica. *Living Bird* 11:129–153.

LACK, D., and D. F. OWEN. 1955. The food of the swift. *Journal of Animal Ecology* 24:120–136.

LAMPREY, H. F. 1963. Ecological separation of the large mammal species in the Tarangive Game Preserve, Tanganyika. *East African Wildlife Journal* 1:63–92.

LANCE, A. N. 1978. Territories and the food plant of individual red grouse. II, Territory size compared with an index of nutrient supply in heather. *Journal of Animal Ecology* 47:307–313.

LANG, J. 1971. Interspecific aggression by scleractinian corals. I, The rediscovery of *Scolymia cubensis* (Milne Edwards and Haime). *Bulletin of Marine Science* 21:952–959.

LANGNER, S. 1973. Zur Biologie des Hochland Kolibris *Oreotrochiles estella* in den Anden Boliviens. *Bonner Zoologische Beiträge* 24:24–47.

LANIER, D. K., D. Q. ESTEP, and D. A. DEWSBURY. 1974. Food hoarding in muroid rodents. *Behavioral Biology* 11:177–187.

LANYON, W. E. 1958. The motivation of sun-bathing in birds. *Wilson Bulletin* 70:280.

LAVIGNE, D. M., and N. A. ØRITSLAND. 1974. Black polar bears. *Nature* 251:218–219.

LAWICK, H. VAN, and J. VAN LAWICK-GOODALL. 1971. *Innocent killers*. Boston: Houghton Mifflin.

LAWICK-GOODALL, J. VAN. 1970. Tool using in primates and other vertebrates. *Advances in the Study of Behavior* 3:195–249.

LAWTON, J. H., J. BEDDINGTON, and R. BONSER. 1974. Switching in invertebrate predators. In M. B. Usher and M. H. Williamson, eds., *Ecological stability*. London: Chapman and Hall, pp. 141–158.

LAYNE, J. N. 1954. The biology of the red squirrel, *Tamiasciurus hudsonicus loquax* (Bangs), in central New York. *Ecological Monographs* 24:227–267.

LAZARUS, J. 1972. Natural selection and the functions of flocking in birds: a reply to Murton. *Ibis* 114:556–558.

LEBOEUF, B. J. 1974. Male-male competition and reproductive success in elephant seals. *American Zoologist* 14:163–176.

LEIN, M. R. 1978. Song variation in a population of chestnut-sided warblers (*Dendroica pensylvanica*): its nature and suggested significance. *Canadian Journal of Zoology* 56:1266–83.

LENTEREN, J. C. VAN, and K. BAKKER. 1975. Discrimination between parasitised and unparasitised hosts in the parasitic wasp *Pseudeucoila bochei:* a matter of learning. *Nature* 254:417–419.

LEVINS, R. 1968. Evolution in changing environments. *Monographs in Population Biology* 2:1–120.

LEWONTIN, R. C. 1979. Fitness, survival and optimality. In D. H. Horn, R. Mitchell, and G. R. Stairs, eds., *Analysis of ecological systems*. Columbus: Ohio State University Press, pp. 3–21.

LEYHAUSEN, P. 1960. *Verhaltensstudien an Katzen*. Berlin: Paul Parey.

LICHT, P., and A. G. BROWN. 1967. Behavioral thermoregulation and its role in the ecology of the red-bellied newt, *Taricha rivularis*. *Ecology* 48:598–611.

LIDICKER, W. Z. 1973. Regulation of numbers in an island population of the California vole, a problem in community dynamics. *Ecological Monographs* 43:271–302.

LIGON, J. D. 1967. Relationships of the cathartid vultures. *Occasional Papers of the University of Michigan Museum of Zoology* 651:1–26.

——. 1968a. Sexual differences in foraging behavior in two species of *Dendrocopos* woodpeckers. *Auk* 85:203–215.

——. 1968b. Observations on Strickland's woodpecker, *Dendrocopos stricklandi. Condor* 70:83–84.

——. 1973. Foraging behavior of the white-headed woodpecker in Idaho. *Auk* 90:862–869.

LIGON, J. D., and S. H. LIGON. 1978. Communal breeding in green woodhoopoes as a case for reciprocity. *Nature* 276:496–498.

LIGON, J. D., and D. J. MARTIN. 1974. Piñon seed assessment by the piñon jay, *Gymnorhinus cyanocephalus. Animal Behaviour* 22:421–429.

LILLYWHITE, H. B. 1970. Behavioral temperature regulation in the bullfrog, *Rana catesbeiana. Copeia* 1970:158–168.

LILLYWHITE, H. B., P. LICHT, and P. CHELGREN. 1973. The role of behavioral thermoregulation in the growth energetics of the toad, *Bufo boreas, Ecology* 54:375–383.

LINCOLN, F. C. 1950. *Migration of birds.* Washington, D.C.: U.S. Government Printing Office.

LINCOLN, G. A. 1974. Predation of incubator birds (*Megapodius freycinet*) by Komodo dragons (*Varanus komodoensis*). *Journal of Zoology, London* 174:419–428.

LIND, H. 1965. Parental feeding in the oystercatcher (*Haematopus o. ostralegus* (L.)). *Dansk Ornithologisk Forenings Tidsskrift* 59:1–31.

LINDAUER, M. 1961. *Communication among social bees.* Cambridge: Harvard University Press.

LISSAMAN, P. B. S., and C. A. SHOLLENBERGER. 1970. Formation flight of birds. *Science* 168:1003–5.

LOCKIE, J. D. 1956. Winter fighting in feeding flocks of rooks, jackdaws, and carrion crows. *Bird Study* 3:180–190.

——. 1966. Territory in small carnivores. *Symposium of the Zoological Society of London* 18:143–165.

LOCKLEY, R. M. 1953. *Puffins.* London: J. M. Dent and Sons.

——. 1961. Social structure and stress in the rabbit warren. *Journal of Animal Ecology* 30:385–423.

LOCKNER, F. 1972. Experimental study of food hoarding in the red-tailed chipmunk, *Eutamias ruficaudus. Zeitschrift für Tierpsychologie* 31:410–418.

LOEHR, K. A., and A. C. RISSER, JR. 1977. Daily and seasonal activity patterns of the Belding ground squirrel in the Sierra Nevada. *Journal of Mammalogy* 58:445–448.

LÖHRL, H. 1955. Schlargewohnheiten der Baumlaufer *Certhia brachydactyla, C. familiaris* und anderer Kleinvogel in kalten Winternachten. *Vogelwarte* 18:71–77.

——. 1959. Zur Frage des Zeitpunktes einer Prägung auf die Heimatregion beim Halsbandschnäpper (*Ficedula albicollis*). *Journal für Ornithologie* 100:132–140.

————. 1977. Zum Brutverhalten des Wiedehopfs *Upupa epops*. *Vogelwelt* 98:41–58.

LORE, R., and K. FLANNELLY. 1977. Rat societies. *Scientific American* 236 (5) : 106–116.

LORENZ, K. 1966. *On aggression*. London: Methuen.

LOW, R. M. 1971. Interspecific territoriality in a pomacentrid reef fish, *Pomacentrus flavicauda* Whitley. *Ecology* 52:648–654.

LOWTHER, J. K. 1961. Polymorphism in the white-throated sparrow. *Canadian Journal of Zoology* 39:281–292.

LOWTHER, J. K., and J. B. FALLS. 1968. *Zonotrichia albicollis* (Gmelin) , white-throated sparrow. *Bulletin of the United States National Museum* 237: 1364–92.

LULL, R. S., and N. E. WRIGHT. 1942. Hadrosaurian dinosaurs of North America. *Geological Society of America, Special Paper* 40:1–242.

LÜSCHER, M. 1961. Air-conditioned termite nests. *Scientific American* 205 (1) : 138–145.

LUSTICK, S. 1969. Bird energetics: effects of artificial radiation. *Science* 163: 387–390.

LUSTICK, S., B. BATTERSBY, and M. KELTY. 1978. Behavioral thermoregulation: orientation toward the sun in herring gulls. *Science* 200:81–83.

LYMAN, C. P. 1954. Activity, food consumption and hoarding in hibernators. *Journal of Mammalogy* 35:545–552.

LYNCH, C. B. 1974. Environmental modification of nest-building in the white-footed mouse, *Peromyscus leucopus*. *Animal Behaviour* 22:402–409.

LYON, D. L., J. CRANDALL, and M. McKONE. 1977. A test of the adaptiveness of interspecific territoriality in the blue-throated hummingbird. *Auk* 94: 448–454.

MACARTHUR, R. H. 1958. Population ecology of some warblers of north-eastern coniferous forests. *Ecology* 39:599–619.

————. 1968. The theory of the niche. In R. C. Lewontin, ed., *Population biology and evolution*. Syracuse: Syracuse University Press, pp. 159–176.

————. 1969. Species packing and what competition minimizes. *Proceedings of the National Academy of Science, U.S.A.* 64:1369–71.

————. 1970. Species packing and competitive equilibrium for many species. *Theoretical Population Biology* 1:1–11.

————. 1972. *Geographical ecology*. New York: Harper and Row.

MACARTHUR, R. H., and R. LEVINS. 1964. Competition, habitat selection, and character displacement in a patchy environment. *Proceedings of the National Academy of Science, U.S.A.* 51:1207–10.

MACARTHUR, R. H., and E. C. PIANKA. 1966. On optimal use of a patchy environment. *American Naturalist* 100:603–609.

MACARTHUR, R. H., and E. O. WILSON. 1967. The theory of island biogeography. *Monographs in Population Biology* 1:1–203.

MCBRIDE, G. 1963. The "teat order" and communication in young pigs. *Animal Behaviour* 11:53–56.

MACDONALD, D. W. 1976. Food caching by red foxes and some other carnivores. *Zeitschrift für Tierpsychologie* 42:170–185.

MacDonald, D. W., and D. G. Henderson. 1977. Aspects of the behaviour and ecology of mixed-species bird flocks in Kashmir. *Ibis* 119:481–493.

MacFarland, C. G., and W. G. Reeder. 1974. Cleaning symbiosis involving Galapagos tortoises and two species of Darwin's finches. *Zeitschrift für Tierpsychologie* 34:464–483.

McKay, G. M. 1973. Behavior and ecology of the Asiatic elephant in southeastern Ceylon. *Smithsonian Contributions to Zoology* 125:1–113.

McLaren, I. A. 1972. Polygyny and breeding territory in birds. *Transactions of the Connecticut Academy of Arts and Sciences* 44:191–210.

MacMillen, R. E., and C. Trost. 1967. Nocturnal hypothermia in the Inca dove, *Scardafella inca*. *Comparative Biochemistry and Physiology* 23:243–253.

McNab, B. K. 1963. Bioenergetics and the determination of home range size. *American Naturalist* 97:133–140.

———. 1971. On the ecological significance of Bergmann's rule. *Ecology* 52:845–854.

McNicholl, M. K. 1975. Larid site tenacity and group adherence in relation to habitat. *Auk* 92:98–104.

MacRoberts, M. H., and B. R. MacRoberts. 1976. Social organization and behavior of the acorn woodpecker in central coastal California. *Ornithological Monographs* 21:1–115.

Maiorana, V. C. 1976. Predation, submergent behavior, and tropical diversity. *Evolutionary Theory* 1:157–177.

Major, P. F. 1977. Predator-prey interactions in schooling fishes during period of twilight: a study of the silverside *Pranesus insularum* in Hawaii. *Fisheries Bulletin* 75:415–426.

———. 1978. Predatory-prey interactions in two schooling fishes, *Caranx ignobilis* and *Stolephorus purpureus*. *Animal Behaviour* 26:760–777.

Malcolm, J. R., and H. van Lawick. 1975. Notes on wild dogs (*Lycaon pictus*) hunting zebras. *Mammalia* 39:231–240.

Mares, M. A., and D. F. Williams. 1977. Experimental support for food particle size resource allocation in heteromyid rodents. *Ecology* 59:1186–90.

Markgren, G. 1963. Studies on wild geese in southernmost Sweden, pt. 1. *Acta Vertebratica* 2:299–418.

Marler, P. 1955. Characteristics of some animal calls. *Nature* 176:6–8.

———. 1956. Behaviour of the chaffinch (*Fringilla coelebs*). *Behaviour, Supplement* 5:1–184.

———. 1957. Specific distinctness in the communication signals of birds. *Behaviour* 11:13–39.

———. 1960. Bird songs and mate selection. In W. E. Lanyon and W. N. Tavolga, eds., *Animal communication*. Washington, D.C.: American Institute of Biological Sciences.

———. 1969. *Colobus guereza*: territoriality and group composition. *Science* 163:93–95.

Marriner, G. R. 1908. *The kea: a New Zealand problem*. Christchurch, New Zealand: Marriner.

Marshall, N. B. 1965. *The life of fishes*. London: Weidenfeld and Nicolson.

MARTIN, A. A. 1970. Parallel evolution in the adaptive ecology of lepto-dactylid frogs of South America and Australia. *Evolution* 24:643–644.

MARTIN, R. D. 1968. Reproduction and ontogeny in tree shrews (*Tupaia belangeri*) with reference to their general behavior and taxonomic relationships. *Zeitschrift für Tierpsychologie* 25:409–495, 505–532.

MASURE, R. H., and W. C. ALLEE. 1934a. The social order in flocks of the common chicken and the pigeon. *Auk* 51:306–327.

———. 1934b. Flock organization of the shell parakeet *Melopsittacus undulatus*. *Ecology* 15:388–398.

MATTHEWS, L. H. 1971. *The life of mammals,* vol. 2. London: Weidenfeld and Nicolson.

MAXWELL, G. R., II, and H. W. KALE II. 1977. Maintenance and anti-insect behavior of six species of ciconiiform birds in South Florida. *Condor* 79:51–55.

MAY, M. L. 1977. Thermoregulation and reproductive activity in tropical dragonflies of the genus *Micrathyria*. *Ecology* 58:787–798.

MAY, R. M., and R. H. MACARTHUR. 1972. Niche overlap as a function of environmental variability. *Proceedings of the National Academy of Science, U.S.A.* 69:1109–13.

MAYNARD SMITH, J. 1964. Kin selection and group selection. *Nature* 201:1145–47.

———. 1977. Parental investment: a prospective analysis. *Animal Behaviour* 25:1–9.

———. 1978. Optimization theory in evolution. *Annual Review of Ecology and Systematics* 9:31–56.

MAYR, E. 1935. Bernard Altum and the territory theory. *Proceedings of the Linnaean Society of New York* 45–46:24–38.

———. 1963. *Animal species and evolution.* Cambridge: Harvard University Press, Belknap Press.

MEADOWS, P. S., and J. I. Campbell. 1972a. Habitat selection and animal distribution in the sea: the evolution of a concept. *Proceedings of the Royal Society of Edinburgh B* 73:145–157.

———. 1972b. Habitat selection by aquatic invertebrates. *Advances in Marine Biology* 10:271–382.

MECH, L. D. 1966. The wolves of Isle Royale. *United States National Park Service, Fauna Series* 7:1–210.

———. 1977. Wolf-pack buffer zones as prey reservoirs. *Science* 198:320–321.

MEINERTZHAGEN, R. 1938. Winter in Arctic Lapland. *Ibis* 1938:754–759.

MENGE, J. L., and B. A. MENGE. 1974. Role of resource allocation, aggression and spatial heterogeneity in coexistence of two competing intertidal starfish. *Ecological Monographs* 44:189–209.

MESERVEY, W. R., and G. F. KRAUS. 1976. Absence of "individual distance" in three swallow species. *Auk* 93:177–178.

MEYER DE SCHAUNSEE, R. 1963. *The birds of Colombia.* Narbeth, Pa.: Livingston Publishing Co.

MICHENER, C. D. 1974. *The social behavior of the bees.* Cambridge: Harvard University Press, Belknap Press.

MICHENER, G. R. 1973. Intraspecific aggression and social organization in ground squirrels. *Journal of Mammalogy* 54:1001–3.

MICHENER, G. R., and D. R. MICHENER. 1977. Population structure and dispersal in Richardson's ground squirrels. *Ecology* 58:359–368.

MILINSKI, M. 1977. Experiments on the selection by predators against spatial oddity of their prey. *Zeitschrift für Tierpsychologie* 43:311–325.

MILLER, A. H. 1931. Systematic revision and natural history of the American shrikes (*Lanius*). *University of California Publications in Zoology* 38:11–242.

MILLER, E. V. 1941. Behavior of the Bewick wren. *Condor* 43:81–99.

MILLER, G. R., and A. WATSON. 1978. Territories and the food plant of individual red grouse. I, Territory size, number of mates and brood size compared with the abundance, production and diversity of heather. *Journal of Animal Ecology* 47:293–305.

MILLER, R. C. 1921. The flock behavior of the coast bush-tit. *Condor* 23:121–127.

————. 1922. The significance of the gregarious habit. *Ecology* 3:122–126.

MILLER, R. S. 1967. Pattern and process in competition. *Advances in Ecological Research* 4:1–74.

MILLIKAN, G. C., and R. I. BOWMAN. 1967. Observations on Galapagos tool-using finches in captivity. *Living Bird* 6:23–41.

MILLS, G. S. 1976. American kestrel sex ratios and habitat separation. *Auk* 93:740–748.

MINOCK, M. E. 1972. Interspecific aggression between black-capped and mountain chickadees at winter feeding stations. *Condor* 74:454–461.

MITCHELL, H. H. 1964. *Comparative nutrition of man and domestic animals,* vol. 1. New York: Academic Press.

MOEHLMAN, P. D. 1978. Jackal helpers and pup survival. *Nature* 277:382–383.

MOLL, E. O., and J. M. LEGLER. 1971. The life history of a neotropical slider turtle, *Pseudemys scripta* (Schoepff) in Panama. *Bulletin of the Los Angeles County Museum, Natural History and Science* 11:1–102.

MOODIE, G. E. E., J. D. McPHAIL, and D. W. HAGEN. 1973. Experimental demonstration of selective predation on *Gasterosteus aculeatus. Behaviour* 47:95–105.

MOORE, P. G. 1975. The role of habitat selection in determining the local distribution of animals in the sea. *Marine Behavior and Physiology* 3:97–100.

MORAN, M. J., and P. F. SALE. 1977. Seasonal variation in territorial response, and other aspects of the ecology of the Australian temperate pomacentrid fish *Parma microlepis. Marine Biology* 39:121–128.

MOREAU, R. E. 1966. *The bird fauna of Africa and its islands.* New York: Academic Press.

————. 1973. *The Palaearctic-African migration systems.* New York: Academic Press.

MOREL, G. 1973. The Sahel Zone as an environment for Palaearctic migrants. *Ibis* 115:413–417.

MOREL, G., and F. BOULIÈRE. 1962. Relations écologiques des avifaunes sedentaire et migratrice dans une savane sahéliene du bas Sénégal. *La Terre et la Vie* 102:371–393.

MOREL, G., and M.-Y. MOREL. 1974. Recherches écologiques sur une savane Sahelienne du Ferlo Sptentrional, Sénégal: influence de la secheresse de l'année 1972–1973 sur l'avifaune. *La Terre et la Vie* 28:95–123.

MORIARTY, D. J. 1976. The adaptive nature of bird flocks: a review. *Biologist* 58:67–79.

MORRIS, D. 1962. The behavior of the green acouchi (*Myoprocta pratti*) with special reference to scatter hoarding. *Proceedings of the Zoological Society of London* 139:701–732.

MORRIS, R. D., and P. R. Grant. 1972. Experimental studies of competitive interaction in a two-species system. IV, *Microtus* and *Clethrionomys* species in a single enclosure. *Journal of Animal Ecology* 41:275–290.

MORRISON, M. L., R. D. SLACK, and E. SHANLEY, JR. 1978. Age and foraging ability relationships of olivaceous cormorants. *Wilson Bulletin* 90:414–422.

MORSE, D. H. 1956. Nighttime activity of the snow bunting. *Maine Field Naturalist* 12:52.

———. 1966. The contexts of songs in the yellow warbler. *Wilson Bulletin* 78:444–455.

———. 1967a. The contexts of songs in the black-throated green and black-burnian warblers. *Wilson Bulletin* 79:62–72.

———. 1967b. Foraging relationships of brown-headed nuthatches and pine warblers. *Ecology* 48:94–103.

———. 1968a. The use of tools by brown-headed nuthatches. *Wilson Bulletin* 80:220–224.

———. 1968b. A quantitative study of foraging of male and female spruce-woods warblers. *Ecology* 49:779–784.

———. 1970a. Territorial and courtship songs of birds. *Nature* 226:659–661.

———. 1970b. Ecological aspects of some mixed-species foraging flocks of birds. *Ecological Monographs* 40:119–168.

———. 1971a. The insectivorous bird as an adaptive strategy. *Annual Review of Ecology and Systematics* 2:177–200.

———. 1971b. The foraging of warblers isolated on small islands. *Ecology* 52:216–228.

———. 1971c. Effects of the arrival of a new species upon habitats utilized by two forest thrushes in Maine. *Wilson Bulletin* 83:57–65.

———. 1972. Habitat utilization of the red-cockaded woodpecker during the winter. *Auk* 89:429–435.

———. 1973a. The foraging of small populations of yellow warblers and American redstarts. *Ecology* 54:346–355.

———. 1973b. Interactions between tit flocks and sparrowhawks *Accipiter nisus*. *Ibis* 115:591–593.

———. 1974. Niche breadth as a function of social dominance. *American Naturalist* 108:818–830.

———. 1975a. Ecological aspects of adaptive radiation in birds. *Biological Reviews* 50:167–214.

———. 1975b. Mourning doves breeding in an unusual habitat: the coastal spruce forest. *Wilson Bulletin* 87:422–424.

———. 1976a. Variables affecting the density and territory size of breeding spruce-woods warblers. *Ecology* 57:290–301.

———. 1976b. Hostile encounters among spruce-woods warblers (*Dendroica,* Parulidae) . *Animal Behaviour* 24:764–771.

———. 1977a. The occupation of small islands by passerine birds. *Condor* 79:399–412.

———. 1977b. Resource partitioning in bumble bees: the role of behavioral factors. *Science* 197:678–680.

———. 1977c. Foraging of bumble bees: the effect of other individuals. *Journal of the New York Entomological Society* 85:237–245.

———. 1977d. Feeding behavior and predator avoidance in heterospecific groups. *Bioscience* 27:332–339.

———. 1978. Structure and foraging patterns of tit flocks in an English woodland. *Ibis* 120:298–312.

———. 1979. Prey capture by the crab spider *Misumena calycina* (Araneae: Thomisidae) . *Oecologia* 39:309–319.

———. 1980. Population limitations: breeding or wintering grounds? In A. Keast and E. S. Morton, eds., *Migrant birds in the American Tropics.* Washington, D.C.: Smithsonian Institution Press, pp. 505–516.

MORTON, E. S. 1973. On the evolutionary advantages and disadvantages of fruit eating in tropical birds. *American Naturalist* 107:8–22.

———. 1975. Ecological sources of selection on avian sounds. *American Naturalist* 109:17–34.

———. 1977. On the occurrence and significance of motivation-structural rules in some bird and mammal sounds. *American Naturalist* 111:855–869.

MORTON, E. S., and M. D. SHALTER. 1977. Vocal response to predators in pair-bonded Carolina wrens. *Condor* 79:222–227.

MORTON, M. L. 1967. The effects of insolation on the diurnal feeding pattern of white-crowned sparrows (*Zonotrichia leucophrys gambeli*) . *Ecology* 49: 690–694.

Moss, R. 1972. Food selection by red grouse (*Lagopus lagopus scoticus* (Lath.)) in relation to chemical composition. *Journal of Animal Ecology* 41:411–428.

Moss, R., G. R. MILLER, and S. E. ALLEN. 1972. Selection of heather by captive red grouse in relation to the age of the plant. *Journal of Applied Ecology* 9:771–781.

MOSTLER, G. 1935. Beobachtungen zur Frage der Wespermimikry. *Zeitschrift für Morphologie und Okologie der Tiere* 29:381–454.

MOUGIN, J. L. 1975. Ecologie comparée des Procellariidae antarctiques et subantarctiques. *Comité National Français des Recherches Antarctiques* 36:1–195.

MOYNIHAN, M. 1962. The organization and probable evolution of some mixed species flocks of neotropical birds. *Smithsonian Miscellaneous Collections* 143 (7) :1–140.

————. 1970. Control, suppression, decay, disappearance and replacement of displays. *Journal of Theoretical Biology* 29:85–112.

MROSOVSKY, M. 1971. *Hibernation and the hypothalamus.* New York: Appleton-Century-Crofts.

MUELLER, H. C. 1968. Prey selection: oddity or conspicuousness? *Nature* 217:92.

————. 1971. Oddity and specific search image more important than conspicuousness in prey selection. *Nature* 233:345–346.

————. 1972. Sunbathing in birds. *Zeitschrift für Tierpsychologie* 30:253–258.

————. 1973. Was oddity conspicuous in prey selection experiments? *Nature* 244:112.

————. 1974a. Food caching behaviour in the American kestrel (*Falco sparverius*). *Zeitschrift für Tierpsychologie* 34:105–114.

————. 1974b. Factors influencing prey selection in the American kestrel. *Auk* 91:705–721.

MULLIGAN, J. A. 1966. Singing behavior and its development in the song sparrow *Melospiza melodia. University of California Publications in Zoology* 81:1–76.

MURDOCH, W. W. 1969. Switching in general predators: experiments on predator specificity and stability of prey populations. *Ecological Monographs* 39:335–354.

MURDOCH, W. W., S. AVERY, and M. E. B. SMITH. 1975. Switching in predatory fish. *Ecology* 56:1094–1105.

MURDOCH, W. W., and J. R. MARKS. 1973. Predation by coccinellid beetles: experiments on switching. *Ecology* 54:160–167.

MURDOCH, W. W., and A. OATEN. 1975. Predation and population stability. *Advances in Ecological Research* 9:1–131.

MURDOCH, W. W., and A. SIH. 1978. Age-dependent interference in a predatory insect. *Journal of Animal Ecology* 47:581–592.

MURRAY, B. G., JR. 1969. A comparative study of the LeConte's and sharp-tailed sparrows. *Auk* 86:199–231.

————. 1971. The ecological consequences of interspecific territorial behavior in birds. *Ecology* 52:414–423.

————. 1976. A critique of interspecific territoriality and character convergence. *Condor* 78:518–525.

MURTON, R. K. 1971a. The significance of a specific search image in the feeding behaviour of the wood-pigeon. *Behaviour* 40:10–42.

————. 1971b. Why do some bird species feed in flocks? *Ibis* 113:534–536.

MURTON, R. K., and A. J. ISAACSON. 1962. The functional basis of some behaviour in the woodpigeon, *Columba palumbus. Ibis* 104:503–521.

MURTON, R. K., A. J. ISAACSON, and N. J. WESTWOOD. 1971. The significance of gregarious feeding behaviour and adrenal stress in a population of woodpigeons *Columba palumbus. Journal of Zoology, London* 165:53–84.

MYERS, J. P. 1978. One deleterious effect of mobbing in the southern lapwing (*Vanellus chilensis*). *Auk* 95:419–420.

MYRBERG, A. A., JR., and R. E. THRESHER. 1974. Interspecific aggression and its relevance to the concept of territoriality in reef fishes. *American Zoologist* 14:81–96.

MYTON, B. 1974. Utilization of space by *Peromyscus leucopus* and other small mammals. *Ecology* 55:277–290.

MYTON, B. A., and R. W. FICKEN. 1967. Seed-size preference in chickadees in relation to ambient temperature. *Wilson Bulletin* 79:319–321.

NEILL, S. R. St. J., and J. M. CULLEN. 1974. Experiments on whether schooling by their prey affects the hunting behaviour of cephalopods and fish predators. *Journal of Zoology, London* 172:549–569.

NEILL, W. E. 1974. The community matrix and interdependence of the competition coefficients. *American Naturalist* 108:399–408.

NELSON, J. B. 1966. The breeding biology of the gannet *Sula bassana* on the Bass Rock, Scotland. *Ibis* 108:584–626.

NELSON, J. B. 1969. The breeding ecology of the red-footed booby in the Galapagos. *Journal of Animal Ecology* 38:181–198.

NELSON, J. F., and R. M. CHEW. 1977. Factors affecting seed reserves in the soil of a Mojave desert ecosystem, Rock Valley, Nye County, Nevada. *American Midland Naturalist* 97:300–320.

NERO, R. W. and J. T. EMLEN. 1951. An experimental study of territorial behavior in breeding red-winged blackbirds. *Condor* 53:105–116.

NEWTON, I. 1967. The adaptive radiation and feeding ecology of some British finches. *Ibis* 109:33–98.

———. 1970. Irruptions of crossbills in Europe. *Symposium of the British Ecological Society* 10:337–357.

———. 1972. *Finches*. London: Collins.

NICE, M. M. 1941. The role of territory in bird life. *American Midland Naturalist* 26:441–487.

———. 1962. Development of behavior in precocial birds. *Transactions of the Linnaean Society of New York* 8:1–211.

NILSSON, N. A. 1955. Studies on the feeding habitat of trout and char in North Swedish lakes. *Report of the Institute of Freshwater Research, Drottningholm* 36:163–225.

NISBET, I. C. T., and LORD MEDWAY. 1972. Dispersion, population ecology and migration of eastern great reed warblers *Acrocephalus orientalis* wintering in Malaysia. *Ibis* 114:451–494.

NOAKES, D. L. G., and G. W. BARLOW. 1973. Ontogeny of parent-contacting in young *Cichlasoma citrinellum* (Pisces, Cichlidae). *Behaviour* 46:221–255.

NOBLE, G. K. 1931. *The biology of the Amphibia*. New York: McGraw-Hill.

———. 1939. The role of dominance on the social life of birds. *Auk* 56:263–273.

NORRIS, K. S. 1967. Color adaptation in desert reptiles and its thermal relationships. In W. W. Milstead, ed., *Lizard ecology: a symposium*. Columbia: University of Missouri Press.

NORTON-GRIFFITHS, M. 1967. Some ecological aspects of the feeding behaviour of the oystercatcher (*Haematopus ostralegus*) on the edible mussel (*Mytilus edulis*). *Ibis* 109:412–424.

———. 1968. The feeding behaviour of the oystercatcher. D. Phil. thesis, Oxford University.

———. 1969. The organization, control and development of parental feeding in the oystercatcher (*Haematopus ostralegus*). *Behaviour* 34:55–114.

NOVIKOV, G. A. 1972. The use of under-snow refuges among small birds of the sparrow family. *Aquilo, Zoology* 13:95–97.

NUDDS, T. D. 1978. Convergence of group size strategies by mammalian social carnivores. *American Naturalist* 112:957–960.

OATEN, A. 1977. Optimal foraging in patches: a case for stochasticity. *Theoretical Population Biology* 12:263–285.

O'BRIEN, W. J., N. A. SLADE, and G. L. VINYARD. 1976. Apparent size as the determinant of prey selection by bluegill sunfish (*Lepomis macrochirus*). *Ecology* 57:1304–10.

O'CONNOR, R. J., and R. A. BROWN. 1977. Prey depletion and foraging strategy in the oystercatcher *Haematopus ostralegus*. *Oecologia* 27:75–92.

ODUM, E. P. 1942. Annual cycle of the black-capped chickadee. *Auk* 59:499–531.

———. 1960. Lipid deposition in nocturnal migrant birds. *Proceedings of the International Ornithological Congress* 12:563–576.

OHMART, R. D. 1973. Observations on the breeding adaptations of the roadrunner. *Condor* 75:140–149.

OHMART, R. D., and R. C. LASIEWSKI. 1970. Roadrunners: energy conservation by hypothermia and absorption of sunlight. *Science* 172:67–69.

OLIVER, W. R. B. 1955. *New Zealand Birds*, 2d ed. Wellington, New Zealand: A. H. and A. W. Reed.

OLSON, F. C. W. 1963. The survival value of fish schooling. *Journal du Conseil, Conseil Permanent International pour l'Exploration de la Mer* 29:115–116.

ORIANS, G. H. 1961. The ecology of blackbird (*Agelaius*) social systems. *Ecological Monographs* 31:285–312.

———. 1969a. The number of bird species in some tropical forests. *Ecology* 50:783–801.

———. 1969b. Age and hunting success in the brown pelican (*Pelecanus occidentalis*). *Animal Behaviour* 17:316–319.

———. 1969c. On the evolution of mating systems in birds and mammals. *American Naturalist* 103:589–603.

———. 1971. Ecological aspects of behavior. In D. S. Farner and J. R. King, eds., *Avian biology*, vol. 1. New York: Academic Press. pp. 513–546.

ORIANS, G. H., and G. COLLIER. 1963. Competition and blackbird social systems. *Evolution* 17:449–459.

ORIANS, G. H., and M. F. WILLSON. 1964. Interspecific territories of birds. *Ecology* 45:736–745.

ØRITSLAND, N. A. 1970. Energetic significance of absorption of solar radiation in polar homeotherms. In M. W. Holdgate, ed., *Antarctic ecology*, vol. 1. New York: Academic Press, pp. 464–470.

OSBORN, H. F. 1929. The titanotheres of ancient Wyoming, Dakota, and Nebraska. *United States Geological Survey Monograph* 55:1–726.

ÖSTERLOF, S. 1966. The migration of the goldcrest (*Regulus regulus*). *Vår Fågelvärld* 25:141–156.

OTTE, D., and A. JOERN. 1975. Insect territoriality and its evolution: population studies of desert grasshoppers on creosote bushes. *Journal of Animal Ecology* 44:29–54.

OWEN, D. F. 1955. The food of the heron, *Ardea cinerea,* in the breeding season. *Ibis* 97:276–295.

OWENS, N. W., and J. D. GOSS-CUSTARD. 1976. The adaptive significance of alarm calls given by shorebirds on their winter feeding grounds. *Evolution* 30:397–398.

PACKARD, A. 1972. Cephalopods and fish: the limits of convergence. *Biological Reviews* 47:241–307.

PAGE, G., and D. F. WHITACRE. 1975. Raptor predation on wintering shorebirds. *Condor* 77:73–83.

PAINE, R. T. 1966. Food web diversity and species diversity. *American Naturalist* 100:65–75.

———. 1976. Size-limited predation: an observational and experimental approach with the *Mytilus-Pisaster* interaction. *Ecology* 57:858–873.

PALMER, R. S. 1949. Maine birds. *Bulletin of the Museum of Comparative Zoology* 102:1–656.

———, ed. 1962. *Handbook of North American birds,* vol. 1. New Haven: Yale University Press.

———, ed. 1976. *Handbook of North American birds,* vol. 3. New Haven: Yale University Press.

PARDI, L. 1948. Dominance order in *Polistes* wasps. *Physiological Zoology* 21:1–13.

PARK, T. 1954. Experimental studies of interspecies competition. II, Temperature, humidity, and competition in two species of *Tribolium. Physiological Zoology* 27:177–238.

PARNELL, J. F. 1969. Habitat relations of the Parulidae during spring migration. *Auk* 86:505–521.

PARTRIDGE, L. 1974. Habitat selection in titmice. *Nature* 247:573–574.

———. 1976. Individual differences in feeding efficiencies and feeding preferences of captive great tits. *Animal Behaviour* 24:230–240.

PATTERSON, T. L., and L. PETRINOVICH. 1978. Territory size in the white-crowned sparrow (*Zonotrichia leucophrys*): measurement and stability. *Condor* 80:97–98.

PAULI, H.-R. 1974. Zur Winterökologie des Birkhuhns *Tetrao tetrix* in den Schweizer Alpen. *Ornithologische Beobachter* 71:247–278.

PEARSON, O. P. 1953. Use of caves by hummingbirds and other species at high altitudes in Peru. *Condor* 55:17–20.

———. 1954. Habits of the lizard *Liolaemus multiformis multiformis* at high altitudes in southern Peru. *Copeia* 1954:111–116.

PEEK, F. W. 1972. An experimental study of the territorial function of vocal and visual displays in the male red-winged blackbird (*Agelaius phoeniceus*). *Animal Behaviour* 20:112–118.

PEITZMEIR, J. 1947. Die Biologie der Misteldrossel (*Turdus v. viscivorus* L) mit besonderer Berucksichtigung der Parklandschaftspopulationen. *Ornithologische Forschungen* 1:42–76.

PERRINS, C. M. 1965. Population fluctuations and clutch-size in the great tit, *Parus major* L. *Journal of Animal Ecology* 34:601–647.

————. 1970. The timing of birds' breeding seasons. *Ibis* 112:242–255.

PERSSON, B. 1971. Habitat selection and nesting of a South Swedish whitethroat *Sylvia communis* Lath. population. *Ornis Scandinavica* 2:119–126.

PETERSON, R. L. 1955. *North American moose.* Toronto: University of Toronto Press.

PHILLIPS, D. W. 1976. The effect of a species-specific avoidance response to predatory starfish on the intertidal distribution of two gastropods. *Oecologia* 23:83–94.

PIANKA, E. R. 1971. Lizard species diversity in the Kalahari Desert. *Ecology* 52:1024–29.

PITCHER, T. 1979. He who hesitates, lives. Is stotting antiambush behavior? *American Naturalist* 113:453–456.

PITELKA, F. A. 1959. Numbers, breeding schedule, and territoriality in pectoral sandpipers of northern Alaska. *Condor* 61:233–264.

PITELKA, F. A., P. Q. TOMICH, and G. W. TREICHEL. 1955. Ecological relations of jaegers and owls as lemming predators near Barrow, Alaska. *Ecological Monographs* 25:85–117.

PITTS, T. D. 1976. Fall and winter roosting habits of Carolina chickadees. *Wilson Bulletin* 88:603–810.

POLES, W. 1955. Animal ways. *Oryx* 3:246–254.

PONTIN, A. J. 1961. Population stabilization and competition between the ants *Lasius flavus* (F.) and *L. niger* (L.). *Journal of Animal Ecology* 32:565–574.

————. 1969. Experimental transplantation of nest-mounds of the ant *Lasius flavus* (F.) in a habitat containing also *L. niger* (L.) and *Myrmica scabrinodis* Nyl. *Journal of Animal Ecology* 38:747–754.

POPHAM, S. J. 1941. The variation in the colour of certain species of *Arctocorixa* (Hemiptera, Corixidae) and its significance. *Proceedings of the Zoological Society of London* 111:135–172.

PORTENKO, L. A. 1948. (The neck pouches in birds.) *Priroda* 37(10):50–54. (In Russian, from Turcek and Kelso, 1968)

PORTER, K. R. 1972. *Herpetology.* Philadelphia: Saunders.

POST, W. 1974. Functional analysis of space-related behavior in the seaside sparrow. *Ecology* 55:564–574.

————. 1978. Social and foraging behavior of warblers wintering in Puerto Rican coastal scrub. *Wilson Bulletin* 90:197–214.

POTTER, E. F., and D. C. HAUSER. 1974. Relationship of anting and sunbathing to molting in wild birds. *Auk* 91:537–563.

POWELL, G. V. N. 1974. Experimental analysis of the social value of flocking by starlings (*Sturnus vulgaris*) in relation to predation and foraging. *Animal Behaviour* 22:501–505.

POWELL, R. W. 1974. Some measures of feeding behavior in captive common crows. *Auk* 91:571–574.

PREVOST, J. 1961. *Ecologie du manchot empereur.* Paris: Herman.

PRICE, M. R. S. 1978. The nutritional ecology of Coke's hartebeest (*Alcelaphus buselaphus cokei*) in Kenya. *Journal of Applied Ecology* 15:33–49.

PROSSER, C. L. 1973. Oxygen: respiration and metabolism. In C. L. Prosser, ed., *Comparative animal physiology,* 3rd ed. Philadelphia: Saunders, pp. 165–211.

PRYS-JONES, O. E. 1973. Interactions between gulls and eiders in St. Andrews Bay, Fife. *Bird Study* 20:311–313.

PULLIAINEN, E. 1970. Winter nutrition of the rock ptarmigan, *Lagopus mutus* (Montin), in northern Finland. *Annales Zoologici Fennici* 7:295–302.

―――. 1974. Winter ecology of the red squirrel (*Sciurus vulgaris* L.) in northeastern Lapland. *Annales Zoologici Fennici* 10:487–494.

PULLIAINEN, E., and L. J. SALO. 1973. Food selection by the willow grouse (*Lagopus lagopus*) in laboratory conditions. *Annales Zoologici Fennici* 10:445–448.

PULLIAM, H. R. 1973. On the advantages of flocking. *Journal of Theoretical Biology* 38:419–422.

―――. 1974. On the theory of optimal diets. *American Naturalist* 108:59–74.

―――. 1975. Diet optimization with nutrient constraints. *American Naturalist* 109:765–768.

PULLIAM, H. R., K. A. ANDERSON, A. MISZTAL, and N. MOORE. 1974. Temperature-dependent social behaviour in juncos. *Ibis* 116:360–364.

PULLIAM, H. R., and M. R. BRAND. 1975. The production and utilization of seeds in plains grassland of southeastern Arizona. *Ecology* 56:1158–66.

PYKE, G. H. 1978. Are animals efficient harvesters? *Animal Behaviour* 26:241–250.

PYKE, G. H., H. R. PULLIAM, and E. L. CHARNOV. 1977. Optimal foraging: a selective review of theory and tests. *Quarterly Review of Biology* 52:137–154.

RABINOWITCH, V. 1969. The role of experience in the development and retention of seed preferences in zebra finches. *Behaviour* 33:222–236.

RADAKOV, D. V. 1958. (On the adaptive significance of shoaling of young coalfish (*Polachia virens* L.) .) *Voprosy Ikhtiologii* 11:69–74. (In Russian)

―――. 1972. *Schooling in the ecology of fish.* New York: Halstead-Wiley.

RALLS, K. 1971. Mammalian scent marking. *Science* 171:443–449.

―――. 1976. Mammals in which females are larger than males. *Quarterly Review of Biology* 51:245–276.

RANDALL, J. A. 1978. Behavioral mechanisms of habitat segregation between sympatric species of *Microtus:* habitat preference and interspecific dominance. *Behavioral Ecology and Sociobiology* 3:187–202.

RAPPOLE, J. H., and D. W. WARNER. 1976. Relationships between behavior, physiology and weather in avian transients at a migration stopover site. *Oecologia* 26:193–212.

RAPPORT, D. J. 1971. An optimization model of food selection. *American Naturalist* 105:575–587.

RAVELING, D. G. 1970. Dominance relationships and agonistic behavior of Canada geese in winter. *Behaviour* 37:291–319.

RAVELING, D. G., W. E. CREWS, and W. D. KLIMSTRA. 1972. Activity patterns of Canada geese during the winter. *Wilson Bulletin* 84:278–295.

RECHER, H. F., and J. A. RECHER. 1969. Comparative foraging efficiency of adult and immature little blue herons (*Florida caerulea*). *Animal Behaviour* 17:320–322.

REESE, E. S. 1963. The behavioural mechanisms underlying shell selection by hermit crabs. *Behaviour* 21:78–126.

REICHMAN, O. J. 1977. Optimization of diets through food preferences by heteromyid rodents. *Ecology* 58:454–457.

RESCIGNO, A., and I. W. RICHARDSON. 1965. On the competitive exclusion principle. *Bulletin of Mathematical Biophysics* 27:85–89.

RHIJN, J. G. VAN. 1973. Behavioural dimorphism in male ruffs, *Philomachus pugnax* (L.). *Behaviour* 47:153–229.

RICE, D. W., and K. V. KENYON. 1962. Breeding cycles and behavior of Laysan and black-footed albatrosses. *Auk* 79:517–567.

RICHARDS, P. W. 1952. *The tropical rain forest.* Cambridge: Cambridge University Press.

RICHARDS, S. M. 1974. The concept of dominance and methods of assessment. *Animal Behaviour* 22:914–930.

RICHDALE, L. E., and J. WARHAM. 1973. Survival, pair bond retention and nest-site tenacity in Buller's mollymawk. *Ibis* 115:257–263.

RICKLEFS, R. E. 1974. Energetics of reproduction in birds. *Publications of the Nuttall Ornithological Club* 15:152–297.

———. 1975. The evolution of cooperative breeding in birds. *Ibis* 117:531–534.

RICKLEFS, R. E., and G. W. COX. 1972. Taxon cycles in the West Indian avifauna. *American Naturalist* 106:195–219.

RIDPATH, M. G. 1972. The Tasmanian native hen, *Tribonyx mortierii. CSIRO Wildlife Research* 17:1–118.

RIECHERT, S. E., and C. R. TRACY. 1975. Thermal balance and prey availability: bases for a model relating web-site characteristics to spider reproductive success. *Ecology* 56:265–284.

RIPLEY, S. D. 1959. Competition between sunbird and honeyeater species in the Moluccan Islands. *American Naturalist* 93:127–132.

———. 1961. Aggressive neglect as a factor in interspecific competition in birds. *Auk* 78:366–371.

RIPPIN, A. D., and D. A. BOAG. 1974a. Recruitment to populations of male sharptailed grouse. *Journal of Wildlife Management* 38:616–621.

———. 1974b. Spatial organization among male sharp-tailed grouse on arenas. *Canadian Journal of Zoology* 52:591–597.

RISDON, D. H. S. 1973. The eating of meat by parrots. *Avicultural Magazine* 79:87–89.

RITCHEY, F. 1951. Dominance-subordination and territorial relationships in the common pigeon. *Physiological Zoology* 24:167–176.

ROBEL, R. J., and W. B. BALLARD. 1974. Lek social organization and reproductive success in the greater prairie chicken. *American Zoologist* 14:121–129.

ROBERTSON, D. R., H. P. A. SWEATMAN, E. A. FLETCHER, and M. G. CLELAND.

1976. Schooling as a mechanism for circumventing the territoriality of competitors. *Ecology* 57:1208–20.

ROBINS, J. D. 1971. Differential niche utilization in a grassland sparrow. *Ecology* 52:1065–70.

RODGERS, W. L. 1967. Specificity of specific hungers. *Journal of Comparative and Physiological Psychology* 64:49–58.

ROEDER, K. D. 1962. The behaviour of free flying moths in the presence of artificial ultrasonic pulses. *Animal Behaviour* 10:300–304.

———. 1975. Neural factors and evitability in insect behavior. *Journal of Experimental Zoology* 194:75–88.

ROGERS, H. E. 1957. An unusual merganser fatality. *Condor* 59:342–343.

ROGERS, Q. P., R. I. TANNOUS, and A. E. HARPER. 1967. Effects of excess leucine on growth and food selection. *Journal of Nutrition* 91:561–572.

ROHWER, S. 1975. The social significance of avian winter plumage variability. *Evolution* 29:593–610.

ROHWER, S., and F. C. ROHWER. 1978. Status signalling in Harris sparrows: experimental deceptions achieved. *Animal Behaviour* 26:1012–22.

ROMER, A. S. 1966. *Vertebrate paleontology*, 3d. ed. Chicago: University of Chicago Press.

ROOKE, I. J., and T. A. KNIGHT. 1977. Alarm calls of honeyeaters with reference to locating sources of sound. *Emu* 77:193–198.

ROOT, R. B. 1967. The niche exploitation pattern of the blue-gray gnatcatcher. *Ecological Monographs* 37:317–350.

ROSE, R. K., and M. S. GAINES. 1976. Levels of aggression in fluctuating populations of the prairie vole, *Microtus ochrogaster,* in eastern Kansas. *Journal of Mammalogy* 57:43–57.

ROSENZWEIG, M. L. 1968. The strategy of body size in mammalian carnivores. *American Midland Naturalist* 80:299–315.

———. 1973. Habitat selection experiments with pair of coexisting heteromyid rodent species. *Ecology* 54:111–117.

ROSENZWEIG, M. L., and P. W. STERNER. 1970. Population ecology of desert rodent communities: body size and seed-husking as bases for heteromyid coexistence. *Ecology* 51:217–224.

ROSENZWEIG, M. L., and J. WINAKUR. 1969. Population ecology of desert rodent communities: habitats and environmental complexity. *Ecology* 50:558–572.

ROTHSTEIN, S. E. 1973. The niche-variation model: is it valid? *American Naturalist* 107:598–620.

ROUGHGARDEN, J. 1974. Niche width: biogeographic patterns among *Anolis* lizard populations. *American Naturalist* 108:429–442.

ROWELL, T. E. 1974. The concept of social dominance. *Behavioral Biology* 11:131–154.

ROWE-ROWE, D. T. 1974. Flight behaviour and flight distances of blesbok. *Zeitschrift für Tierpsychologie* 34:208–211.

ROWLEY, I. 1965. The life history of the superb blue wren. *Emu* 64:251–297.

———. 1978. Communal activities among white-winged choughs *Corcorax melanorhamphus. Ibis* 120:178–197.

ROYAMA, T. 1966a. Factors governing feeding rates, food requirement and brood size of nestling great tits *Parus major. Ibis* 108:315–347.

———. 1966b. A re-interpretation of courtship feeding. *Bird Study* 13:116–129.

———. 1970. Factors governing the hunting behaviour and selection of food by the great tit *(Parus major* L.). *Journal of Animal Ecology* 39:619–668.

ROZIN, P. 1967. Specific aversions as a component of specific hungers. *Journal of Comparative and Physiological Psychology* 64:237–242.

ROZIN, P., and W. ROGERS. 1967. Novel diet preferences in vitamin deficient rats and rats recovered from vitamin deficiency. *Journal of Comparative and Physiological Psychology* 63:421–428.

RUDEBECK, G. 1950–51. The choice of prey and modes of hunting of predatory birds with special reference to their selective effect. *Oikos* 2:65–88, 3:200–231.

RUDNAI, J. 1974. The pattern of lion predation in Nairobi Park. *East African Wildlife Journal* 12:213–225.

RUSH, W. M. 1932. *Northern Yellowstone elk study.* Missoula: Montana Fish and Game.

RYDER, J. P. 1967. The breeding biology of Ross' goose in the Perry River region, Northwest Territories. *Canadian Wildlife Service, Report Series* 3:1–56.

SABINE, W. S. 1955. The winter society of the Oregon junco: the flock. *Condor* 57:88–111.

———. 1959. The winter society of the Oregon junco: intolerance, dominance, and the pecking order. *Condor* 61:110–135.

SADE, D. S. 1967. Determinants of dominance in a group of free-ranging rhesus monkeys. In S. A. Altman, ed., *Social communication among primates.* Chicago: University of Chicago Press, pp. 99–114.

SALE, P. F. 1968. Influence of cover availability on depth preference of the juvenile manini, *Acanthurus triostegus sandvichensis. Copeia* 1968:802–807.

———. 1969. A suggested mechanism for habitat selection by the juvenile manini *(Acanthurus triostegus sandvichensis* Streets). *Behaviour* 35:27–44.

———. 1972. Influence of corals in the dispersion of the pomacentrid fish, *Dascyllus aruanus. Ecology* 53:741–744.

SALOMONSEN, F. 1968. The moult migration. *Wildfowl* 19:5–24.

SALT, G. W. 1967. Predation in an experimental protozoan population *(Woodruffia-Paramecium). Ecological Monographs* 37:113–144.

SAMSON, F. B. 1976. Territory, breeding density, and fall departure in Cassin's finch. *Auk* 93:477–497.

———. 1977. Social dominance in winter flocks of Cassin's finch. *Wilson Bulletin* 89:57–66.

SANTOS, M. A. 1976. Prey selectivity and switching response of *Zetzellia mali. Ecology* 57:390–394.

SCHALLER, G. B. 1963. *The mountain gorilla.* Chicago: University of Chicago Press.

———. 1967. *The deer and the tiger.* Chicago: University of Chicago Press.

————. 1972. *The Serengeti lion.* Chicago: University of Chicago Press.

SCHJELDERUP-EBBE, T. 1922. Beiträge zur Sozialpsychologie des Haushuhns. *Zeitschrift für Psychologie* 88:225–252.

SCHLEIDT, W. M. 1961. Reaktionen von Truthühnern auf fliegende Raubvögel und Versuche zur Analyse ihrer AAM's. *Zeitschrift für Tierpsychologie* 18:534–560.

SCHOENER, T. W. 1965. The evolution of bill size differences among sympatric congeneric species of birds. *Evolution* 19:189–213.

————. 1967. The ecological significance of sexual dimorphism in size in the lizard *Anolis conspersus. Science* 155:474–477.

————. 1968a. The *Anolis* lizards of Bimini: resource partitioning in a complex fauna. *Ecology* 49:704–726.

————. 1968b. Sizes of feeding territories among birds. *Ecology* 49:123–141.

————. 1969a. Models of optimal size for solitary predators. *American Naturalist* 103:277–313.

————. 1969b. Optimal size and specialization in constant and fluctuating environments: an energy time approach. *Brookhaven Symposium in Biology* 22:103–114.

————. 1971. The theory of foraging strategies. *Annual Review of Ecology and Systematics* 2:369–404.

————. 1974. Resource partitioning in ecological communities. *Science* 185:27–39.

SCHULTZ, A. 1969. *The life of primates.* London: Weidenfeld and Nicolson.

SEARCY, W. A. 1978. Foraging succession in three age classes of glaucous-winged gulls. *Auk* 95:586–588.

SEGHERS, B. H. 1974. Schooling behavior in the guppy (*Poecilia reticulata*): an evolutionary response to predation. *Evolution* 28:486–489.

SEITZ, A. 1940. Die Paarbildung bei einigen Cichliden. I, Die Paarbildung bei *Astatotilapia atrigigena* Pfeffer. *Zeitschrift für Tierpsychologie* 1:40–84.

SELANDER, R. K. 1965. On mating systems and sexual selection. *American Naturalist* 99:129–141.

————. 1966. Sexual dimorphism and differential niche utilization in birds. *Condor* 68:113–151.

SERVENTY, D. L. 1971. Biology of desert birds. In D. S. Farner and J. R. King, eds., *Avian biology,* vol. 1. New York: Academic Press, pp. 287–339.

SEYMOUR, R. S. 1974. Convective and evaporative cooling in sawfly larvae. *Journal of Insect Physiology* 20:2447–57.

SHALTER, M. D. 1975. Lack of spatial generalization in habituation tests of fowl. *Journal of Comparative and Physiological Psychology* 89:258–262.

————. 1978a. Localization of passerine seeet and mobbing calls by goshawks and pygmy owls. *Zeitschrift für Tierpsychologie* 46:260–267.

————. 1978b. Effect of spatial context on the mobbing reaction of pied flycatchers to a predator model. *Animal Behaviour* 26:1219–21.

SHALTER, M. D., and W. M. SCHLEIDT. 1977. The ability of barn owls *Tyto alba* to discriminate and localize avian alarm calls. *Ibis* 119:22–27.

SHAW, E. 1970. Schooling in fishes: critique and review. In L. R. Aronson, E. Tobach, D. S. Lehrman, and J. S. Rosenblatt, eds., *Development and evolution of behavior: essays in memory of T. C. Schneirla*. San Francisco: Freeman, pp. 452–480.

SHEPPARD, D. H., P. H. KLOPFER, and H. OELKE. 1968. Habitat selection: differences in stereotypy between insular and continental birds. *Wilson Bulletin* 80:452–457.

SHERMAN, P. W. 1977. Nepotism and the evolution of alarm calls. *Science* 197:1246–53.

SHORT, L. L., JR. 1961. Interspecies flocking of birds of montane forest in Oaxaca, Mexico. *Wilson Bulletin* 73:341–347.

SIBLEY, C. G. 1955. Behavioral mimicry in titmice (Paridae) and certain other birds. *Wilson Bulletin* 67:128–132.

SIEGFRIED, W. R. 1971. Communal roosting of the cattle egret. *Transactions of the Royal Society of South Africa* 39:419–443.

SIEGFRIED, W. R., and L. G. UNDERHILL. 1975. Flocking as an anti-predator strategy in doves. *Animal Behaviour* 23:504–508.

SIKES, S. K. 1971. The natural history of the African elephant. London: Weidenfeld and Nicolson.

SIMMONS, K. E. L. 1951. Interspecific territorialism. *Ibis* 92:407–413.

———. 1952. The nature of the predator-reactions of breeding birds. *Behaviour* 4:161–172.

SIMON, C. A. 1975. The influence of food abundance on territory size in the iguanid lizard *Sceloporus jarrovi*. *Ecology* 56:993–998.

SIMPSON, G. G. 1953. *The major features of evolution*. New York: Columbia University Press.

SINCLAIR, A. R. E. 1978. Factors affecting the food supply and breeding season of resident birds and movements of palearctic migrants in a tropical savannah. *Ibis* 120:480–497.

SKUTCH, A. F. 1949. Do tropical birds rear as many young as they can nourish? *Ibis* 91:430–458.

———. 1961. Helpers among birds. *Condor* 63:198–226.

———. 1976. *Parent birds and their young*. Austin: University of Texas Press.

SLANEY, P. A., and T. G. NORTHCOTE. 1974. Effects of prey abundance on density and territorial behavior of young rainbow trout (*Salmo gairdneri*) in laboratory stream channels. *Journal of the Fisheries Research Board of Canada* 31:1201–09.

SLOAN, N. F., and G. A. SIMMONS. 1973. Foraging behavior of the chipping sparrow in response to high populations of the jack pine budworm. *American Midland Naturalist* 90:210–215.

SLOBODKIN, L. B. 1961. *Growth and regulation of animal populations*. New York: Holt, Rinehart, and Winston.

SLOBODKIN, L. B., F. E. Smith, and N. L. Hairston. 1967. Regulation in terrestrial ecosystems, and the implied balance of nature. *American Naturalist* 101:109–124.

SLUD, P. 1960. The birds of Finca "La Selva," Costa Rica: a tropical wet forest locality. *Bulletin of the American Museum of Natural History* 121: 49–148.

————. 1964. The birds of Costa Rica. *Bulletin of the American Museum of Natural History* 128:1–430.

SMIGEL, B. W., and M. L. ROSENZWEIG. 1974. Seed selection in *Dipodomys merriami* and *Perognathus penicillatus*. *Ecology* 55:329–339.

SMITH, C. C. 1968. The adaptive nature of social organization in the genus of tree squirrels *Tamiasciurus*. *Ecological Monographs* 38:31–63.

SMITH, D. G., and D. H. HOLLAND. 1974. Mobbing red-winged blackbirds force American kestrel into water. *Auk* 91:843–844.

SMITH, J. N. M. 1974a. The food searching behaviour of two European thrushes. I, Description and analysis of search paths. *Behaviour* 48:276–302.

————. 1974b. The food searching behaviour of two European thrushes. II, The adaptiveness of the search patterns. *Behaviour* 49:1–61.

————. 1978. Division of labour by song sparrows feeding fledged young. *Canadian Journal of Zoology* 56:187–191.

SMITH, J. N. M., and R. DAWKINS. 1971. The hunting behaviour of individual great tits in relation to spatial variations in their food density. *Animal Behaviour* 19:695–706.

SMITH, J. N. M., and H. P. A. SWEATMAN. 1974. Food-searching behavior of titmice in patchy environments. *Ecology* 55:1216–32.

SMITH, N. G. 1968. The advantage of being parasitized. *Nature* 219:690–694.

————. 1969. Provoked release of mobbing—a hunting technique of *Micrastur* falcons. *Ibis* 111:241–243.

SMITH, S. M. 1972. Roosting aggregations of bushtits in response to cold temperatures. *Condor* 74:478–479.

————. 1975. Innate recognition of coral snake patern by a possible avian predator. *Science* 187:759–760.

————. 1977. Coral-snake pattern recognition and stimulus generalisation by naive great kiskadees (Aves: Tyrannidae). *Nature* 265:535–536.

————. 1978. The "underworld" in a territorial sparrow: adaptive strategy for floaters. *American Naturalist* 112:571–582.

SMITH, W. I., and S. ROSS. 1953. The hoarding behavior of the mouse. I, The role of previous feeding experience. *Journal of Genetical Psychology* 82:279–297.

SMYTHE, N. 1970a. Ecology and behavior of the agouti (*Dasyprocta punctata*) and related species on Barro Colorado Island, Panama. Ph.D. thesis, University of Maryland.

————. 1970b. On the existence of "pursuit invitation" signals in mammals. *American Naturalist* 104:491–494.

SNOW, B. L., and D. W. SNOW. 1972. Feeding niches of hummingbirds in a Trinidad valley. *Journal of Animal Ecology* 41:471–485.

SNOW, D. W. 1952. The winter avifauna of Lapland. *Ibis* 94:137–143.

————. 1962a. A field study of the black-and-white manakin, *Manacus manacus*, in Trinidad. *Zoologica* 47:65–104.

————. 1962b. A field study of the golden-headed manakin, *Manacus erythrocephala*, in Trinidad. *Zoologica* 47:183–198.

————. 1971. Evolutionary aspects of fruit-eating by birds. *Ibis* 113:194–202.

SOLOMON, M. E. 1949. The natural control of animal populations. *Journal of Animal Ecology* 18:1–35.

SOULÉ, M., and B. R. STEWART. 1970. The "niche-variation" hypothesis: a test and alternatives. *American Naturalist* 104:85–97.

SOUTHERN, H. N. 1970. The natural control of a population of tawny owls (*Strix aluco*). *Journal of Zoology, London* 162:197–285.

SPRINGER, S. 1957. Some observations on the behavior of schools of fishes in the Gulf of Mexico and adjacent waters. *Ecology* 38:166–171.

STALLCUP, J. A., and G. E. WOOLFENDEN. 1978. Family status and contributions to breeding by Florida scrub jays. *Animal Behaviour* 26:1144–56.

STAMPS, J. A. 1977. The relationship between resource competition, risk, and aggression in a tropical territorial lizard. *Ecology* 58:349–358.

STANFORD, J. K. 1947. Bird parties in forest in Burma. *Ibis* 89:507–509.

STANLEY, S. M. 1974. Relative growth of the titanothere horn: a new approach to an old problem. *Evolution* 28:447–457.

STATES, J. B. 1976. Local adaptations in chipmunk (*Eutamias amoenus*) populations and evolutionary potential at species' borders. *Ecological Monographs* 46:221–256.

STEIN, R. A., and J. J. MAGNUSON. 1976. Behavioral response of crayfish to a fish predator. *Ecology* 57:751–761.

STEINIGER, K. 1950. Beitrage zur Soziologie und sonstigen Biologie der Wanderratte. *Zeitschrift für Tierpsychologie* 7:356–379.

STENGER, J., and J. B. FALLS. 1959. The utilized territory of the ovenbird. *Wilson Bulletin* 71:125–140.

STENSETH, N. C., and L. HANSSON. 1979. Optimal food selection: a graphic model. *American Naturalist* 113:373–389.

STIMSON, J. 1970. Territorial behavior of the owl limpet, *Lottia gigantea*. *Ecology* 51:113–118.

STOECKER, R. E. 1972. Competitive relations between sympatric populations of voles (*Microtus montanus* and *M. pennsylvanicus*). *Journal of Animal Ecology* 41:311–329.

STORER, R. W. 1966. Sexual dimorphism and food habits in three North American accipiters. *Auk* 83:423–436.

———. 1971. Adaptive radiation in birds. In D. S. Farner and J. R. King, eds., *Avian biology*, vol. 1. New York: Academic Press, pp. 149–188.

STORER, R. W., W. R. SIEGFRIED, and J. KINAHAN. 1975. Sunbathing in grebes. *Living Bird* 14:45–57.

STRESEMANN, E. 1917. Über gemischte Vogelschwärme. *Verhandlungen der Ornithologischen Gessellschaft in Bayern* 13:127–151.

STURMAN, W. A. 1968. The foraging ecology of *Parus atricapillus* and *Parus rufescens* in the breeding season, with comparisons with other species of *Parus*. *Condor* 70:309–332.

SULKAVA, S. 1969. On small birds spending the night in the snow. *Aquilo, Zoology* 7:33–37.

SUZUKI, A. 1971. Carnivory and cannibalism observed among forest-living chimpanzees. *Journal of the Anthropological Society of Nippon (Tokyo)* 79:30–48.

SVÄRDSON, G. 1949. Competition and habitat selection in birds. *Oikos* 1:157–174.

———. 1950. Swift (*Apus apus*) movement in summer. *Proceedings of the International Ornithological Congress* 10:335–338.

SVENDSEN, G. E. 1974. Behavioral and environmental factors in the spatial distribution and population dynamics of a yellow-bellied marmot population. *Ecology* 55:760–771.

SWANBERG, P. O. 1951. Food storage, territory and song in the thick-billed nutcracker. *Proceedings of the International Ornithological Congress* 10:545–554.

SWINGLAND, I. R. 1975. The influence of weather and individual interactions on the food intake of captive rooks (*Corvus frugilegus*). *Physiological Zoology* 48:295–302.

———. 1977. The social and spatial organization of winter communal roosting in rooks (*Corvus frugilegus*). *Journal of Zoology, London* 182:509–528.

SWYNNERTON, C. F. M. 1915. Mixed bird parties. *Ibis* 1915:346–354.

TAMARIN, R. H., and C. J. KREBS. 1969. *Microtus* population biology. II, Genetic changes at the transferrin locus in fluctuating populations of two vole species. *Evolution* 23:183–211.

TAYLOR, L. R. 1961. Aggregation, variance and the mean. *Nature* 189:732–735.

TELEKI, G. 1973. *The predatory behavior of wild chimpanzees.* Lewisburg, Pa.: Bucknell University Press.

TEMPLETON, J. 1970. Reptiles. In G. C. Whittow, ed., *Comparative physiology of thermoregulation*, vol. 1. New York: Academic Press, pp. 167–221.

TENER, J. S. 1965. *Muskoxen in Canada.* Ottawa: Queen's Printer.

TERBORGH, J., and J. S. WESKE. 1969. Colonization of secondary habitats by Peruvian birds. *Ecology* 50:765–782.

TEVIS, L. 1958. Germination and growth of ephemerals induced by sprinkling a sandy desert. *Ecology* 39:681–687.

THIOLLAY, J. M. 1977. Les migrations de rapaces en Afrique occidentale: adaptations ecologiques aux fluctuations saisonnierres de production des ecosystemes. *La Terre et la Vie* 32:89–133.

THOMAS, G. 1974. The influence of encountering a food object on the subsequent searching behaviour in *Gasterosteus aculeatus* L. *Animal Behaviour* 22:941–952.

THOMPSON, W. A., I. Vertinsky, and J. R. Krebs. 1974. The survival value of flocking in birds: a simulation model. *Journal of Animal Ecology* 43:785–820.

THOMSON, A. L., ed. 1964. *A new dictionary of birds.* London: Nelson.

THORNEYCROFT, H. B. 1975. A cytogenetic study of the white-throated sparrow, *Zonotrichia albicollis* (Gmelin). *Evolution* 29:611–621.

THORPE, W. H. 1943. A type of insight learning in birds. *British Birds* 37:29–31.

———. 1963. *Learning and instinct in animals*, 2d ed. London: Methuen.

———. 1972. Duetting and antiphonal song in birds, its extent and significance. *Behaviour, Supplement* 18:1–197.

THRESHER, R. E. 1976. Field experiments on species recognition by the threespot damselfish, *Eupomacentrus planifrons* (Pisces: Pomacentridae). *Animal Behaviour* 24:562–569.

THUROW, G. 1976. Aggression and competition in eastern *Plethodon* (Amphibia, Urodela, Plethodontidae). *Journal of Herpetology* 10:277–291.

TINBERGEN, J. M. 1976. How starlings (*Sturnus vulgaris* L) apportion their foraging time in a virtual single-prey situation on a meadow. *Ardea* 64: 155–170.

TINBERGEN, L. 1946. Die Sperwer als roofvijand van Zangvogels. *Ardea* 34:1–213.

———. 1949. Bosvogels en insecten. *Nederlandsch Boschbouw Tijdschrift* 4:91–105.

———. 1960. The natural control of insects in pinewoods. I, Factors influencing the intensity of predation by songbirds. *Archives Néerlandaises de Zoologie* 13:265–336.

TINBERGEN, N. 1951. *The study of instinct.* Oxford: Clarendon Press.

———. 1953a. *The herring gull's world.* London: Collins.

———. 1953b. *Social behaviour in animals.* London: Methuen.

———. 1957. The functions of territory. *Bird Study* 4:14–27.

TINBERGEN, N., M. IMPEKOVEN, and D. FRANK. 1967. An experiment on spacing out as defence against predation. *Behaviour* 28:307–321.

TOMPA, F. S. 1962. Territorial behavior: the main controlling factor of a local song sparrow population. *Auk* 79:687–697.

———. 1971. Catastrophic mortality and its population consequences. *Auk* 88:753–759.

TOPINSKI, P. 1974. The role of antlers in establishment of the red deer herd hierarchy. *Acta Theriologica* 19:509–514.

TRAVIS, S. E., and K. B. ARMITAGE. 1972. Some quantitative aspects of the behavior of marmots. *Transactions of the Kansas Academy of Science* 75: 308–321.

TREISMAN, M. 1975. Predation and the evolution of gregariousness. I, Models for concealment and evasion. *Animal Behaviour* 23:779–800.

TRIMBLE, S. A. 1976. Galapagos mockingbird pecks at sea lion mouth. *Condor* 78:566.

TRIVERS, R. L. 1971. The evolution of reciprocal altruism. *Quarterly Review of Biology* 46:35–57.

———. 1972. Parental investment and sexual selection. In B. Campbell, ed., *Sexual selection and the descent of man 1871–1971.* Chicago: Aldine, pp. 136–179.

———. 1974. Parent-offspring conflict. *American Zoologist* 14:249–264.

TRIVERS, R. L., and H. HARE. 1976. Haplodiploidy and the evolution of the social insects. *Science* 191:249–263.

TUCKER, V. A. 1965. The relation between the torpor cycle and heat exchange in the California pocket mouse *Perognathus californicus*. *Journal of Cellular and Comparative Physiology* 65:405–414.

———. 1966. Diurnal torpor and its relation to food consumption and weight changes in the California pocket mouse *Perognathus californicus*. *Ecology* 47:245–252.

TURCEK, F. J., and L. KELSO. 1968. Ecological aspects of food transportation and storage in the Corvidae. *Communications in Behavioral Biology A* 1:277–297.

TURNER, F. B. 1970. The ecological efficiency of consumer populations. *Ecology* 51:741–742.

UDVARDY, M. D. F. 1954. Summer movements of black swifts in relation to weather conditions. *Condor* 56:261–267.

VANCE, R. R. 1972. The role of shell adequacy in behavioral interactions involving hermit crabs. *Ecology* 53:1075–83.

VAN TYNE, J., and A. J. BERGER. 1959. *Fundamentals of ornithology.* New York: Wiley.

VAN VALEN, L. 1965. Morphological variation and width of ecological niche. *American Naturalist* 99:377–390.

VARLEY, G. C., G. R. GRADWELL, and M. P. HASSELL. 1973. *Insect population ecology.* Oxford: Blackwell.

VAUGIEN, L. 1952. Sur le comportement sexuel singulier de la Peruch ondulée, maintenue à l'obscurité. *Comptes Rendus Hebdomodaires des Séances de l'Academie des Sciences D* 234:1489.

———. 1953. Sur l'apparition de la maturité sexuelle des jeunes perruches ondulées mâls soumises à diverses conditions d'éclairement: Le développe-testiculaire est plus rapide dans l'obscurité complète. *Bulletin Biologique de la France et de la Belgique* 87:274–286.

VEHRENCAMP, S. L. 1977. Relative fecundity and parental effort in communally nesting anis, *Crotophaga sulcirostris. Science* 197:403–405.

VERBEEK, N. A. M. 1973. The exploitation system of the yellow-billed magpie. *University of California Publications in Zoology* 99:1–57.

———. 1977. Comparative feeding behavior of immature and adult herring gulls. *Wilson Bulletin* 89:415–421.

VERNER, J. 1965. Time budget of the male long-billed marsh wren during the breeding season. *Condor* 67:125–139.

———. 1975. Avian behavior and habitat management. In *Proceedings of the symposium on management of forest and range habitats for nongame birds.* Washington, D.C.: USDA Forest Service, pp. 39–58.

———. 1977. On the adaptive significance of territoriality. *American Naturalist* 111:769–775.

VERNER, J., and M. F. WILLSON. 1966. The influence of habitats on mating systems of North American passerine birds. *Ecology* 47:143–147.

WARD, P. 1965a. Feeding ecology of the black-faced dioch *Quelea quelea* in Nigeria. *Ibis* 107:173–214.

———. 1965b. The breeding biology of the black-faced dioch *Quelea quelea* in Nigeria. *Ibis* 107:326–349.

———. 1971. The migration patterns of *Quelea quelea* in Africa. *Ibis* 113:275–297.

WARD, P., and A. ZAHAVI. 1973. The importance of certain assemblages of birds as "information-centres" for food-finding. *Ibis* 115:517–534.

WARE, D. 1974. Predation by rainbow trout (*Salmo gairdneri*): the influence of hunger, prey density and prey size. *Journal of the Fisheries Research Board of Canada* 31:1193–1201.

———. 1975. Growth, metabolism and optimal swimming speed in a pelagic fish. *Journal of the Fisheries Research Board of Canada* 32:33–41.

WATSON, A., and R. MOSS. 1970. Dominance, spacing behaviour and aggression in relation to population limitation in vertebrates. *Symposium of the British Ecological Society* 10:167–220.

WATSON, J. R. 1970. Dominance-subordination in caged groups of house sparrows. *Wilson Bulletin* 82:268–278.

WATTS, C. R., and A. W. STOKES. 1971. The social order of turkeys. *Scientific American* 224 (6):112–118.

WEBER, W. C. 1972. Birds in cities: a study of populations, foraging ecology and nest-sites of urban birds. M.S. thesis, University of British Columbia.

WEBSTER, F. A., and D. R. GRIFFIN. 1962. The role of flight membranes in insect capture by bats. *Animal Behaviour* 10:332–340.

WECKER, S. C. 1963. The role of early experience in habitat selection by the prairie deer mouse, *Peromyscus maniculatus bairdi. Ecological Monographs* 33:307–325.

———. 1964. Habitat selection. *Scientific American* 211 (4):109–116.

WEIHS, D. 1973. Hydrodynamics of fish schooling. *Nature* 241:290–291.

WELLS, K. D. 1977. The social behaviour of anuran amphibians. *Animal Behaviour* 25:666–693.

WELSH, D. A. 1975. Savannah sparrow breeding and territoriality on a Nova Scotia dune beach. *Auk* 92:235–251.

WELTY, J. C. 1934. Experiments in group behavior of fishes. *Physiological Zoology* 7:85–128.

———. 1962. *The life of birds.* New York: Knopf.

WERNER, E. E. 1974. The fish size, prey size, handling time relation in several sunfishes and some implications. *Journal of the Fisheries Research Board of Canada* 31:1531–36.

———. 1977. Species packing and niche complementarity in three sunfishes. *American Naturalist* 111:553–578.

WERNER, E. E., and D. J. HALL. 1974. Optimal foraging and the size selection of prey by the bluegill sunfish (*Lepomis macrochirus*). *Ecology* 55:1042–52.

———. 1977. Competition and habitat shift in two sunfishes (Centrarchidae). *Ecology* 58:869–876.

WEST, G. C. 1960. Seasonal variation in the energy balance of the tree sparrow in relation to migration. *Auk* 77:306–329.

———. 1968. Bioenergetics of captive willow ptarmigan under natural conditions. *Ecology* 49:1035–45.

WHITE, F. N., G. A. BARTHOLOMEW, and T. R. HOWELL. 1975. The thermal significance of the nest of the sociable weaver *Philetairus socius:* winter observations. *Ibis* 117:171–179.

WHITE, F. N., and J. L. KINNEY. 1974. Avian incubation. *Science* 186:107–115.

WICKLER, W. 1968. *Mimicry.* London: Weidenfeld and Nicolson.

WIENS, J. A. 1966. On group selection and Wynne-Edwards' hypothesis. *American Scientist* 54:273–287.

————. 1970. Effects of early experience on substrate pattern selection in *Rana aurora* tadpoles. *Copeia* 1970:543–548.

————. 1972. Anuran habitat selection: early experience and substrate selection in *Rana cascadae* tadpoles. *Animal Behaviour* 20:218–220.

WIESE, J. H., and R. L. CRAWFORD. 1974. Joint "leap-frog" feeding by ardeids. *Auk* 91:836–837.

WILBUR, H. W. 1969. The breeding biology of Leach's petrel, *Oceanodroma leucorhoa*. *Auk* 86:433–442.

WILEY, R. H. 1973. Territoriality and non-random mating in sage grouse, *Centrocercus urophasianus*. *Animal Behavior Monographs* 6:85–169.

————. 1974. Evolution of social organization and life-history patterns among grouse. *Quarterly Review of Biology* 49:201–227.

WILEY, R. H., and D. G. RICHARDS. 1978. Physical constraints on acoustic communication in the atmosphere: implications for the evolution of animal vocalizations. *Behavioral Ecology and Sociobiology* 3:69–94.

WILKINSON, P. F. 1974. Behaviour and domestication of the musk ox. In V. Geist and F. Walther, eds., *The behaviour of ungulates and its relation to management*. Morges, Switzerland: International Union for the Conservation of Nature, pp. 909–920.

WILKINSON, P. F., and C. C. SHANK. 1976. Rutting-fight mortality among musk oxen on Banks Island, Northwest Territories, Canada. *Animal Behaviour* 24:756–758.

WILLIAMS, G. C. 1964. Measurement of consociation among fishes and comments on the evolution of schooling. *Publications of the Museum, Michigan State University, Biological Series* 2:349–384.

WILLIAMS, G. C. 1966. *Adaptation and natural selection*. Princeton: Princeton University Press.

WILLIAMSON, P. 1971. Feeding ecology of the red-eyed vireo (*Vireo olivaceus*) and associated foliage-gleaning birds. *Ecological Monographs* 41:129–152.

WILLIS, E. O. 1966a. The role of migrant birds at swarms of army ants. *Living Bird* 5:187–231.

————. 1966b. Competitive exclusion and birds at fruiting trees in western Colombia. *Auk* 83:479–480.

————. 1967. The behavior of bicolored antbirds. *University of California Publications in Zoology* 79:1–127.

————. 1968. Studies of lunulated and Salvin's antbirds. *Condor* 70:128–148.

————. 1972. The behavior of spotted antbirds. *Ornithological Monographs* 10:1–162.

————. 1973a. Local distribution of mixed flocks in Puerto Rico. *Wilson Bulletin* 85:75–77.

————. 1973b. Do birds flock in Hawaii, a land without predators? *California Birds* 3:1–8.

WILLIS, E. O., and Y. ONIKI. 1978. Birds and army ants. *Annual Review of Ecology and Systematics* 9:243–264.

WILLSON, M. F. 1969. Avian niche size and morphological variation. *American Naturalist* 103:531–542.

————. 1971. Seed selection in some North American finches. *Condor* 73: 415–429.

WILLSON, M. F., and J. C. HARMESON. 1973. Seed preferences and digestive efficiency of cardinals and song sparrows. *Condor* 75:225–234.

WILLSON, M. F., J. R. KARR, and R. R. ROTH. 1975. Ecological aspects of avian bill-size variation. *Wilson Bulletin* 87:32–44.

WILSON, D. P. 1970. Additional observations on larval growth and settlement of *Sabellaria alveolata*. *Journal of the Marine Biological Association, United Kingdom* 50:1–31.

WILSON, D. S. 1975. The adequacy of body size as a niche difference. *American Naturalist* 109:769–784.

WILSON, E. O. 1970. Chemical communication within animal species. In E. Sondheimer and J. B. Simeone, eds., *Chemical ecology*. New York: Academic Press, pp. 133–155.

————. 1971a. *The insect societies*. Cambridge: Harvard University Press, Belknap Press.

————. 1971b. Competitive and aggressive behavior. In J. F. Eisenberg and W. Dillon, eds., *Man and beast: comparative social behavior*. Washington, D.C.: Smithsonian Institution Press, pp. 183–217.

————. 1975. *Sociobiology*. Cambridge: Harvard University Press, Belknap Press.

WINKEL, W., and D. WINKEL. 1973. Höhlenschafen bei Kohlmeisen (*Parus major*) zur Zeit der Brut und Mauser. *Vogelwelt* 94:50–60.

WINSTANLEY, D., R. SPENCER, and K. WILLIAMSON. 1974. Where have all the whitethroats gone? *Bird Study* 21:1–14.

WINTERBOTTOM, J. M. 1943. On woodland bird parties in Northern Rhodesia. *Ibis* 1943:437–442.

————. 1949. Mixed bird parties in the tropics, with special reference to Northern Rhodesia. *Auk* 66:258–263.

WOOD-GUSH, D. G. M. 1971. *The behaviour of the domestic fowl*. London: Heineman.

WOOLFENDEN, G. E. 1975. Florida scrub jay helpers at the nest. *Auk* 92:1–15.

WOOLFENDEN, G. E., S. C. WHITE, R. L. MUMME, and W. B. ROBERTSON, JR. 1976. Aggression among starving cattle egrets. *Bird-banding* 47:48–53.

WYNNE-EDWARDS, V. C. 1962. *Animal dispersion in relation to social behaviour*. Edinburgh: Oliver and Boyd.

WYRWOLL, T. 1977. The hunting tendency of the goshawk (*Accipiter gentilis*) in relation to nesting place. *Journal für Ornithologie* 118:21–34.

YEATON, R. I., and M. L. CODY. 1974. Competitive release in island song sparrow populations. *Theoretical Population Biology* 5:42–58.

YOM-TOV, Y. 1974. The effect of food and predation on breeding density and success, clutch size and laying date of the crow (*Corvus corone* L.). *Journal of Animal Ecology* 43:479–498.

YOM-TOV, Y., A. IMBER, and J. OTTERMAN. 1977. The microclimate of winter roosts of the starling *Sturnus vulgaris*. *Ibis* 119:366–368.

YOUNG, A. M. 1971. Wing coloration and reflectance in *Morpho* butterflies as related to reproductive behavior and escape from avian predators. *Oecologia* 7:209–222.

ZACH, R. 1978. Selection and dropping of whelks by northwestern crows. *Behaviour* 67:134–148.

———. 1979. Shell dropping: decision-making and optimal foraging in northwestern crows. *Behaviour* 68:106–117.

ZACH, R., and J. B. FALLS, 1976. Do ovenbirds (Aves: Parulidae) hunt by expectation? *Canadian Journal of Zoology* 54:1894–1903.

———. 1978. Prey selection by captive ovenbirds (Aves: Parulidae). *Journal of Animal Ecology* 47:929–943.

ZAHAVI, A. 1971. The function of pre-roost gatherings and communal roosts. *Ibis* 113:106–109.

———. 1974. Communal nesting by the Arabian babbler. *Ibis* 116:84–87.

ZIMEN, E. 1976. On the regulation of pack size in wolves. *Zeitschrift für Tierpsychologie* 40:300–341.

ZUMETA, D. C., and R. T. HOLMES. 1978. Habitat shift and roadside mortality of scarlet tanagers during a cold wet New England spring. *Wilson Bulletin* 90:575–586.

Index

373